# TOPICS IN MOLECULAR MEDICINE
*Volume 1*

---

# Adhesion Molecules and Cell Signaling:
## Biology and Clinical Applications

TOPICS IN MOLECULAR MEDICINE
*Volume 1*

# Adhesion Molecules and Cell Signaling:

## Biology and Clinical Applications

Editors

**Wolfgang Siess, M.D.**
*University of Munich*
*Institut für Prophylaxe und Epidemiologie der Kreislaufkrankheiten*
*Munich, Germany*

**Reinhard Lorenz, M.D.**
*University of Munich*
*Institut für Prophylaxe und Epidemiologie der Kreislaufkrankheiten*
*Munich, Germany*

**Peter C. Weber, M.D.**
*University of Munich*
*Institut für Prophylaxe und Epidemiologie der Kreislaufkrankheiten*
*Munich, Germany*

Raven Press New York

**Raven Press, Ltd., 1185 Avenue of the Americas, New York, New York 10036**



Made in the United States of America

Adhesion molecules and cell signaling: biology and clinical applications / editors, Wolfgang Siess, Reinhard Lorenz, and Peter C. Weber.

p. cm.—(Topics in molecular medicine; v. 1)

Proceedings of an international symposium, held in Prien, Germany, June 1994.

Includes bibliographical references and index.

ISBN 0-7817-0323-9

1. Cell adhesion molecules—Congresses. 2. Cell adhesion molecules—Pathophysiology—Congresses. 3. Cellular signal transduction—Congresses. I. Siess, Wolfgang. II. Series.

[DNLM: 1. Cell Adhesion—congresses. 2. Cell Adhesion Molecules—physiology—congresses. 3. Cell Communication—congresses. 4. Signal Transduction—physiology—congresses. QU 55 A2347 1995]

QP552.C42A34 1995

574.87′6—dc20

DNLM/DLC

for Library of Congress 95-178

The material contained in this volume was submitted as previously unpublished material, except in the instances in which credit has been given to the source from which some of the illustrative material was derived.

Great care has been taken to maintain the accuracy of the information contained in the volume. However, neither Raven Press nor the authors can be held responsible for errors or for any consequences arising from the use of the information contained herein.

9 8 7 6 5 4 3 2 1

# Preface

This book is the first volume of the new Raven Press Review Series "Topics in Molecular Medicine." The aim of the new series is to enhance the transfer of new insights of molecular mechanisms of physiologic and pathophysiologic processes into clinical investigation and application. The volume contains the proceedings of an International Symposium on "Adhesion Molecules and Cell Signalling: Biology and Clinical Applications," held in Prien, Lake Chiemsee, Germany, on June 8–12, 1994. The purpose of the symposium was to bring together leading basic and clinical scientists working on adhesion molecules.

Physiologic and pathophysiologic cell–cell interactions in the circulation, the lymphatic system, and extravascular space are regulated via specific adhesion molecules. As the understanding of the role and signaling pathways of these molecules increases, their possible impact on medicine becomes obvious. It was surprising to see how much this frontier has already penetrated clinical science. Sections I and II consider the structure and regulation of the main classes of adhesion molecules: integrins, selectins, and ICAMs. Sections III and IV detail postangioplasty reactions, ischemia–reperfusion injury, or metastasis, either by directly blocking specific adhesion molecules or by inhibiting signaling mechanisms leading to the surface expression of adhesion molecules. Section V outlines the involvement of adhesion molecules in immunologic responses and inflammation, such as asthma and transplant rejection.

The symposium was limited to 120 participants from academia and industry. The generous support by Bayer AG helped to create optimal conditions for the meeting. It was generally felt that the goal of increasing communications among the invited scientists and the external participants was achieved. We hope that this volume serves a similar useful function for a broader audience.

To facilitate reading, a table containing the names, synonyms, and interactions of the various adhesion molecules has been added on page xvii.

*Wolfgang Siess*
*Reinhard Lorenz*
*Peter C. Weber*

# Acknowledgment

On behalf of all participants of the International Symposium on Topics in Molecular Medicine titled "Adhesion Molecules and Cell Signalling: Biology and Clinical Applications" held in Prien, Lake Chiemsee, Germany, June 8–12, 1994, we thank Bayer AG for making the symposium possible.

# Contents

# Contributors and Chairmen

**Walter Birchmeier** *Max-Delbrück-Centrum, für Molekulare Medizin, Robert-Rössle-Str. 10, 13125 Berlin-Buch, Germany*

**W. F. Bodmer** *Imperial Cancer Research Fund, Lincoln's Inn Fields, London WC2A 3PX, England*

**Michael R. Buchanan** *Department of Pathology, McMaster University Health Sciences Centre, 1200 Main Street, Hamilton, Ontario, Canada L8N 3Z5*

**Barry S. Coller** *Department of Medicine, Mount Sinai Medical Center, One Gustave L. Levy Place, New York, New York 10029-6574*

**Myron I. Cybulsky** *Departments of Pathology, Brigham and Women's Hospital, Harvard Medical School, 221 Longwood Avenue (LMRC-4), Boston, Massachusetts 02115*

**Helmut E. Feucht** *Department of Internal Medicine, Klinikum Innenstadt, and Institute of Immunology, University of Munich, Ziemssenstrasse 1, D-80336 Munich, Germany*

**Bruce Furie** *Division of Hematology–Oncology, New England Medical Center, Departments of Medicine and Biochemistry, Tufts University School of Medicine, 750 Washington Street, Boston, Massachusetts 02111*

**A. J. Gearing** *British Bio-technology Plc, Watlington Road, Cowley, OX4 5LY, England*

**J. A. Glomset** *Howard Hughes Medical Institute, University of Washington, HSB J-611F SL-15, Seattle, Washington 98195*

**Robert H. Gundel** *Preclinical Research, Miles, Inc., 400 Morgan Lane, West Haven, Connecticut 06516*

**Stephen Haskill** *Lineberger Comprehensive Cancer Center, University of North Carolina at Chapel Hill, Chapel Hill, North Carolina 27599-7295*

**Per Hedqvist** *Karolinska Institutet, Physiology and Pharmacology, S-17177 Stockholm, Sweden*

**Martin E. Hemler** *Dana-Farber Cancer Institute, Harvard Medical School, 44 Binney Street, Boston, Massachusetts 02115*

**Bernhard Holzmann** *Institute of Medical Microbiology and Hygiene, Technical University, Trogerstrasse 4a, D-81675 Munich, Germany*

**Martin J. Humphries** *School of Biological Sciences, University of Manchester, Oxford Road, Manchester, M13 9PT, England*

**Mitsuaki Isobe** *The First Department of Internal Medicine, Shinshu University School of Medicine, 3-1-1 Asahi, Matsumoto 390, Japan*

**Judith P. Johnson** *Institute for Immunology, University of Munich, Goethestrasse 31, 80336 Munich, Germany*

**Laurence A. Lasky** *Department of Immunobiology, Gentech. Inc., 460 Point San Bruno Blvd., South San Francisco, California 94080*

**Reinhard Lorenz** *Institut für Kreislaufprophylaxe, Pettenkoferstr. 9, D-80336 München, Germany*

**Rodger P. McEver** *Departments of Medicine and Biochemistry, W. K. Warren Medical Research Institute, University of Oklahoma Health Sciences Center, 825 North East 13th Street, Oklahoma City, Oklahoma 73104*

**Stephen M. Prescott** *Program in Human Molecular Biology and Genetics, Eccles Institute of Human Genetics, University of Utah, Salt Lake City, Utah 84112*

**Marek W. Radomski** *Departments of Gynecology, Obstetrics, and Pharmacology, Perinatal Research Center, Heritage Medical Research Centre, University of Alberta, Edmonton, Alberta, Canada T6G 2S2*

**G. Riethmüller** *Vorstand des Instituts für Immunologie der LMU, Goethestr. 31, 80336 München, Germany*

**Lewis H. Romer** *Departments of Pediatrics, and Cell Biology and Anatomy, University of North Carolina at Chapel Hill, Chapel Hill, North Carolina 27599-7220*

**Erkki Ruoslahti** *Cancer Research Center, La Jolla Cancer Research Foundation, 109 North Torrey Pines Road, La Jolla, California 92037*

**Martin R. Schneider** *Experimental Oncology, Schering AG, Mullerstrasse 170-178, 13342 Berlin, Germany*

**Wolfgang Siess** *Institut für Kreislaufprophylaxe, Pettenkoferstr. 9, D-80336 München, Germany*

**David L. Simmons** *Cell Adhesion Laboratory, Institute of Molecular Medicine, John Radcliffe Hospital, University of Oxford, Headington, Oxford, OX3 9DU, England*

**Gustav Steinhoff** *Department of Cardiovascular Surgery, Christian Albrechts University of Kiel, Arnold Hellerstrasse 7, 24105 Kiel, Germany*

**Nicholas Vedder** *Department of Surgery, University of Washington, Harborview Medical Center, ZA-16, Seattle, Washington 98104*

**Denisa Wagner** *The Center for Blood Research, Harvard Medical School, 800 Huntington Avenue, Boston, Massachusetts 02115*

**Peter A. Ward** *Department of Pathology, University of Michigan Medical School, 1301 Catherine Road, Ann Arbor, Michigan 48109-0602*

**Margot Zöller** *Department of Tumor Progression and Immune Defense, German Cancer Research Center, Im Neuenheimer Feld 506, 69120 Heidelberg, Germany*

**Peter C. Weber** *Institut für Prophylaxe und Epidemiologie, der Kreislaufkrankheiten Ludwig-Maximilians-Universität München, Pettenkoferstrasse 9, 80336 Munich, Germany*

Adhesion molecules

| Receptors | Ligands/Counter-receptors |
|---|---|
| Integrins | |
| $\alpha_2\beta_1$ (glycoprotein Ia/IIa, VLA-2) | Collagen type I |
| $\alpha_{IIb}\beta_3$ (glycoprotein IIb/IIIa) | Fibrinogen, fibronectin, von Willebrand factor, vitronectin |
| $\alpha_v\beta_3$ (CD51, vitronectin receptor) | Vitronectin, fibrinogen, von Willebrand factor, thrombospondin |
| $\alpha_L\beta_2$ (CD11a/CD18, LFA-1) | ICAM-1 (CD54), ICAM-2 (CD102), ICAM-3 (CD50) |
| $\alpha_M\beta_2$ (CD11b/CD18, Mac-1, Mo1, CR3) | ICAM-1 (CD54) |
| $\alpha_X\beta_2$ (CD11c/CD18, p150/95) | Complement component iC3b, fibrinogen |
| $\alpha_4\beta_1$ (CD49d/CD29, VLA-4) | VCAM-1, fibronectin |
| $\alpha_4\beta_7$ (LPAM-1, CD49d/CD$^-$) | MAdCAM-1, VCAM-1, fibronectin |
| $\alpha_{IEL}\beta_7$ | |
| Selectins | |
| P-Selectin (CD62P, GMP-140, PADGEM) | PSGL-1 |
| E-Selectin (CD62E, ELAM-1) | PSGL-1 |
| L-Selectin (CD62L, LECAM-1) | GlyCAM-1, CD34 |

TOPICS IN MOLECULAR MEDICINE
*Volume 1*

---

# Adhesion Molecules and Cell Signaling:
## Biology and Clinical Applications

*Topics in Molecular Medicine, Volume 1,*
edited by Wolfgang Siess, Reinhard Lorenz,
and Peter C. Weber. Raven Press, Ltd.,
New York © 1995.

# 1

# Comparison of Minimum α-Chain Cytoplasmic Tail Sequences Needed to Support Integrin Adhesion and Localization Functions

Martin E. Hemler, Paul D. Kassner, and Satoshi Kawaguchi

*The Dana Farber Cancer Institute, Harvard Medical School, Boston, Massachusetts 02115*

Integrin-mediated adhesion is involved in a wide variety of biological processes, including hematopoietic differentiation (1–3), leukocyte trafficking (4–6), wound healing (7), tumor cell metastasis (8), and muscle development (9). The adhesive activity of integrin heterodimers is highly controlled, as exemplified by T-cell activation (10–12), keratinocyte differentiation (13), and platelet activation (14–16). Regulation of integrin function proceeds by an unknown mechanism, termed inside-out signaling, that may cause conformational changes that alter the affinity of integrin for ligand (17). The cytoplasmic domains of both α- and β-chains are candidates for translation of signals from cytosolic elements into conformational changes within the extracellular domains.

The cytoplasmic domains of both α- (18–22) and β- (23–26) chains influence integrin-mediated adhesion, presumably by altering ligand binding affinity (18). Deletion of either the $\alpha^4$ (19), $\alpha^2$ (20), or $\alpha^6$ (22) cytoplasmic domain causes loss of both constitutive adhesive activity and phorbol ester-stimulated activity mediated by VLA-4, VLA-2, or VLA-6 in several different cellular environments. Exchange of the $\alpha^2$ or $\alpha^4$ cytoplasmic sequences with those from the $\alpha^2$, $\alpha^4$, or $\alpha^5$ sequences did not alter the levels of integrin-mediated adhesion (19,20,27), suggesting that these domains play a similar positive role in the regulation of integrin activity. Because the $\alpha^2$, $\alpha^4$, and $\alpha^5$ cytoplasmic domains have no obvious shared sequence motif, it has not been clear how they could make similar contributions to cell adhesion. In contrast to results seen with $\alpha^2$, $\alpha^4$, and $\alpha^6$, deletion of $\alpha^L$ (24), $\alpha^5$ (28), and $\alpha^1$ (29) cytoplasmic domains appeared to have no impact on integrin-mediated adhesion. Furthermore, deletion of $\alpha^{IIb}$ (18) caused an increase in adhesion mediated by the $\alpha^{IIb}\beta_3$ integrin. Therefore on the basis of these diverse results it has been difficult to derive general concepts applicable to most integrins.

Integrin α-chain tails also play key roles in regulating the outside-in signaling that occurs after ligand binding. For example, the type of α-chain tail can variably influence cell migration and collagen gel contraction (27). One post-ligand binding event that occurs for many integrins is the organization of focal adhesion complexes (FACs) (30). The integrin $\beta_1$-chain, by itself, has the capability to be recruited into FACs, whereas the α-chain alone does not (31). Instead, the $\alpha^{IIb}$ (32) and $\alpha^1$ (29) cytoplasmic domains appear to act as negative regulators, preventing integrin recruitment into pre-formed FACs. The critical regions within α-chains that confer negative regulatory activity have not yet been identified.

In this study we analyzed the mutated cytoplasmic domains of $\alpha^4$ and $\alpha^2$ and determined that relatively short cytoplasmic tail sequences are sufficient to enable positive regulation of cell adhesion. In addition, we utilized a series of $\alpha^2$ cytoplasmic tail mutants to identify the critical residues needed for regulation of ligand-independent recruitment into FACs. Thus, we were able to determine that there may be a relationship between the regulation of localization into FACs and the regulation of cell adhesion.

## RESULTS

### Analysis of Adhesion Mediated by $\alpha^4$ Cytoplasmic Deletion Mutants in MIP101 Cells

To identify specific residues involved in positive regulation of adhesion by the $\alpha^4$ cytoplasmic tail, wild-type $\alpha^4$ and multiple-truncated $\alpha^4$ constructs (Table 1, top) were expressed in the MIP101 cell line at comparable levels, as determined by flow cytometry (not shown). In cell adhesion assays, MIP101 cells transfected with the $\alpha^4$-974 construct (0 amino acids after GFFKR) displayed no attachment to CS1–BSA coated at 3 μg/ml (Fig. 1). In addition, no attachment to sVCAM-1 was observed at a coating concentration of 2.5 μg/ml (Fig. 1). In contrast, $\alpha^4$-980, $\alpha^4$-984, $\alpha^4$-990, and $\alpha^4$-997 transfectants (with 6, 10, 16, or 23 amino acids after GFFKR) showed pronounced adhesion to CS1–BSA and sVCAM-1, comparable to that seen with $\alpha^4$ wild-type transfectants (25 amino acids after GFFKR). Complete inhibition by the anti-$\alpha^4$ monoclonal antibody (MAb) HP1/2 indicated that adhesion was $\alpha^4$-dependent (not shown). On the basis of these results, it is clear that only five to ten amino acids following the GFFKR motif were sufficient to restore almost completely the constitutive adhesive function of $\alpha^4$ and that amino acids at positions 10–25 (after the GFFKR) did not cause substantial further changes in adhesive activity.

**TABLE 1.** *List of $\alpha^2$ and $\alpha^4$ mutants*

| $\alpha^4$ Constructs | Cytoplasmic tail sequence | CHO | MIP |
|---|---|---|---|
| X4C4 (wt) | KAGFFKRQYKSILQEENRRDSWSYINSKSNDD | + | + |
| X4C0 (974) | KAGFFKR | + | + |
| 976 | KAGFFKRQY | + | |
| 977 | KAGFFKRQYK | + | |
| 979 | KAGFFKRQYKSI | + | |
| 980 | KAGFFKRQYKSIL | + | + |
| 984 | KAGFFKRQYKSILQEEN | | + |
| 990 | KAGFFKRQYKSILQEENRRDSWS | | + |
| 997 | KAGFFKRQYKSILQEENRRDSWSYINSKSN | | + |
| $\alpha^2$ Constructs | Cytoplasmic tail sequence | CHO | |
| X2C2 (wt) | KLGFFKRKYEKMTKNPDEIDETTELSS | + | |
| X2C0 (1132) | KLGFFKR | + | |
| 1133 | KLGFFKRK | + | |
| 1134 | KLGFFKRKY | + | |
| 1136 | KLGFFKRKYEK | + | |
| 1139 | KLGFFKRKYEKMTK | + | |
| 1146 | KLGFFKRKYEKMTKNPDEIDE | + | |

## Fine Mapping of $\alpha^4$ Cytoplasmic Tail Sequences Involved in Regulation of Adhesion

To analyze further the functional importance of the membrane-proximal region in the $\alpha^4$ tail, additional $\alpha^4$ mutants containing termination codons within the 6-amino acid QYKSIL sequence were generated. These additional mutants (listed in Table 1) were expressed in CHO cells at similar levels as indicated by flow cytometry (not shown).

Whereas adhesion by $\alpha^4$-transfected MIP101 cells was analyzed in the presence of only a single dose of divalent cations, $\alpha^4$-transfected CHO cells were analyzed over a range of seven or eight different divalent cation concentrations. This approach enabled the determination of divalent cation $ED_{50}$ values, equivalent to the level of divalent cation needed for one-half maximal adhesion. Notably, $ED_{50}$ values for $\alpha^4$-CHO-974 were two- to threefold greater than those of CHO-X4C4 on CS1–BSA or on sVCAM-1, respectively. Thus, CHO-974 cells, lacking an $\alpha^4$ cytoplasmic domain, utilized divalent cations less efficiently to support adhesion. The $\alpha^4$-CHO-976 and CHO-977 cells showed intermediate $ED_{50}$ values, indicating that only two or three additional amino acids were able to cause partial recovery of cation usage during cell adhesion to CS1–BSA and to sVCAM-1 mediated by $\alpha^4\beta_1$. The presence of only five or six $\alpha^4$ amino acids after the GFFKR region restored $\alpha^4$-mediated adhesion to nearly its full activity, yielding $ED_{50}$ values almost as low as those observed for X4C4 adhesion to CS1 peptide and sVCAM-1. These results obtained with CHO cells (summarized in Fig. 2) are consistent with those seen with MIP-101 cells (Fig. 1).

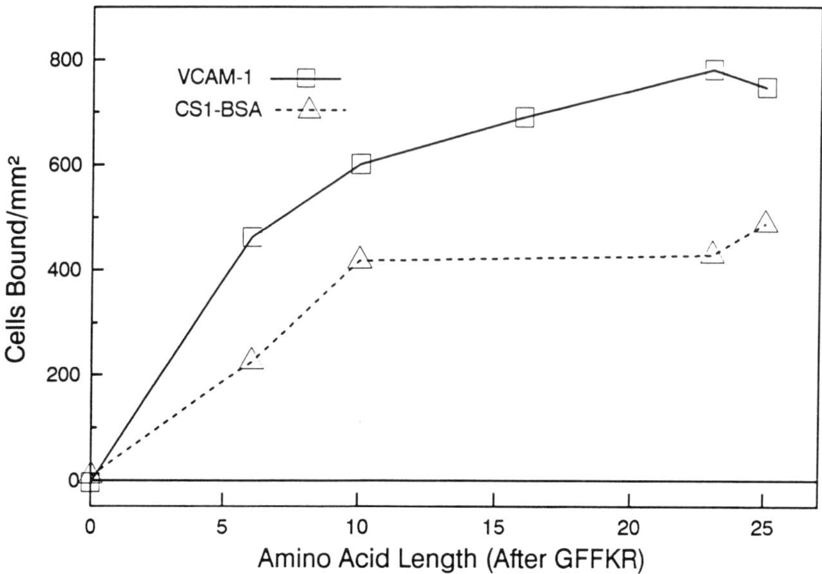


**FIG. 1.** Adhesion of MIP101 cells containing $\alpha^4$ cytoplasmic truncation mutants to CS1–BSA and sVCAM-1. Transfected cells were obtained and adhesion assays were carried out as described (37), using RPMI medium and 0.1% BSA. MIP101 cells expressing the $\alpha^4$ truncation mutants were assayed for their adhesion to CS1–BSA coated at 3 μg/ml or to sVCAM-1 coated at 2.5 μg/ml. The means of triplicate determinations are indicated, with an SD of 5% to 15% (not shown).

For $\alpha^4$-transfected CHO cells, adhesion to sVCAM-1 and CS1–BSA was entirely dependent on $\alpha^4$ expression because neotransfectants showed no adhesion to either substrate (not shown). Furthermore, adhesion to sVCAM-1 and CS1–BSA was inhibited (by 85% and 92%, respectively) by the $\alpha^4$-specific MAb HP1/2, even when a very high concentration of $Mg^{2+}$ (30 m*M*) was used (data not shown).

Although the deletion results suggested that the QYKSIL region in $\alpha^4$ may

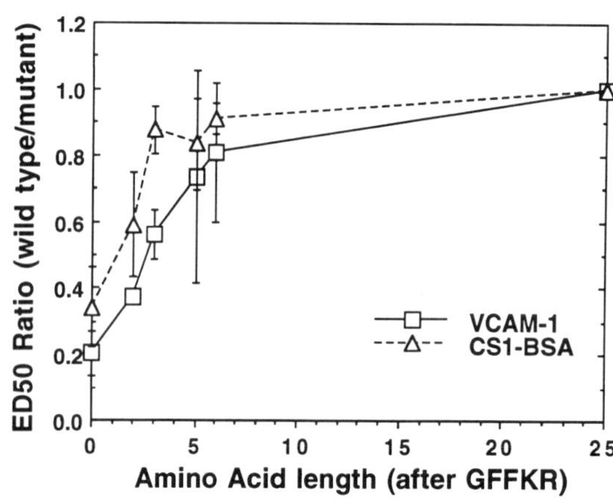


**FIG. 2.** Adhesion as a function of length of $\alpha^4$ chain cytoplasmic tail. For each mutant and wild-type $\alpha^4$-CHO transfectant, adhesion to CS1–BSA (2.5 μg/ml) and sVCAM-1 (2.0 μg/ml) was determined using 7–8 $Mg^{2+}$ concentrations, and then divalent cation $ED_{50}$ values were determined as previously described (37). Each data point represents the mean ± SD determined from at least three separate $ED_{50}$ ratio (wild-type: mutant) calculations.

be a critical motif needed for adhesion, point mutations in that region (Y976A, Y976F, K977A, or S978A) did not cause a significant loss of adhesive activity, as evidenced by $ED_{50}$ values similar to wild-type $\alpha^4$ $ED_{50}$ values (not shown). Furthermore, replacement of the entire $\alpha^4$ tail with that of $\alpha^2$ (to give an X4C2 construct) did not cause a marked alteration in $ED_{50}$ values.

### Analysis of $\alpha^2$ Cytoplasmic Tail Sequences Involved in Regulation of Adhesion

To determine if another integrin α-subunit would behave like $\alpha^4$, CHO cells were also transfected with integrin $\alpha^2$ constructs containing cytoplasmic domain deletions of various lengths (Table 1). Each construct was expressed in CHO cells at comparable cell surface levels (not shown), and adhesion to collagen was assayed over a range of $Mg^{2+}$ concentrations. In seven experiments, the $ED_{50}$ values for CHO-1132 cells (also called CHO-X2CO) were elevated by an average of 11-fold compared to CHO-X2C2.

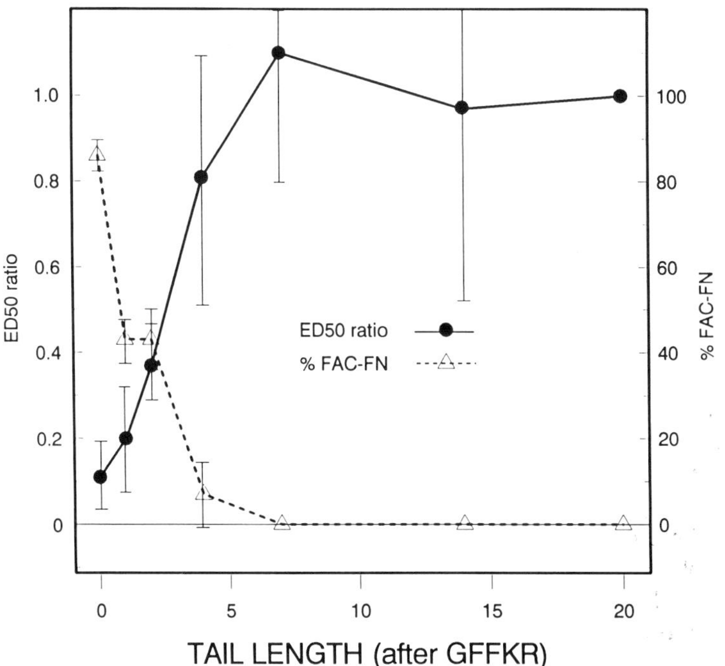


**FIG. 3.** Inverse correlation between the adhesive activity of $\alpha^2$ mutants and their localization in FACs on fibronectin. The relative adhesive activities of the various mutant $\alpha^2$-CHO cells (reported as wild-type:mutant $ED_{50}$ ratios) and the percentage of each of the mutant $\alpha^2$-CHO cells that localized in FACs on fibronectin are shown relative to the number of $\alpha^2$ amino acids present after GFFKR.

Notably, $ED_{50}$ values obtained with CHO-1133 and CHO-1134 cells were intermediate between CHO-X2C2 and CHO-X2C0 cells. Thus, only one or two amino acids after the GFFKR motif were sufficient to cause partial recovery of adhesive activity. The presence of only four amino acids after GFFKR ($\alpha^2$-1136) was sufficient to cause a further decrease in the $ED_{50}$ value, and seven ($\alpha^2$-1139) or 14 ($\alpha^2$-1146) amino acids following the GFFKR motif yielded adhesion results (and $ED_{50}$ values) not significantly different from X2C2 (Fig. 3). Replacement of the entire $\alpha^2$ cytoplasmic domain with that of $\alpha^4$ or $\alpha^5$ (X2C4 and X2C5 constructs) did not cause a major change in $ED_{50}$ values (not shown).

Adhesion of CHO-$\alpha^2$ transfectants to collagen was entirely $\alpha^2$-dependent, since CHO-p901 (mock transfectant) showed no adhesion to collagen, and adhesion of CHO-X2C2 was completely inhibited by the anti-$\alpha^2$ MAb 5E8 under conditions of maximal adhesion (data not shown).

## $\alpha^2$ Cytoplasmic Domain Effects on Integrin Localization into Focal Adhesion Complexes

After cell spreading of $\alpha^2$-CHO transfectants on collagen, both wild-type $\alpha^2$ and fully deleted $\alpha^2$-1132 showed similar localization into typical focal adhesion complexes (FACs; not shown). To determine whether the cytoplasmic domain of $\alpha^2$ plays a role in the regulation of indiscriminate recruitment of VLA-2 into focal adhesion sites formed by other integrins, we examined wild-type $\alpha^2$ (X2C2) and fully deleted $\alpha^2$ (X2C0) in CHO cells spread on fibronectin. In this case, wild-type VLA-2 in CHO cells failed to localize into focal adhesion sites and remained diffuse, whereas X2C0 ($\alpha^2$-1132) obviously localized into FACs formed on fibronectin. Although the distributions of mutant and wild-type $\alpha^2$ were strikingly different, CHO cells in both cases were able to form FACs, as indicated by staining with an anti-phosphotyrosine antibody (not shown). In another experiment, the X2C0 protein, but not wild-type $\alpha^2$, was recruited into FACs formed when CHO cells spread on vitronectin (not shown). Therefore, the $\alpha^2$ subunit cytoplasmic domain prevents VLA-2 from being recruited into focal adhesion sites formed by other integrins.

To identify the specific residues in the $\alpha^2$ cytoplasmic domain responsible for regulation of VLA-2 localization on fibronectin, we evaluated each of the $\alpha^2$ cytoplasmic domain mutants shown in Table 1. The $\alpha^2$-1133 and $\alpha^2$-1134 mutants did localize into FACs in CHO cells spread on fibronectin, but when four amino acids were present after the GFFKR motif (as in $\alpha^2$-1136), immunoreactivity of anti-$\alpha^2$ MAb at focal adhesion sites was markedly decreased. Other deletion mutants with more amino acids, such as $\alpha^2$-1139 and $\alpha^2$-1146, also were excluded from FACs, even though such sites were clearly present, as indicated by staining of phosphotyrosine. The percentage of

CHO cells showing $\alpha^2$ or mutant $\alpha^2$ localization into fibronectin FACs is indicated in Fig. 3. Notably, there is a striking inverse correlation between localization into FACs and regulation of cell adhesion.

## DISCUSSION

We have demonstrated here that only a short $\alpha^4$ or $\alpha^2$ cytoplasmic domain sequence of five to seven amino acids (immediately following the GFFKR motif) is required for a positive effect on cell adhesion. The results from deletion, point mutation, and chimera experiments indicate that diverse amino acids can contribute, and these results argue against a highly specific positive regulatory motif within α-chain tails. Elsewhere (19,20,27), the diverse cytoplasmic domain sequences of $\alpha^2$, $\alpha^4$, and $\alpha^5$ supported similar levels of constitutive adhesive activity, and the $\alpha^{6A}$ and $\alpha^{6B}$ cytoplasmic tails exerted comparable positive effects on adhesion (22,33,34), despite having totally different sequences after the GFFKR motif. Similarly, only seven amino acids after the GFFKR region in the $\alpha^V$ sequence were sufficient to support almost maximal adhesion (35), even though those seven amino acids (VRPPQEE) bear no resemblance to the functionally active sequences after GFFKR in $\alpha^4$ (QYKSILQ) or in $\alpha^2$ (KYEKMTK).

Our results also argue against an essential role for integrin α-chain phosphorylation in the regulation of cell adhesion. Although sometimes correlated with increased adhesive activity, phosphorylation is clearly not essential, because mutation of potential sites within the critical $\alpha^4$ QYKSIL sequence had no effect on adhesion.

We hypothesize that five- to seven-α-chain amino acids are sufficient to block a negative regulatory region in the $\beta_1$ tail. In this regard, the length of α chain tail needed to exclude $\alpha^2$ from ligand-independent focal adhesion sites correlates closely with the tail length that makes a positive contribution toward cell adhesion. Thus, the same regulatory site on the β-chain that is exposed by α chain deletion could conceivably cause both diminished cell adhesion and ligand-independent focal adhesion formation. For example, the same cytoskeletal interactions that lead to FAC formation may alter either the conformation or the distribution of the integrin, thus resulting in diminished cell adhesion. In this regard, it was recently shown that pre-existing variations in integrin cell surface distribution may determine its level of constitutive adhesive activity (36).

## ACKNOWLEDGMENT

We thank Dr. Roy Lobb (Biogen, Cambridge, MA) for providing sVCAM-1 and Dr. Tadashi Shimo-Oka (Iwaki Glass Co., Tokyo, Japan) for donating CS1-BSA. This work was supported by National Institutes of

Health Grant GM46526 and by funding from the DFCI-Sandoz Drug Discovery program.

## REFERENCES


1. Miyake K, Weissman IL, Greenberger JS, Kincade PW. Evidence for a role of the integrin VLA-4 in lympho-hemopoiesis. *J Exp Med* 1991;173:599–607.
2. Utsumi K, Sawada M, Narumiya S, et al. Adhesion of immature thymocytes to thymic stromal cells through fibronectin molecules and its significance for the induction of thymocyte differentiation. *Proc Natl Acad Sci USA* 1991;88:5685–89.
3. Roldán, García-Pardo A, Brieva JA. VLA-4-fibronectin interaction is required for the terminal differentiation of human bone marrow cells capable of spontaneous and high rate immunoglobulin secretion. *J Exp Med* 1992;175:1739–47.
4. Issekutz TB, Wyrkretowicz A. Effect of a new monoclonal antibody, TA-2, that inhibits lymphocyte adherence to cytokine stimulated endothelium in the rat. *J Immunol* 1991;147: 109–16.
5. Baron JL, Madri JA, Ruddle NH, Hashim G, Janeway CA. Surface expression of α4 integrin by CD4 T cells is required for their entry into brain parenchyma. *J Exp Med* 1993;177:57–68.
6. Weg VB, Williams TJ, Lobb RR, Nourshargh S. A monoclonal antibody recognizing very late activation antigen-4 inhibits eosinophil accumulation in vivo. *J Exp Med* 1993;177: 561–6.
7. Schiro J, Chan BMC, Roswit WT, et al. Integrin α2β1 (VLA-2) mediates reorganization and contraction of collagen matrices by human cells. *Cell* 1991;67:403–10.
8. Chan BMC, Matsuura N, Takada Y, Zetter BR, Hemler ME. In vitro and in vivo consequences of VLA-2 expression on rhabdomyosarcoma cells. *Science* 1991;251:1600–2.
9. Rosen GD, Sanes JR, LaChance R, Cunningham JM, Roman J, Dean DC. Roles for the integrin VLA-4 and its counter receptor VCAM-1 in myogenesis. *Cell* 1992;69:1107–19.
10. Dustin ML, Springer TA. T-cell receptor cross-linking transiently stimulates adhesiveness through LFA-1. *Nature* 1989;341:619–24.
11. Van Kooyk Y, Van DeWiel-Van Kemenade P, Weder P, Kuijpers TW, Figdor CG. Enhancement of LFA-1-medaited cell adhesion by triggering through CD2 or CD3 on T lymphocytes. *Nature* 1989;342:811–3.
12. Shimizu Y, Van Seventer GA, Horgan KJ, Shaw S. Regulated expression and binding of three VLA (β1) integrin receptors on T cells. *Nature* 1990;345:250–3.
13. Adams JC, Watt FM. Changes in keratinocyte adhesion during terminal differentiation: reduction in fibronectin binding precedes α5β1 integrin loss from the cell surface. *Cell* 1990;63:425–35.
14. Marguerie GA, Plow EF, Edgington TS. Human platelets possess an inducible and saturable receptor specific for fibrinogen. *J Biol Chem* 1979;254:5357–63.
15. Bennett JS, Vilaire G. Exposure of platelet fibrinogen receptors by ADP and epinephrine. *J Clin Invest* 1979;64:1393–1401.
16. Du X, Plow EF, Frelinger AL, O'Toole TE, Loftus JC, Ginsberg MH. Ligands "activate" integrin $\alpha_{IIb}\beta_3$ (platelet GPIIb-IIIa). *Cell* 1991;65:409–16.
17. Ginsberg MH, Du X, Plow EF. Inside-out integrin signalling. *Curr Opin Cell Biol* 1992;4: 766–71.
18. O'Toole TE, Mandelman D, Forsyth J, Shattil SJ, Plow EF, Ginsberg MH. Modulation of the affinity of integrin αIIbβ3 (GPIIb-IIIa) by the cytoplasmic domain of αIIb. *Science* 1991;254:845–7.
19. Kassner PD, Hemler ME. Interchangeable alpha chain cytoplasmic domains play a positive role in control of cell adhesion mediated by VLA-4, a $\beta_1$-integrin. *J Exp Med* 1993;178: 649–60.
20. Kawaguchi S, Hemler ME. Role of the α subunit cytoplasmic domain in regulation of adhesive activity mediated by the integrin VLA-2. *J Biol Chem* 1993;268:16279–85.
21. Rabb H, Michishita M, Sharma CP, Brown D, Arnaout MA. Cytoplasmic tails of human

complement receptor type 3 (CR3, CD11b/CD18) regulate ligand avidity and the internalization of occupied receptors. *J Immunol* 1993;151:990–1002.
22. Shaw LM, Mercurio AM. Regulation of α6β1 integrin laminin receptor function by the cytoplasmic domain of the α6 subunit. *J Cell Biol* 1993;123:1017–25.
23. Hayashi Y, Haimovich B, Reszka A, Boettiger D, Horwitz A. Expression and function of chicken integrin beta-1 subunit and its cytoplasmic domain mutants in mouse NIH 3T3 cells. *J Cell Biol* 1990;110:175–84.
24. Hibbs ML, Xu H, Stacker SA, Springer TA. Regulation of adhesion to ICAM-1 by the cytoplasmic domain of LFA-1 integrin beta subunit. *Science* 1991;251:1611–3.
25. Hibbs ML, Jakes S, Stacker SA, Wallace RW, Springer TA. The cytoplasmic domain of the integrin lymphocyte function-associated antigen 1 β subunit: sites required for binding to intercellular adhesion molecule 1 and the phorbol ester-stimulated phosphorylation site. *J Exp Med* 1991;174:1227–38.
26. Chen YP, Djaffar I, Pidard D, Steiner B, Cieutat AM, Caen JP, Rosa JP. Ser-752 → Pro mutation in the cytoplasmic domain of integrin $\beta_3$ subunit and defective activation of platelet integrin $\alpha_{IIb}\beta_3$ (glycoprotein IIb-IIIa) in a variant of Glanzmann thrombasthenia. *Proc Natl Acad Sci USA* 1992;89:10169–73.
27. Chan BMC, Kassner PD, Schiro JA, Byers HR, Kupper TS, Hemler ME. Distinct cellular functions mediated by different VLA integrin α subunit cytoplasmic domains. *Cell* 1992; 68:1051–60.
28. Bauer JS, Varner J, Schreiner C, Kornberg L, Nicholas R, Juliano RL. Functional role of the cytoplasmic domain of the integrin α5 subunit. *J Cell Biol* 1993;122:209–21.
29. Briesewitz R, Kern A, Marcantonio EE. Ligand-dependent and -independent integrin focal contact localization: the role of the α chain cytoplasmic domain. *Mol Biol Cell* 1993;4:593–604.
30. Turner CE, Burridge K. Transmembrane molecular assemblies in cell-extracellular matrix interactions. *Curr Opin Cell Biol* 1991;3:849–53.
31. LaFlamme SE, Akiyama SK, Yamada KM. Regulation of fibronectin receptor distribution. *J Cell Biol* 1992;117:437–47.
32. Ylänne J, Chen Y, O'Toole TE, Loftus JC, Takada Y, Ginsberg MH. Distinct functions of integrin α and β subunit cytoplasmic domains in cell spreading and formation of focal adhesions. *J Cell Biol* 1993;122:223–33.
33. Delwel GO, Hogervorst F, Kuikman I, Paulsson M, Timpl R, Sonnenberg A. Expression and function of the cytoplasmic variants of the integrin α6 subunit in transfected K562 cells. *J Biol Chem* 1993;268:25865–75.
34. Shaw LM, Lotz MM, Mercurio AM. Inside-out integrin signalling in macrophages: analysis of the role of the α6Aβ1 and α6Bβ1 integrin variants in laminin adhesion by cDNA expression in an α6 integrin-deficient macrophage cell line. *J Biol Chem* 1993;268:11401–8.
35. Filardo EJ, Cheresh DA. A β turn in the cytoplasmic tail of the integrin αv subunit influences conformation and ligand binding of αvβ3. *J Biol Chem* 1994;269:4641–7.
36. Van Kooyk Y, Weder P, Heije K, Figdor CG. Extracellular $Ca^{2+}$ modulates leukocyte function-associated antigen-1 cell surface distribution on T lymphocytes and consequently affects cell adhesion. *J Cell Biol* 1994;124:1061–70.
37. Kassner PD, Kawaguchi S, Hemler ME. Minimum α chain sequence needed to support integrin-mediated adhesion. *J Biol Chem* 1994;269:19859–67.

*Topics in Molecular Medicine, Volume 1,* edited by Wolfgang Siess, Reinhard Lorenz, and Peter C. Weber. Raven Press, Ltd., New York © 1995.

# 2

# The Role of Intercellular Adhesion Molecules 1, 2, and 3 in Cell Recognition and Signaling

David L. Simmons

*Cell Adhesion Laboratory, Institute of Molecular Medicine, John Radcliffe Hospital, Headington, Oxford, OX3 9DU, England*

## THE CELLULAR CONTEXT

The professional career of leukocytes involves the development of multiple but brief interactions with many different cell types. Leukocytes are the global business travelers of the cellular world. By contrast, adherent cells comprising the mass of solid tissues—fibroblasts, muscle cells, and epithelial cells—lead quiet and sedentary lives, apart from the volcanic eruptions of embryonic morphogenesis. In the adult, the cells in solid tissues contact only a small number of other cells during their lives. For example, colon epithelial cells are the progeny of crypt stem cells and then differentiate and migrate towards the tip of the villus before being sloughed off. Again, keratinocytes originate from basal stem cells and then migrate upwards through suprabasal layers to eventual terminal differentiation in epidermal entombment. Epithelial cells largely converse with other epithelial cells. Leukocytes, however, get to converse with everyone. Through constant circulation, adhesion to vascular endothelium, transmigration, movement over and around all cell types in tissues, re-entry into lymphatics, and homing to specialized lymphoid tissues, leukocytes are exposed to a cellular metropolis.

Given this lifestyle, it is not surprising that leukocytes have evolved with an extensive armamentarium of surface proteins whose job includes cell recognition, adhesion, and signaling. There are two key elements to leukocyte behavior: recognize who you are talking to and act on that knowledge, preferably very quickly.

The development of monoclonal antibodies (MAbs) expedited the study of the phenotype of the leukocyte surface. These allowed not only structural definition but functional assessment of individual proteins. There are now almost 130 defined surface proteins on leukocytes (CD1 to CD130, barring a few gaps). A surprising number of these play a role in intercellular adhesion; surprising, as the prevailing paradigm in the mid-1980s was that cytokines would be the major influences on leukocyte responses. It was not anticipated that so many of the leukocyte surface proteins would bind other surface proteins.

## INTERCELLULAR ADHESION MOLECULES: THE HISTORICAL CONTEXT

Early work by Springer's group (1) led to the definition of a set of proteins involved in interleukocyte adhesion and function, the so-called leukocyte function-associated antigens LFA-1 (CD11a/CD18), LFA-2 (CD2), and LFA-3 (CD58). MAbs to these antigens blocked a wide range of interleukocyte and leukocyte–target cell interactions in an antigen- and HLA-independent manner. Through the work of Rothlein and Springer (2), it became clear that LFA-1 did not function as a homotypic adhesin, as LFA-1$^+$ cells could bind to LFA-1$^-$ cells (derived from LAD-I patients congenitally deficient in $\beta_2$ integrin expression) in an in vitro assay for cell aggregation induced by phorbol esters. This was the first indication that LFA-1 bound to a distinct protein. Springer's lab (3) set about defining this protein by systematically screening for non–LFA-1 MAbs that blocked LFA-1–dependent adhesion. The prototypic MAb RR1 defined a new molecule called intercellular adhesion molecule 1 or ICAM-1 (CD54). This was clearly a major ligand for LFA-1, and the cDNAs encoding it were isolated in 1988 (4,5).

However, the seeds of future work were sown in the original paper in 1986 (3), as there were cell lines, especially SKW3, whose LFA-1–dependent adhesion processes could not be inhibited by the anti–ICAM-1 MAb. This was an early indication of additional, non–ICAM-1 ligand(s) for LFA-1. Work in the related field of leukocyte–endothelial cell adhesion had pointed to the existence of non–ICAM-1 ligands for LFA-1, as LFA-1$^+$ leukocytes could bind via LFA-1 to resting endothelium that did not express ICAM-1. Springer defined this second path, this time by employing an innovative development of the transient expression/panning technology introduced by Brian Seed in 1987. Staunton et al. (6) cloned this non–ICAM-1 ligand by screening for LFA-1–dependent adhesion in the presence of the prototypic blocking anti–ICAM-1 MAb RR1, thus preventing the recloning of this ligand. Thus, ICAM-2 (CD102) was born in 1989 (6).

With the development of adhesion-blocking anti–ICAM-2 MAbs in 1991 (7), it became clear that there remained an anti–ICAM-1/anti–ICAM-2-resistant component of LFA-1–dependent adhesion. Springer, again, set up LFA-1 adhesion blockade screens to identify MAbs to this remaining element. MAbs were isolated that defined a third ICAM, ICAM-3 (8). Now, for the first time, using a range of cell lines, combinations of anti–ICAM-1, anti–ICAM-2, and anti–ICAM-3 MAbs could completely block all LFA-1–dependent adhesion (8). In 1992 and 1993, three labs employing completely different strategies [transient expression (9), degenerate PCR searching for ICAM-1–related molecules (10), and purification of the ICAM-3 protein and microsequencing (11)] independently cloned ICAM-3. Since that time no new ICAMs have been cloned or hinted at. However, in politics, never say "never," and in biology, never say you understand anything fully; it would be foolish to say that there will be no additional ICAMs or additional ligands for LFA-1. Indeed, a recent report by Linsley and Clark (12) indicates that the IgSF member CTLA-4, known to be a ligand for B7.1 and B7.2, can also act as a ligand for LFA-1.

## MOLECULAR BIOLOGY OF THE ICAMs

### Gene and Protein Organization

All three ICAMs are members of the Ig superfamily. ICAM-1 consists of five Ig domains, ICAM-2 of two domains, and ICAM-3 of five domains. All the ICAMs consist of the short C2-type domain consisting of seven β strands (Fig. 1).

The ICAM-1 and ICAM-2 genes have a relatively simple organization. Each Ig domain is encoded in a single exon, with signal peptide and transmembrane domains and cytoplasmic domains in separate exons (13,14). The genes are medium sized (25 to 30 kb). The ICAM-3 gene structure has not been determined, but preliminary data indicate that the overall exon organization and even the sizes of some of the introns are conserved be-

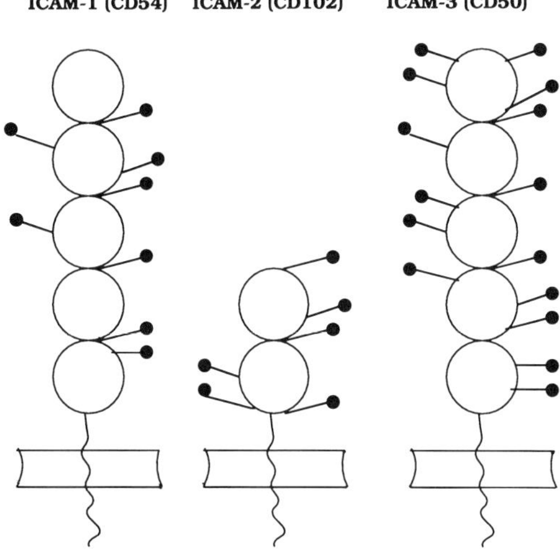


**FIG. 1.** Molecular structure of ICAM-1 (CD54), ICAM-2 (CD102), and ICAM-3 (CD50). ●, Potential N-linked glycan site.

tween ICAM-1 and ICAM-3 (D. Bossy and D. Simmons, unpublished observations).

cDNAs for the ICAMs have been cloned from a number of species, reflecting the interest in the development of animal models for in vivo therapies. ICAM-1 has been cloned from humans (4,5), chimpanzees (15), mice (16), rats (17), and dogs (15). ICAM-2 has been cloned from humans (6) and mice (14). ICAM-3 has been cloned from humans (9–11), mice, and rats (M. Gallatin, unpublished observations).

ICAM-1 and ICAM-3 are more closely related than either is to ICAM-2, indicating that these arose from a more recent gene duplication and that ICAM-2 arose somewhat earlier. Indeed, ICAMs 1 and 3 map very close together on chromosome 19p13.3-13.2 (18,19), whereas ICAM-2 maps to chromosome 17q23-25 (20). Furthermore, the genes for ICAM-1 and ICAM-3 map within only a few hundred kb of each other (D. Bossy and D. Simmons, unpublished observations).

The most closely related domains are the N-terminal ones. There is a gradient of homology, with decreasing similarity toward the C terminus. The LFA-1 binding site has been mapped to domain 1, so this may not seem unusual. However, between ICAM-1 and ICAM-3 there is higher than expected homology between domain 2 and half of domain 3. Thus far these have not been implicated in LFA-1 engagement, so it is intriguing that they should remain so conserved in an apparently nonfunctional area. One exciting postulation is that this region does in fact define a functional binding site for either LFA-1 or an as yet uncharacterized ligand(s). It is possible that the ICAMs bind other molecules in addition to leukocyte integrins LFA-1 and Mac-1. We are currently screening for additional ICAM ligands using ICAM-1, -2, and -3 Fc chimeric fusion proteins to biopan a range of cDNA expression libraries.

The cytoplasmic tails of the ICAMs are the most divergent region of the molecules. On this basis, it was suggested that the ICAMs might have evolved to subserve very different signaling or cytoskeletal roles. Indeed, some evidence indicates that the tail of ICAM-1 interacts with the actin-containing cytoskeleton and α-actinin (21). To date, no evidence exists for proteins interacting with the cytoplasmic tail of ICAM-2. However, many recent reports allude to and demonstrate a role for ICAM-3 in a number of signaling events. These are described more fully below.

The ICAMs are all heavily glycosylated proteins with up to one-half of the mass being due to oligosaccharides; most of this is N-linked. ICAM-3 is the most heavily N-glycan decorated. Diamond et al. (22) reported in 1991 that the N-linked glycans in domain 3 of ICAM-1 negatively regulated the ability of ICAM-1 to bind to Mac-1. Elimination of these glycans by mutagenesis of the N-X-S/T consensus sites or total N-glycan removal by enzymatic digestion or chemical inhibition resulted in increased Mac-1/ICAM-1 adhesion. Preliminary results (23) indicate that removal of the N-linked glycans from

ICAM-3 enables ICAM-3 to bind Mac-1. However, in contrast to ICAM-1, for which some degree of Mac-1 adhesion to the native fully glycosylated protein was detected, there was no binding of fully glycosylated ICAM-3 to Mac-1. Therefore, regulation of the glycosylation status of ICAM-3 could very tightly control the ability of this ICAM to bind to Mac-1; even a small degree of N-glycan decoration could reduce Mac-1 binding. This is particularly relevant to the control of neutrophil adhesion, as these cells are rich in both Mac-1 and a heavily and heterogeneously glycosylated form of ICAM-3. It will be interesting to see if the N-linked glycans are modified as leukocytes are activated, i.e., in a kinetically rapid manner, so that this may be a further way of controlling ICAM/integrin adhesion.

## Molecular Aspects of ICAM Gene Expression

Despite the high degree of homology at the level of protein sequence and overall domain organization, the expression profiles of the ICAMs are very different, suggesting that the enhancer/promoter sequences have evolved at a faster rate than the coding regions. This is especially true for ICAM-1 and ICAM-3, which seem to have arisen from a very recent gene duplication but have very different expression patterns. ICAM-1 is expressed at very low levels in resting leukocytes and at virtually undetectable levels on vascular endothelium and other cell types. However, ICAM-1 is rapidly upregulated by inflammatory cytokines such as interferon-$\gamma$, IL-1-$\beta$ and TNF-$\alpha$ on a wide range of cell types, including leukocytes, endothelium, keratinocytes, epithelial cells, and fibroblasts. Therefore, ICAM-1 can be viewed as a rapid-response ICAM, present at low levels in quiescent states but able to be induced rapidly under appropriate circumstances. In contrast, ICAM-2 is expressed at low levels by leukocytes, including platelets, and also by vascular endothelium, but is not responsive to cytokine stimulation. It is therefore a constitutively expressed ICAM.

ICAM-3 is similar to ICAM-2 in its constitutive profile of expression. However, two features make ICAM-3 worthy of special comment. First, ICAM-3 is expressed at very high levels on all resting leukocyte populations. Second, ICAM-3 expression is restricted to leukocytes and does not have the broader profiles of ICAM-1 and ICAM-2. The only exception to this limited profile, to date, is the intriguing report of ICAM-3 expression on vascular endothelium in greater than 50% of cases of lymphoma and myeloma (24). This may be due to two factors. The neovascularization occurring around the tumors could be of a different phenotype, yielding a novel expression of ICAM-3. Alternatively, as yet uncharacterized secreted factors deriving from the lymphoma or myeloma, e.g., a novel cytokine, could lead to specific expression of ICAM-3 by the new vessels developing in and around these tumors. These alternatives are presently under investigation.

Significant effort is being focused on understanding the regulation of ICAM-1 expression with a view to developing specific novel therapeutic targets (13). NF-κB has been identified as a key controlling transcription factor, but this is too broad in its uses as a controller of cytokine-regulated gene expression to be useful as a specific anti–ICAM-1 target. ICAM-1 has acquired a broader expression profile than the other ICAMs, so a comparison of the promoter sequences and transcription factor binding patterns of ICAM-1 and ICAM-3, which is leukocyte-specific, could yield valuable information about the basis of leukocyte-restricted gene expression.

## Molecular Dissection of the ICAM Binding Site(s)

Considerable effort has been focused on a definition of the binding surface(s) on ICAM-1, -2, and -3 for LFA-1 (25–27). In addition, for ICAM-1, the binding sites for Mac-1 (22), rhinovirus (25,28) and *Plasmodium falciparum*-infected erythrocytes (29,30) have also been dissected. The dominant contact points for ICAM-1, -2, and -3 with LFA-1 seem to reside in domain 1. Similarly, for ICAM-1 the key interactions with rhinovirus and *P. falciparum* are also located in the N-terminal domain. However, Mac-1 is thus far the exception in two respects. First, the key domain is domain 3. Second, binding is negatively regulated by the presence of N-linked glycans. The presence of N-linked glycans in domain 1 does not appear to affect LFA-1 interactions.

Returning to LFA-1, a common theme emerging from all the mutagenesis screens, MAb epitope maps, and more limited peptide inhibition studies is that residues located in a flat surface consisting of the C, F, and G β strands of the Ig fold contribute to the LFA-1 binding site (Fig. 2). In addition, at least for ICAM-1 and ICAM-3, a key residue in the F to G loop is also involved (25,27). Concurrent work on the interactions between the IgSF member VCAM and the integrin VLA-4 has also implicated residues in the CFG surface of VCAM domain 1 as the binding site (31,32). Furthermore, there appears to be a common, short linear motif (L D/E V) in the C′ loop that is an essential component of IgSF–integrin interactions (32). As recently argued, this may form the kernel of how the Ig domain binds to the active site of integrins, but other residues in different regions must provide the specificity, so that the ICAMs bind only to LFA-1 and VCAM binds only to VLA-4 (32). Binding of either the ICAMs or VCAMs to their cognate integrin receptors may involve a sequential two-stage process: an initial low-affinity docking interaction involving the common exposed C-loop motif (LDV), followed by a higher-affinity interaction bringing in other residues that impart specificity to the entire process. The definition of these "specificity" residues is a key area for research to allow the development of discriminatory peptide therapeutics targeted at either LFA-1 or VLA-4 (Fig. 3).

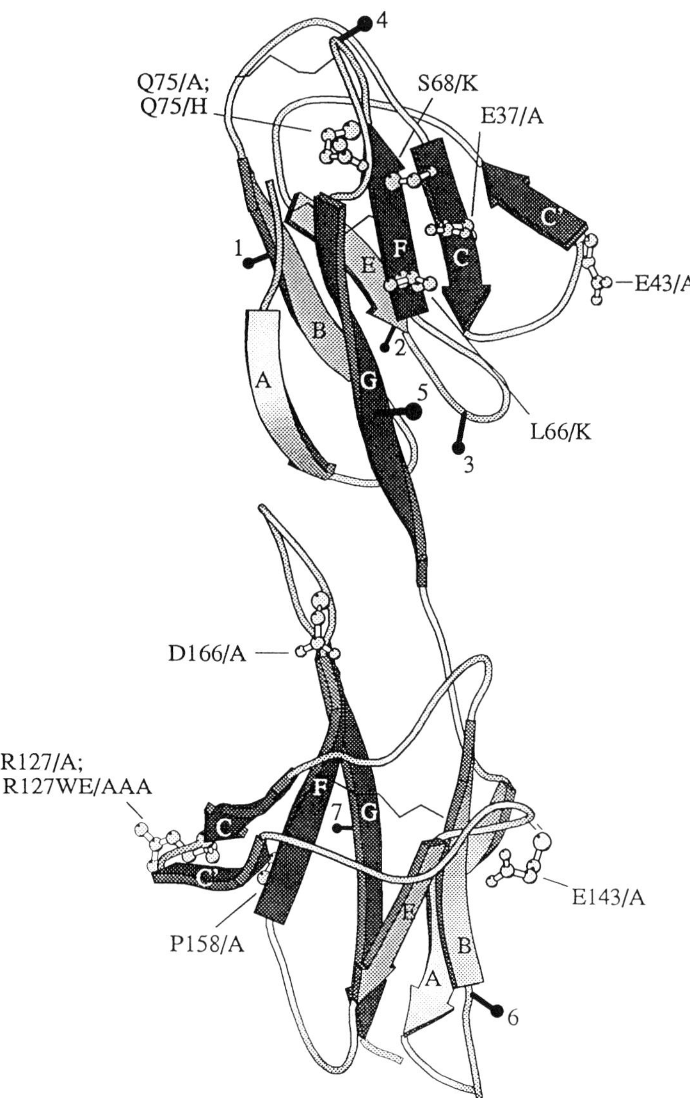


**FIG. 2.** Molecular model of domains 1 and 2 of ICAM-3 with key residues involved in LFA-1 binding indicated.

The resting state affinity of the ICAMs for LFA-1 is very low, estimated at $10^{-5}$ to $10^{-6}$ *M* (33). This resting affinity state is increased by regulating two different aspects of LFA-1 (33,34): first, the distribution of the integrin in the membrane, increasing the degree of clustering or aggregation and thus increasing avidity; and second, a conformational change in the integrin itself, leading to a higher intrinsic affinity for ligand engagement. The cellular

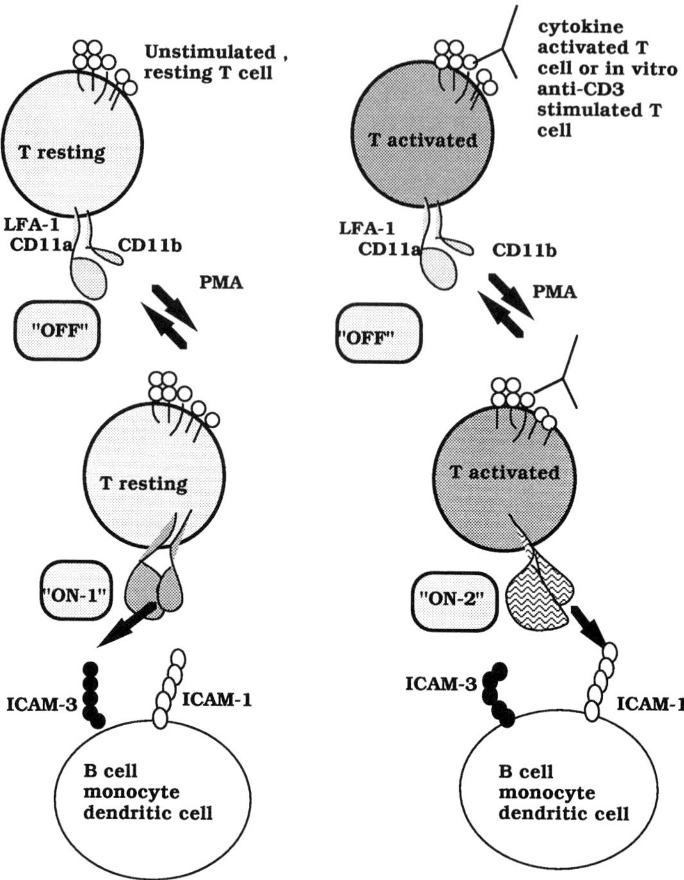


**FIG. 3.** Model for the differential roles of ICAM-1 and ICAM-3 on resting versus activated T cells. The model proposes that LFA-1 can exist in at least three affinity states: a low-affinity state, the "OFF" state; a high-affinity state that preferentially binds ICAM-3, the "ON-1" state; and a high-affinity state that preferentially binds ICAM-1, the "ON-2" state.

regulators of these events are by no means understood at present, but they can be modeled by a number of in vitro tools. These tools consist of two distinct types. First, the so-called "outside-inside" signals are thought to act by stabilizing a shape change or transition state of the extracellular domain of LFA-1. To date, outside-inside signalers are all anti–LFA-1 MAbs (e.g., MEM83, KIM185, NKI-L16) (35). In the near future, it is highly likely that small peptide agonists selected from peptide phage display libraries will emerge as equally effective mimetics. The second class of in vitro tools are so-called "inside-outside" signalers. This class includes activators of intracellular signal transduction pathways such as phorbol esters (34). We can mimic the process of LFA-1 activation with our in vitro tools, but we do not know the true nature of the normal regulators of this key leukocyte integrin. A major goal of ICAM/LFA research at present is the definition of the normal cellular correlates of each of these classes of regulators.

A recent study has found that the binding sites for ICAM-1 and ICAM-3 on LFA-1 may be subtly different (36). A screen of nearly 30 anti–LFA-1 MAbs for their effects on adhesion to either ICAM-1 or ICAM-3 revealed that two of the MAbs (YTH81.5 and 122.2A5) blocked binding to ICAM-3 but did not affect binding to ICAM-1. No MAbs blocked ICAM-1 but not ICAM-3. This differential blocking effect was not due to the steric effects of glycosylation. It could be argued that because ICAM-3 is much more heavily glycosylated than ICAM-1 the oligosaccharide chains affected accessibility to the binding site in LFA-1. However, the two MAbs still blocked the binding of fully deglycosylated ICAM-3 to LFA-1, and still did not affect the binding of fully deglycosylated ICAM-1 to LFA-1. Therefore, ICAM-1 and ICAM-3 can bind differentially to LFA-1, and it is possible that LFA-1 exists in different activation states that enable it to discriminate between the two ICAM ligands (see below).

## CELLULAR FUNCTIONS OF THE ICAMs

### Adhesion

Through the development of blocking MAbs to both LFA-1 and the ICAM-1, -2, and -3, it has been overwhelmingly established that the LFA-1/ICAM interaction plays a pivotal role in a wide range of leukocyte interactions, including those between antigen-presenting cells and T and B cells, T-helper cells and B cells, cytotoxic T cells and their targets, natural killer cells and their targets, and antibody-dependent cell-mediated cytotoxicity. In addition, ICAM-1 is utilized as a major adhesion path in the multistep cascade of leukocyte interaction with vascular endothelium. In leukocyte cell–cell interactions, the reciprocal binding of ICAM to LFA-1 provides key co-stimulatory signals that allow engagement of cellular effector programs, such as cytokine secretion and mitogenesis by $CD4^+$ helper T cells, or unleashing of killing functions by cytotoxic T cells.

Of course, the ICAM/LFA interaction is only one of a series of ligand/receptor interactions, including CD28 and CTLA4, which interact with B7.1 and B7.2, and CD2/LFA-3 and VCAM-1/VLA-4, which act in combination and synergistically to yield effective cellular function (37).

### Signaling

Since the mid-1980s the paradigm to rationalize the role of the ICAMs in cell–cell interactions has been that they serve solely as adhesion molecules, strengthening otherwise weak intercellular recognition events mediated by either B- or T-cell receptor complexes recognizing specific antigen or processed antigen in the context of class I and class II HLA complexes. The adhesion role is clearly established and is patently important to the successful outcome of each of these cell recognition events. However, it has only recently been appreciated that the ICAMs could also play a role as signal transducers as well as mediators of cell adhesion.

Recent work has shown that cross-linking of ICAM-1 can deliver a signal to neutrophils to yield an oxidative burst (38) and to T cells to activate expression of surface proteins (39). In addition, data are accumulating for a very strong role of ICAM-3 as a potent signal transducer. Cross-linking ICAM-3 produced a significant $Ca^{2+}$ flux and also association with tyrosine kinases $p56^{lck}$ and $p59^{fyn}$ (40).

Sanchez-Madrid's group has established that ICAM-3 can signal to LFA-1 to upregulate binding to ICAM-1 (41) and that ICAM-3 can also act as a

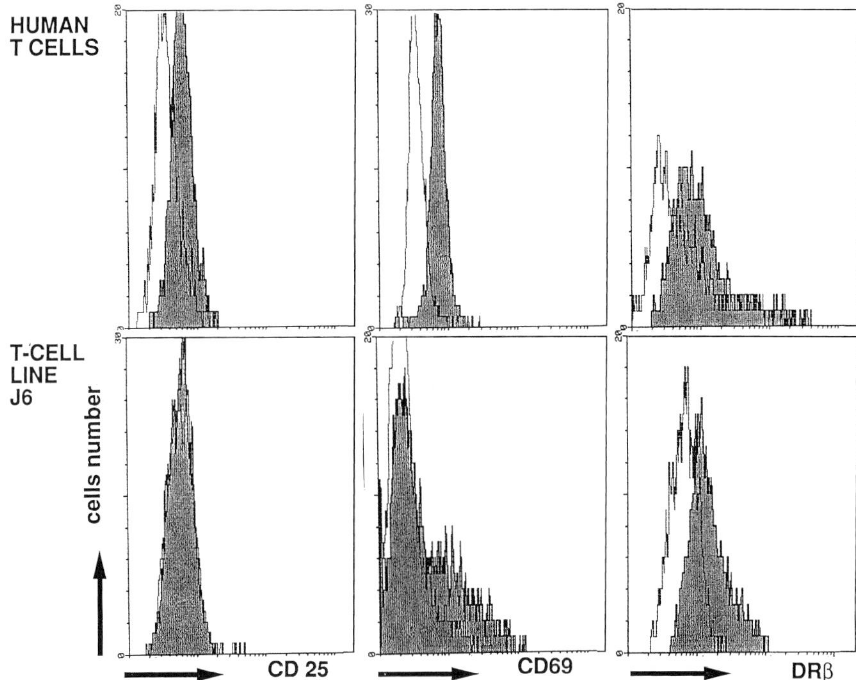


**FIG. 4.** Effect of anti–ICAM-3 monoclonal antibodies on the expression of surface proteins CD25 and CD69. Peripheral blood T cells **(top row)** and Jurkat (J6) cells **(bottom row)** were stimulated with immobilized anti–ICAM-3 MAb CH3.3, and the surface expressions of CD25, CD69 and class DRβ are indicated in shaded profiles. The expression profiles of cells stimulated with isotype-matched anti–ICAM-1 and VCAM-1 MAbs are indicated in open profiles. Anti–ICAM-1 MAbs had no stimulatory effects on CD25, CD69, or class II expression.

potent signaling molecule to upregulate expression of CD25 and CD69 in combination with submitogenic concentrations of anti-CD3 MAbs (42). Our own work has shown that quite different anti–ICAM-3 MAbs (both blockers of LFA-1 adhesion and nonblocking MAbs) can induce expression of both T-cell and B-cell activation markers in the presence of anti-CD3 MAbs (43) (Fig. 4).

### ICAM-3 and Initiation of Immune Responses

The recent cloning and biochemical characterization of ICAM-3 has led to the notion that this particular ICAM may play a key role as the primary ligand for LFA-1 when immune responses to foreign antigen are initiated. The experimental data supporting this claim are not at present definitive, but some data are intriguing and very suggestive.

Both ICAM-1 and -2 are present at low levels on resting leukocytes, whereas ICAM-3 is constitutively and abundantly expressed. In addition, ICAM-3 is expressed by "professional" antigen-presenting cells, such as Langherans cells in the skin and dendritic cells in lymphoid organs (44). ICAM-3 is expressed by these cells even in the resting, "unchallenged" state. In contrast, ICAM-1 either is not expressed or is expressed at much lower levels on these cells until the immune response has been initiated and cytokine release by T cells and monocytes has led to upregulation of ICAM-1 gene expression. Second, functional studies have shown that on tonsil dendritic cells, even though ICAM-1 and ICAM-2 are expressed, it is ICAM-3 that actually serves as the functional ligand for LFA-1 (45). In addition, resting T cells use ICAM-3 rather than ICAM-1 as the ligand for LFA-1 (8,41,42). Activated T-cell blasts use ICAM-1 in preference to ICAM-3. MAbs to ICAM-3 are potent activators of cell responses, both as inducers of homotypic cell–cell aggregation and as inducers of activation markers such as CD25, CD69, class II, and B7 (43). Cross-linking ICAM-1 has not been shown to be as potent an activation signal.

Following the cloning of ICAM-3 and the observation that the cytoplasmic tails of the three ICAMs were the most divergent domains of the molecules, it has become clear that the ICAMs serve different roles in intracellular programs. ICAM-1 has previously been shown to link to the cytoskeleton via α-actinin. As already mentioned, ICAM-1 can also converse with the cellular activation programs that lead to increased expression of CD25. Thus far, there are no data on the structural or functional activities of ICAM-2. In contrast, the recent work described above points to an exciting role for ICAM-3 in leukocyte activation.

In summary, the expression and functional data strongly support a role for ICAM-3 as the key ligand for leukocyte LFA-1 in initiating immune responses. ICAM-1 may take over later when the response is actively in progress.

## ICAMs AND DISEASE

### ICAMs and Inflammation

Because the ICAM/LFA interaction is a key element of successful interleukocyte recognition and activation, it greatly contributes to the initiation and development of inflammatory processes. The clinical corollaries of these biologic roles are now being explored by attempts to block the interaction, initially by using MAbs to both ICAM-1 and LFA-1 and, at a more experimental phase, anti-sense oligonucleotides to ICAM-1. There are some encouraging indications that anti-CD18 and anti–ICAM-1 MAb therapies will both prove effective in controlling inflammatory phases of allograft rejection, but full results from phase II trials are still not available. Given the recent discovery of ICAM-3, its role in initiating responses, and the differential binding activities of ICAM-1 and ICAM-3 for LFA-1, specific anti–ICAM-3-targeted therapies will be eagerly awaited.

### ICAMs and Cancer

The ICAMs are important in the development of cancer in two distinct ways. First, as key weapons in the leukocyte armory, they are involved in the successful interdiction of tumor targets. Second, however, the ICAMs could play a negative role, aiding tumor cell dissemination. In 1988, ICAM-1 was independently discovered as a surface molecule whose expression was elevated as melanocytes transformed to melanomas and continued to increase as they progressed to the metastatic state (46). In addition, in murine models the level of soluble ICAM-1 shed from colorectal carcinomas has been found to be a reliable indicator of overall tumor burden (47). The role of ICAM expressed by these tumors is not established. In one sense it is counterproductive to the tumor, as increased ICAM-1 expression should lead to increased attack by cytotoxic T cells and natural killer cells. However, it has been proposed but still not definitively proven that increased ICAM-1 expression by tumor cells allows leukocyte to attach to them, and thus the tumor cells can hitch a ride on the leukocytes as they adhere to vascular endothelium and diapedese, enabling them to spread to secondary sites. It is very important that we find out whether this mechanism is in fact used, so that anti-ICAM therapies for cancer can be tested and developed.

## SUMMARY

The ICAM family consists of three members. All are ligands for LFA-1, and ICAM-1 can also bind Mac-1. However, the complete repertoire of other ligands has not yet been defined. The ICAMs are not functionally redundant, as they have quite distinct expression profiles and can bind LFA-1 differentially. ICAM-3 is constitutively and abundantly expressed by all leukocytes, whereas ICAM-1 and ICAM-2 are expressed at very low levels. ICAM-1 is rapidly upregulated by inflammatory cytokines. A current working model for the interplay of the ICAMs is that ICAM-3 is the major ligand for LFA-1 in the initiation of immune responses and that ICAM-1 takes over after these early stages. ICAM-3 not only functions as an adhesion molecule but is clearly emerging as a potent signaling molecule.

## ACKNOWLEDGMENT

I thank all members of my laboratory: Dr. David Bossy, Dr. Christopher Buckley, Dr. Claire Holness, and Amanda Littler for published data, unpublished data, and discussions; Dr. Derek Hart for dendritic cell studies; Dr. Nancy Hogg and Dr. Clive Landis (Leukocyte Adhesion Lab), and Dr. Paul Bates (Biomolecular Modeling Lab, ICRF, London) for ICAM-3 mutagenesis and ICAM-1, 3/LFA-1 binding studies. Work in my laboratory is supported by the Imperial Cancer Research Fund. Dr. Christopher Buckley is supported by the Wellcome Trust.

## REFERENCES

1. Sanchez–Madrid F, Krensky AM, Ware CF, et al. Three distinct antigens associated with T lymphocyte-mediated cytolysis: LFA-1, LFA-2 and LFA-3. *Proc Natl Acad Sci USA* 1982;79:7489–93.
2. Rothlein R, Springer TA. The requirement for lymphocyte function associated antigen 1 in homotypic leukocyte adhesion stimulated by phorbol esters. *J Exp Med* 1986;161:1132–8.
3. Rothlein R, Dustin ML, Marlin SD, Springer TA. A human intercellular adhesion molecule (ICAM-1) distinct from LFA-1. *J Immunol* 1986;137:1270–6.
4. Simmons DL, Makgoba MW, Seed B. ICAM-1, an adhesion ligand of LFA-1, is homologous to the neural cell adhesion molecule NCAM. *Nature* 1988;331:624–7.
5. Staunton DE, Marlin SD, Stratowa C, Dustin ML, Springer TA. Primary structure of intercellular adhesion molecule 1 (ICAM-1) demonstrates interaction between members of the immunoglobulin and integrin supergene families. *Cell* 1988;52:925–33.
6. Staunton DE, Dustin ML, Springer TA. Functional cloning of ICAM-2, a cell adhesion ligand for LFA-1 homologous to ICAM-1. *Nature* 1989;339:361–4.
7. de Fougerolles AR, Stacker SA, Schwarting R, Springer TA. Characterisation of ICAM-2 and evidence for a third counter-receptor for LFA-1. *J Exp Med* 1991;174:253–63.
8. de Fougerolles AR, Springer TA. Intercellular adhesion molecule 3 a third counter-receptor for leukocyte function associated antigen 1 on resting lymphocytes. *J Exp Med* 1992;175: 185–90.
9. Fawcett J, Holness CLL, Needham LA, et al. Molecular cloning of ICAM-3, a third ligand for LFA-1, constitutively expressed on resting leukocytes. *Nature* 1992;360:481–4.
10. Vazeux R, Hoffman PA, Tomita JK, et al. Cloning and characterisation of a new intercellular adhesion molecule ICAM-R. *Nature* 1992;360:485–8.
11. de Fougerolles AR, Klicstein LB, Springer TA. Cloning and expression of ICAM-3 reveals strong homology to other Ig family counter-receptors for LFA-1. *J Exp Med* 1993;177: 1187–90.
12. Linsley P, Clark E. CTLA-4 is a ligand for LFA-1. *J Cell Biochem* 1994;18D:A120.
13. Voraberger G, Schafer R, Stratowa C. Cloning of the human gene for intercellular adhesion

molecule 1 and analysis of its 5′-regulatory region: induction by cytokines and phorbol ester. *J Immunol* 1991;147:2777–84.
14. Xu H, Tong IL, de Fougerolles AR, Springer TA. Sequence and gene organisation of murine ICAM-2. *J Immunol* 1992;149:2650–5.
15. Hammond L, McClelland A. *Genbank Release 80.0. National Center for Biotechnology Information*. Bethesda, MD; National Library of Medicine, NIH, 1993.
16. Horley KJ, Carpenito C, Baker B, Takei F. Molecular cloning of murine intercellular adhesion molecule (ICAM-1). *EMBO J* 1989;8:2889–96.
17. Kita Y, Takashi T, Iigo Y, Tamatari T, Miyasaka M, Horiuchi T. Sequence and expression of rat ICAM-1. *Biochim Biophys Acta* 1992;1131:108–10.
18. Ropers HH, Mohrenweiser H. Report of the committee on the genetic constitution of chromosome 19. *Genet Prior Rep* 1993;1:524–47.
19. Bossy D, Mattei MG, Simmons DL. The luminar intercellular adhesion molecule 3 (ICAM 3) gene is located in the 19p13.2–p13.3 region, close to the ICAM-1 gene. *Genomics* 1994;23:712–3.
20. Samsom D, Borrow J, Solomon E, Trowsdale J. The ICAM-2 gene maps to 17q23-25. *Genomics* 1991;11:462–4.
21. Carpen O, Pallai P, Staunton DE, Springer TA. Association of ICAM-1 with actin-containing cytoskeleton and α-actinin. *J Cell Biol* 1992;118:1223–34.
22. Diamond MS, Staunton DE, Marlin SD, Springer TA. Binding of the integrin Mac-1 (CD11b/CD18) to the third Ig-like domain of ICAM-1 (CD54) and its regulation by glycosylation. *Cell* 1991;65:961–71.
23. Littler AJ, Holness CL, Simmons DL. N-linked glycans negatively regulate binding of ICAM-3 to Mac-1. In preparation.
24. Doussis-Anagnostopoulou I, Kaklamanis L, Cordell J, et al. ICAM-3 expression on endothelium in lymphoid malignancy. *Am J Pathol* 1993;143:1040–3.
25. Staunton DE, Dustin ML, Erickson HP, Springer TA. The arrangement of the immunoglobulin-like domains of ICAM-1 and the binding sites for LFA-1 and rhinovirus. *Cell* 1990;61:243–54.
26. Li R, Nortamo P, Valmu L, et al. A peptide from ICAM-2 binds to the leukocyte integrin CD11a/CD18 and inhibits endothelial cell adhesion. *J Biol Chem* 1993;268:17513–8.
27. Holness CL, Bates PA, Littler AJ, et al. Analysis of the binding site on intercellular adhesion molecule 3 for the leukocyte integrin lymphocyte function associated antigen 1. *J Biol Chem* 1995;270:877–84.
28. Staunton DE, Merluzzi VJ, Rothlein R, Barton R, Marlin SD, Springer TA. A cell adhesion molecule ICAM-1, is the major surface receptor for rhinovirus. *Cell* 1989;56:849–53.
29. Berendt AR, McDowall A, Craig AG, et al. The binding site on ICAM-1 for Plasmodium falciparum-infected erythrocytes overlaps, but is distinct from, the LFA-1 binding site. *Cell* 1992;68:71–81.
30. Ockenhouse CF, Betageri R, Springer TA, Staunton DE. Plasmodium falciparum-infected erythrocytes bind ICAM-1 at a site distinct from LFA-1, Mac-1, and human rhinovirus. *Cell* 1992;68:63–9.
31. Osborn L, Vassallo C, Griffiths Browning B, et al. Arrangement of domains, and amino acid residues required for binding of vascular cell adhesion molecule-1 to its counter-receptor VLA-4 (α4β1). *J Cell Biol* 1994;124:601–8.
32. Vonderheide RH, Tedder TF, Springer TA, Staunton DE. Residues within a conserved amino acid motif of domains 1 and 4 of VCAM-1 are required for binding to VLA-4. *J Cell Biol* 1994;125:215–22.
33. Lollo BA, Chan KWH, Hanson EM, Moy VT, Brian AA. Direct evidence for two affinity states for lymphocyte function-associated antigen 1 on activated cells. *J Biol Chem* 1993; 268:21693–700.
34. Dransfield I, Cabanas C, Barrett J, Hogg N. Interaction of leukocyte integrins with ligand is necessary but not sufficient for function. *J Cell Biol* 1992;116:1527–35.
35. Andrew D, Shock A, Ball E, Ortlepp S, Bell J, Robinson M. KIM185, a monoclonal antibody to CD18 which induces a change in the conformation of CD18 and promotes a change in the conformation of CD18 and promotes both LFA-1 and CR3-dependent adhesion. *Eur J Immunol* 1993;23:2217–22.
36. Landis C, McDowall A, Holness CL, Littler A, Simmons DL, Hogg N. Involvement of the

"I" domain of LFA-1 in selective binding to ligands ICAM-1 and ICAM-3. *J Cell Biol* [*in press*].
37. Springer TA. Adhesion receptors of the immune system. *Nature* 1990;346:425–34.
38. Rothlein R, Kishimoto TK, Mainolfi E. Cross-linking of ICAM-1 induces co-signalling of an oxidative burst from mononuclear leukocytes. *J Immunol* 1994;152:2488–95.
39. Poudrier J, Owens T. CD54/Intercellular adhesion molecule 1 and major histocompatibility complex II signalling induces B cells to express interleukin 2 receptors and complements help provided through CD40 ligation. *J Exp Med* 1994;179:1417–27.
40. Juan M, Vinas O, Pino-Otin MR, et al. CD50 (intercellular adhesion molecule 3) stimulation induces calcium mobilisation and tyrosine phosphorylation through p59fyn and p56lck in Jurkat T cell line. *J Exp Med* 1994;179:1747–56.
41. Campanero MR, del Pozo MA, Arroyo AG, et al. ICAM-3 interacts with LFA-1 and regulates the LFA-1/ICAM-1 cell adhesion pathway. *J Cell Biol* 1993;123:1007–16.
42. Hernandez-Caselles T, Rubio G, Campanero MR, et al. ICAM-3, the third LFA-1 counter-receptor, is a co-stimulatory molecule for both resting and activated T lymphocytes. *Eur J Immunol* 1993;23:2799–806.
43. Bossy D, Buckley C, Littler AJ, et al. Epitope mapping and functional properties of anti-intercellular adhesion molecule 3 (CD50) monoclonal antibodies. *Eur J Immunol* 1995 [*in press*].
44. Acavedo A, del Pozo MA, Arroyo AG, Sanchez-Mateos P, Gonzalez-Amaro R, Sanchez-Madrid F. Distribution of ICAM-3 bearing cells in normal human tissues. *Am J Pathol* 1993;143:774–83.
45. Hart D, Simmons DL. ICAM-3 is a functional ligand for LFA-1 on tonsil dendritic cells. *Eur J Immunol* 1995 [*in press*].
46. Johnson JP, Stade BG, Holzmann B, Schwable W, Riethmüler G. De novo expression of ICAM-1 in melanoma correlates with increased risk of metastasis. *Proc Natl Acad Sci USA* 1989;86:641–4.
47. Gearing AJH, Newman W. Circulating adhesion molecules in disease. *Immunol Today* 1993;14:506–12.

*Topics in Molecular Medicine, Volume 1,*
edited by Wolfgang Siess, Reinhard Lorenz,
and Peter C. Weber. Raven Press, Ltd.,
New York © 1995.

# 3

# The LDV Peptide Motif: Its Role in Integrin–Ligand Binding

Martin J. Humphries

*School of Biological Sciences, University of Manchester, Manchester, England*

Cell–cell and cell–extracellular matrix adhesive interactions play a key role in regulating cell positioning. In some situations, adhesion is used by cells to translocate from one site to another, and in others it retains them in a stable location. In normal circumstances, therefore, adhesion is used to maintain tissue compartment boundaries, allow repair of damaged tissue, and permit trafficking of migratory cells. Each of these processes is highly regulated and precisely coordinated. In disease, however, normal adhesive interactions can contribute to disease progression by allowing cells to adhere at inappropriate times in inappropriate locations. Most of the major human diseases, including cardiovascular diseases, cancers, and autoimmune diseases, are characterized by abnormal cell trafficking. In these cases, if cell adhesive interactions could be prevented or controlled through the use of inhibitory agents, then disease progression might be arrested.

The major molecular targets for the generation of adhesion inhibitors are adhesion receptors, of which integrins are the principal class. Integrins are a family of $\alpha,\beta$ heterodimeric glycoproteins currently known to contain 21 different members in vertebrates (1). Each dimer possesses a unique ligand specificity, suggesting that different receptors exhibit different functions in vivo and implying that specific integrin antagonists could have many applications (2). Integrins recognize extracellular matrix macromolecules such as fibronectin, laminins, and collagens, but also bind to cell surface proteins, including several members of the immunoglobulin superfamily. The diversity of integrin ligands has prompted interest in the mechanisms they employ to mediate binding, as this information should help to explain ligand specificity and thereby facilitate the synthesis of specific antagonists with limited side effects.

The first step in the rational design of adhesion antagonists requires information about the structure of adhesion molecules. Because many of these molecules are large, modular proteins and because some, like integrins, are intercalated into the plasma membrane, progress by crystallization of the intact molecules has been slow. As an alternative, the subdomains of extracellular matrix macromolecules that confer activity have been mapped by progressive truncation. Large fragments of fibronectin, laminin, and the interstitial collagens have been isolated and shown to retain similar levels of cell-binding activity to their parent molecules. More recently, biochemical separation of these fragments into their smaller modular compartments, and ultimately to their minimal active sequences, has been performed. A key concept that has emerged from truncation-based approaches is that the majority of integrin ligands use very short protein sequences as receptor recognition motifs (2,3). Indeed, in initial studies using fibronectin as a model, adhesive domains were mapped using proteolytic fragments to the point at which it was possible to reproduce activity in the form of synthetic peptides. Subsequently, a key minimal active sequence was identified as the tripeptide RGD (4). Subsequent studies revealed that other adhesion molecules, including fibrinogen, von Willebrand factor, vitronectin, and osteopontin, also employ this motif, albeit with different receptor specificity profiles (2). Apart from the core tripeptide, there is little similarity in the flanking sequences, except for a preference for hydrophobic amino acids immediately after the aspartate (Fig. 1).

The ability to reproduce adhesive activity in synthetic peptide form was a key strategic transition, because analogues of these sequences were essentially lead compounds for drug design (5). The first integrin–ligand interaction to be targeted for drug development was the RGD-dependent binding of fibrinogen to the platelet integrin $\alpha_{IIb}\beta_3$, an interaction that plays a key role in thrombosis. This work has now led to the development of RGD mimics that exhibit potency, specificity, and a capability for oral delivery (5). Current indications from clinical trials with these antagonists are encouraging.

Since the discovery of the RGD recognition sequence, the molecular basis of two other integrin–ligand interactions has been partly resolved; again, activity can be described by synthetic peptides. First, proteolytic cleavage

| Protein | Sequence | | |
|---|---|---|---|
| FN | VTYAVTG | RGD | SPASSKP |
| FG (1) | TNIMEIL | RGD | FSSANNR |
| FG (2) | TSSTSYN | RGD | STFESKS |
| VN | ECKPQVT | RGD | VFTMPED |
| VWF | VVTGSPE | RGD | QSSWKSV |
| OP | TVDTYDG | RGD | SVVYGLR |

**FIG. 1.** Sequence alignments of active RGD sequences. FN, fibronectin; FG, fibrinogen; VN, vitronectin; VWF, von Willebrand factor; OP, osteopontin. Human sequences are used.

of fibrinogen has led to the identification of an $\alpha_{IIb}\beta_3$-binding peptide from the C-terminus of the γ-chain with a minimal sequence QAGDV (6). Second, a peptide-based approach has localized the major integrin $\alpha_4\beta_1$-binding site in fibronectin to the tripeptide LDV (7). There appears to be a functional analogy between these integrin-binding motifs, as QAGDV and RGD peptides bind to mutually exclusive binding sites on $\alpha_{IIb}\beta_3$ (8) and LDV and RGD have the same relationship for binding $\alpha_4\beta_1$ (9). This suggests that the concepts learned from studies of RGD may be applicable to other integrin-binding motifs.

## INTEGRIN $\alpha_4\beta_1$ LIGANDS

In recent years, work in this laboratory has focused on the integrin $\alpha_4\beta_1$ and its two ligands, the extracellular matrix glycoprotein fibronectin (10–12) and the cytokine-inducible endothelial cell-surface protein VCAM-1 (13). Results from several studies have suggested that $\alpha_4\beta_1$–ligand interactions are promigratory. $\alpha_4\beta_1$ tends to be expressed by highly motile cells such as leukocytes and neuroectodermal cells (14), expression of $\alpha_4\beta_1$ by tumor cells correlates with a malignant phenotype (15), and cells expressing chimeric integrins containing the $\alpha_4$ cytoplasmic domain display enhanced migratory activity (16). In addition, anti-$\alpha_4$ monoclonal antibodies have been shown to inhibit trafficking of leukocytes in a number of acute and chronic inflammatory conditions in vivo. It therefore appears that $\alpha_4\beta_1$ contributes to leukocyte extravasation and is an important therapeutic target (17,18). We have therefore been interested in determining the molecular basis of $\alpha_4\beta_1$–ligand binding with a view to using this information to develop specific antagonists.

VCAM-1 is a member of the immunoglobulin superfamily, consisting of six or seven immunoglobulin repeats. Domain deletion and chimera experiments have shown that the sites in VCAM-1 recognized by $\alpha_4\beta_1$ lie within immunoglobulin domains 1 and 4 (19). In the six-domain form of VCAM-1, domain 4 is removed by alternative splicing. $\alpha_4\beta_1$ binding to fibronectin occurs via the HepII/IIICS region. HepII is the principal proteoglycan-binding domain of fibronectin, and the IIICS is one of three alternatively spliced sites in the molecule. Three $\alpha_4\beta_1$-binding sites have been identified within the HepII/IIICS (20–22). Two of these sites (contained within the peptides CS1 and CS5) are present in independently spliced segments of the IIICS (20,21), and the other site (contained within the peptide H1) is found in the constitutively expressed HepII region (22). The CS1 peptide contains the tripeptide LDV as its minimal active site (7), H1 contains a related motif, IDA, and CS5 employs a variant of RGD, REDV (9,20). On the basis of studies of the relative activities of the peptides, CS1 is approximately 20-fold more active than CS5, and H1 is similar to CS5 (9,22).

## FUNCTIONAL IMPORTANCE OF ALTERNATIVE SPLICING

The location of integrin-binding sites in relatively small, alternatively spliced segments of both $\alpha_4\beta_1$ ligands suggests that splicing may regulate function. To test this possibility, we adopted the approach of generating recombinant versions of the HepII/IIICS splice variants. Four variants have been studied which contain the four possible combinations of the three active sites, CS1, CS5, and H1. The H/120 variant contains all three known sites, H/89 contains CS1 and H1, H/95 contains CS5 and H1, and H/0 contains H1 alone. When tested in cell attachment and spreading assays, the relative activities of the different variants reflected the relative activities of CS1, CS5, and H1 peptides, i.e., the CS1-containing variants were most active (23). A difference in activity between H/95 and H/0 implies that the CS5 site is active in the H/95 variant. Splicing also regulates the integrin-binding activity of VCAM-1, because a molecule containing domains 1 and 4 possesses approximately twice the adhesive activity of a molecule containing only domains 1 and 2 (Mould et al., manuscript in preparation).

The major factor that determines the adhesive activity mediated by different spliced variants of fibronectin and VCAM-1 appears to be the inherent affinity of their interactions with $\alpha_4\beta_1$. Measurements of receptor-binding affinities for the different recombinant proteins in a solid-phase binding assay in the presence of 1 m*M* $MnCl_2$ showed that VCAM-1 bound to $\alpha_4\beta_1$ with approximately four-fold higher affinity than H/120 and H/89 (which were both of similar affinity). H/120 exhibited approximately 10-fold higher binding affinity than H/95, which in turn exhibited approximately twofold higher affinity than H/0 (24). These results therefore match the observed differences between the ligands in adhesion assays. An unexpected finding from kinetic measurements of $\alpha_4\beta_1$–ligand binding was that changes in ligand affinity were due principally to changes in the rate of ligand–receptor association ($k_1$) rather than ligand–receptor dissociation ($k_{-1}$). This suggests that receptor–ligand engagement requires changes in ligand conformation (which may be small for sequences in VCAM-1, but very large for low-affinity sequences, such as in the HepII region of fibronectin), whereas, in contrast, receptor–ligand dissociation appears to require changes in receptor conformation. These studies suggest that the conformations of the active site peptide sequences in fibronectin and VCAM-1 endow them with a differential ability to bind receptor. The significance of differences in affinity could be related to receptor clustering, and therefore to variably stable interactions of integrins with the cytoskeleton, or to the transduction of different signals to the cell interior. In vivo, the high-affinity binding of VCAM-1 to $\alpha_4\beta_1$ may promote leukocyte arrest on the surface of endothelial cells and/or diapedesis across the endothelial monolayer. In vitro, H/120, H/89, and VCAM-1 elicited a biphasic promotion of cell migration as a function of ligand concentration. This dual stimulation and suppression of migration suggests that a threshold level of adhesiveness is needed to initiate cell migration but that high levels of adhesion cause migration to decline owing to an inability of the cells to break contact with the substrate. In vivo, the density of ligand active sites and their orientation with respect to the cell surface are key factors in determining whether a cell moves or stays still.

## ACTIVE SITES IN VCAM-1

As described above, VCAM-1–mediated leukocyte extravasation is an important therapeutic target for the development of anti-inflammatory agents. Recently, we showed that fibronectin and VCAM-1 act as competitive inhibitors of each other's binding to $\alpha_4\beta_1$ (25). This suggests, first, that the fibronectin- and VCAM-1-binding sites on $\alpha_4\beta_1$ are either identical or spatially very close and, second, that fibronectin and VCAM-1 may employ similar peptide active sites. The key active sequence within the $\alpha_4\beta_1$-binding domain of fibronectin is the tripeptide LDV (7), and both active immunoglobulin domains of VCAM-1 (domains 1 and 4) contain a related sequence, I(39 and 327)DSP. Homology modeling of VCAM-1 domain 1 using immunoglobulin crystal structures predicts that this sequence resides in a flexible loop region that is potentially accessible to $\alpha_4\beta_1$.

A number of mutations within the IDSP-containing loop and within other sequences predicted to be surface-accessible were screened for their effects on adhesive activity. Using a simplified VCAM-1 construct lacking domains 4–6, mutation of the I(39)DSP sequence essentially abrogated $\alpha_4\beta_1$–VCAM-1 binding (26). The profile of monoclonal antibody binding to mutated VCAM-1 molecules was not affected by the mutation, indicating that the effects on adhesion were a direct consequence of perturbing integrin recognition. To examine the contribution of the IDSP sequence in domain 4 of VCAM-1, I(327)DSP and/or I(39)DSP were mutated in the context of full-length VCAM-1 and examined for adhesive activity at 37°C, when both domains are functional, or at 4°C, when domain 4 function is compromised (27). The domain 1 mutation only partially reduced adhesion at 37°C, suggesting that domain 4 compensated for its loss (26), but at 4°C almost all binding was abolished. The domain 4 mutation had no significant effect at either assay temperature, indicating that $\alpha_4\beta_1$ preferentially binds domain 1 over domain 4. Simultaneous mutation of both IDSP sequences in domains 1 and 4 produced a molecule that was unable to support adhesion at either temperature. Taken together, these data suggest that $\alpha_4\beta_1$ binds domain 1 and domain 4 via a common mechanism mediated by the IDSP motif. However, if binding to one domain is compromised, binding can still take place via the other domain.

As would be predicted from studies of integrin binding to fibronectin, a synthetic peptide containing the IDSP sequence was found to perturb cell adhesion to VCAM-1 in a dose-dependent manner (26). Interestingly, the peptide possessed activity similar to that of the fibronectin CS1 peptide. These results confirm those obtained from VCAM-1 mutagenesis and, importantly, suggest that the mechanism of integrin binding by immunoglobulin ligands is related to that employed by most extracellular matrix ligands.

Although much structure–function work remains, the IDSP motif in VCAM-1 is now the third LDV-like sequence to be identified as a functional integrin-binding site. The first is the LDV(P) site in CS1 and the second the IDA(P) sequence in the fibronectin HepII peptide H1. A series of other reports suggest that there may be many more examples of the use of similar sequences and that this is, in fact, a second common integrin-binding motif. The QAGDV peptide from the carboxyl terminus of the γ-chain of fibrinogen (6) bears some homology to LDV and, like LDV (21), is a competitive inhibitor of RGD function (8). It is most striking, however, that all other integrin-binding immunoglobulin ligands contain either an aspartate or a glutamate residue in their distal membrane domains in a position analogous to the IDSP of VCAM-1 (Fig. 2). There is now considerable evidence that each of these residues contributes to function: mutation of E34 within ICAM-1 domain 1 (28) and of E37 within ICAM-3 (D. Simmons, personal communication) substantially inhibits LFA-1 ($\alpha_L\beta_2$) binding; mutations in the region of D41 of MAdCAM-1 inhibits $\alpha_4\beta_1$ binding (M. Briskin, personal communication); and a peptide spanning the L(39)ETP site of ICAM-2 is an effective inhibitor of $\alpha_L\beta_2$ binding and is capable of a direct interaction with $\alpha_L\beta_2$ (29). Therefore, a consensus integrin-binding motif on immunoglobulin cell adhesion molecules can be defined as aliphatic—aspartate or glutamate—serine or threonine—proline or hydrophilic. The use of a glutamate

| Protein | Sequence | | |
|---|---|---|---|
| FN (CS1) | PEI | LDVP | STV |
| FN (H1) | STA | IDAP | SNL |
| FG (γ-chain) | KQA | GDV. | |
| VCAM-1 (D1) | RTQ | IDSP | LNG |
| VCAM-1 (D4) | RTQ | IDSP | LSG |
| MAdCAM-1 (D1) | WRG | LDTS | LGS |
| ICAM-1 (D1) | LLG | IETP | LPK |
| ICAM-2 (D1) | VGG | LETS | LNK |
| ICAM-3 (D1) | KIA | LETS | LSK |

**FIG. 2.** Sequence homology in the domain 1 C-D loop region of integrin-binding CAMs and comparison with integrin-binding sequences in fibronectin (FN) and fibrinogen (FG). "LDVP" homology is shaded. Where possible, human sequences are used.

rather than an aspartate residue by ICAMs contrasts with the absolute dependence of RGD sequences on aspartate, but might reflect a slightly deeper active site pocket in $\beta_2$ integrins. In the future, it will be of pressing importance to test the ability of LDV-like ("LDV") peptides to perturb each of these integrin–ligand interactions using sensitive assays for adhesion, as this would have implications for the design of novel classes of antiadhesive agents. After this, the role of amino acids flanking the acidic residue in determining the activity and specificity of these sites needs to be examined.

## PEPTIDE POTENCY AND SPECIFICITY

The fact that both RGD and "LDV" are used as a common recognition signals provides a partial explanation for the ligand-binding promiscuity of integrins and, for the purpose of this article, explains how $\alpha_4\beta_1$ can bind both fibronectin and VCAM-1. Some integrins, however, exhibit exquisite ligand-binding specificity, implying that either the integrins, the ligands, or both have structural features that generate this specificity. Two possible explanations are that the peptides are held in different conformations in different ligands, or that other regions of the ligands contribute to receptor binding. There is some evidence supporting the latter view (30,31), but it is also clear that structural alteration of RGD peptides can alter receptor selectivity. This therefore implies that flanking sequences in the ligand may constrain RGD conformation. As yet, in the absence of crystal structure data, there are few indications as to how this is achieved. What is apparent thus far from "LDV" structure–function studies is that the motif has greater tolerance of amino acid substitutions in the residues flanking the key aspartate. This offers hope that peptide engineering might produce selective inhibitors for CAM-binding integrins. Studies are under way to introduce conformational restrictions into these peptides.

Tertiary structure determinations of RGD sites in fibronectin, tenascin, kistrin, echistatin and, recently, in the foot-and-mouth disease VP1 coat protein have in all cases localized these sites to flexible surface loops (32–36). Interestingly, our prediction is that "LDV" active sites in integrin-binding immunoglobulins are also localized to loops. This location may be functionally important in endowing a relatively high on-rate for ligand binding. Elucidation of the conformational changes that take place in both ligands and receptors during binding and dissociation and how this varies among ligand isoforms remain fundamentally important questions. In addition, further investigation of the molecular basis and functional consequences of ligand binding by the integrin $\alpha_4\beta_1$ promises to answer a number of other key questions. It is possible that the family of "LDV"-containing proteins is much larger than that discussed here, and further putative active sites warrant study. In addition, the essential structural features of an active

"LDV" sequence are unknown, as is the functional role of the short aspartate- or glutamate-containing peptides employed by integrin ligands.

## ACKNOWLEDGMENT

The work performed in the author's laboratory described in this article was supported by grants from the Wellcome Trust.

## REFERENCES

1. Hynes RO. Integrins: versatility modulation and signaling in cell adhesion. *Cell* 1992;69: 11–25.
2. Humphries MJ. The molecular basis and specificity of integrin-ligand interactions. *J Cell Sci* 1990;97:585–92.
3. Yamada KM. Adhesive recognition sequences. *J Biol Chem* 1991;266:12809–12.
4. Pierschbacher MD, Ruoslahti E. The cell attachment activity of fibronectin can be duplicated by small synthetic fragments of the molecule. *Nature* 1984;309:30–3.
5. Humphries MJ, Doyle PM, Harris CJ. Integrin antagonists as modulators of adhesion. *Expert Opin Therapeutic Patents* 1994;4:227–35.
6. Kloczewiak M, Timmons S, Lukas TJ, Hawiger J. Platelet receptor recognition site on human fibrinogen. Synthesis and structure-function relationship of peptides corresponding to the carboxy-terminal segment of the γ-chain. *Biochemistry* 1984;23:1767–74.
7. Komoriya A, Green LJ, Mervic M, Yamada SS, Yamada KM, Humphries MJ. The minimal essential sequence for a major cell type-specific adhesion site CS1 within the alternatively spliced type III connecting segment domain of fibronectin is leucine-aspartic acid-valine. *J Biol Chem* 1991;266:15075–9.
8. Santoro SA, Lawing WJ. Competition for related but nonidentical binding sites on the glycoprotein IIb-IIIa complex by peptides derived from platelet adhesive proteins. *Cell* 1987;48:867–73.
9. Mould AP, Komoriya A, Yamada KM, Humphries MJ. The CS5 peptide is a second site in the IIICS region of fibronectin recognized by the integrin $\alpha_4\beta_1$. Inhibition of $\alpha_4\beta_1$ function by RGD peptide homologues. *J Biol Chem* 1991;266:3579–85.
10. Wayner EA, Garcia-Pardo A, Humphries MJ, McDonald JA, Carter WG. Identification and characterization of the T lymphocyte adhesion receptor for an alternative cell attachment domain CS-1 in plasma fibronectin. *J Cell Biol* 1989;109:1321–30.
11. Guan JL, Hynes RO. Lymphoid cells recognize an alternatively spliced segment of fibronectin via the integrin receptor $\alpha_4\beta_1$. *Cell* 1990;60:53–61.
12. Mould AP, Wheldon LA, Komoriya A, Wayner EA, Yamada KM, Humphries MJ. Affinity chromatographic isolation of the melanoma adhesion receptor for the IIICS region of fibronectin and its identification as the integrin $\alpha_4\beta_1$. *J Biol Chem* 1990;265:4020–4.
13. Elices MJ, Osborn L, Takada Y, et al. VCAM-1 on activated endothelium interacts with the leukocyte integrin VLA-4 at a site distinct from the VLA-4/fibronectin binding site. *Cell* 1990;60:577–84.
14. Hemler ME, Elices MJ, Parker C, Takada Y. Structure of the integrin VLA-4 and its cell-cell and cell-matrix adhesion functions. *Immunol Rev* 1990;114:45–65.
15. Albelda SM, Mette SA, Elder DE, et al. Integrin distribution in malignant melanoma: association of the β3 subunit with tumor progression. *Cancer Res* 1990;50:6757–64.
16. Chan BMC, Kassner PD, Schiro JA, Byers HR, Kupper TS, Hemler ME. Distinct cellular functions mediated by different VLA integrin α subunit cytoplasmic domains. *Cell* 1992; 68:1051–60.
17. Yednock TA, Cannon C, Fritz LC, Sanchez-Madrid F, Steinman L, Karin N. Prevention of experimental autoimmune encephalomyelitis by antibodies against $\alpha_4\beta_1$ integrin. *Nature* 1992;356:63–6.
18. Weg VB, Williams TJ, Lobb RR, Nourshargh S. A monoclonal antibody recognizing very

late activation antigen-4 inhibits eosinophil accumulation *in vivo*. *J Exp Med* 1993;177:561–6.

19. Vonderheide RH, Springer TA. Lymphocyte adhesion through very late antigen 4: evidence for a novel binding site in the alternatively spliced domain of vascular cell adhesion molecule 1 and an additional α4 integrin counterreceptor on stimulated endothelium. *J Exp Med* 1992;175:1433–42.
20. Humphries MJ, Akiyama SK, Komoriya A, Olden K. Yamada KM. Identification of an alternatively spliced site in human plasma fibronectin that mediates cell type-specific adhesion. *J Cell Biol* 1986;103:2637–47.
21. Humphries MJ, Komoriya A, Akiyama SK, Olden K, Yamada KM. Identification of two distinct regions of the type III connecting segment of human plasma fibronectin that promote cell type-specific adhesion. *J Biol Chem* 1987;262:6886–92.
22. Mould AP, Humphries MJ. Identification of a novel recognition sequence for the integrin $\alpha_4\beta_1$ in the carboxy-terminal heparin-binding domain of fibronectin. *EMBO J* 1991;10:4089–95.
23. Askari JA, Craig SE, Garratt AN, Clements J, Humphries MJ. Integin α4β1-mediated melanoma cell adhesion and migration on vascular cell adhesion molecule-1 (VCAM-1) and the alternatively spliced IIICS region of fibronectin. *J Biol Chem* 1994;269:27224–30.
24. Dudgeon TJ, Bottomley MJ, Driscoll PC, Humphries MJ, Mould AP, Wingfield GI, Clements JM. Expression and characterisation of a very-late antigen-4 (α4β1) integrin-binding fragment of vascular cell adhesion molecule-1. *Eur J Biochem* 1994;226:517–23.
25. Makarem R, Newham P, Askari JA, et al. Competitive binding of vascular cell adhesion molecule-1 and the HepII/IIICS domain of fibronectin to the integrin $\alpha_4\beta_1$. *J Biol Chem* 1994;269:4005–11.
26. Clements JM, Newham P, Shepherd M, et al. Identification of a key integrin-binding sequence in VCAM-1 homologous to the LDV active site in fibronectin. *J Cell Sci* 1994;107:2127–35.
27. Needham LA, Van Dijk S, Pigott R, et al. Activation dependent and independent binding sites on vascular cell adhesion molecule-1. *Cell Adhesion Commun* 1994;2:87–99.
28. Staunton DE, Dustin ML, Erickson HP, Springer TA. The arrangement of the immunoglobulin-like domains of ICAM-1 and the binding sites for LFA-1 and rhinovirus. *Cell* 1990;61:243–54.
29. Li R, Nortamo P, Valmu L, et al. A peptide form ICAM-2 binds to the leukocyte integrin CD11a/CD18 and inhibits endothelial cell adhesion. *J Biol Chem* 1993;268:17513–8.
30. Obara M, Kang MS, Yamada KM. Site-directed mutagenesis of the cell-binding domain of human fibronectin: separable synergistic sites mediate adhesive function. *Cell* 1988;53:649–57.
31. Bowditch RD, Hariharan M, Tominna EF, et al. Identification of a novel integrin binding site in fibronectin. Differential utilization by β3 integrins. *J Biol Chem* 1994;269:10856–63.
32. Adler M, Lazarus RA, Dennis MS, Wagner G. Solution structure of kistrin, a potent platelet aggregation inhibitor and GP IIb-IIIa antagonist. *Science* 1991;253:445–8.
33. Chen Y, Pitzenberger SM, Garsky VM, Lumma PK, Sanyal G, Baum J. Proton NMR assignments and secondary structure of the snake venom protein echistatin. *Biochemistry* 1991;30:11625–36.
34. Leahy DJ, Hendrickson WA, Aukhil I, Erickson HP. Structure of a fibronectin type III domain from tenascin phased by MAD analysis of the selenomethionyl protein. *Science* 1992;258:987–91.
35. Main AL, Harvey TS, Baron M, Boyd J, Campbell ID. The three-dimensional structure of the tenth type III module of fibronectin: an insight into RGD-mediated interactions. *Cell* 1992;71:671–8.
36. Logan D, Abu-Ghazaleh R, Blakemore W, et al. Structure of a major immunogenic site on foot-and-mouth disease virus. *Nature* 1993;362:566–8.

*Topics in Molecular Medicine, Volume 1,*
edited by Wolfgang Siess, Reinhard Lorenz,
and Peter C. Weber. Raven Press, Ltd.,
New York © 1995.

# 4

# Integrin-Mediated Tyrosine Phosphorylation of Focal Adhesion Proteins and Organization of the Cytoskeleton

Lewis H. Romer

*Departments of Pediatrics (Division of Critical Care), and Cell Biology and Anatomy, University of North Carolina at Chapel Hill, Chapel Hill, North Carolina 27599-7220*

## CELL ADHESION: COMMUNICATION FROM THE CELL SURFACE

Cell adhesion is mediated by a diverse group of receptors including integrins, cadherins, and members of the immunoglobulin superfamily (1–3). These adhesion receptors contribute to tissue organization and structural integrity during embryonic development and in the maintenance of homeostatic function (3). Signaling across the plasma membrane via these receptors provides vital information about the cellular environment. Coordinated responses to these signals ensure optimal cellular adaptation through changes in growth and differentiation (4–6).

Adhesion to extracellular matrix (ECM) proteins profoundly affects the morphology and motility of many cell types (7–9). Cell–ECM adhesion is mediated by the integrin family of adhesion receptors. Integrins are heterodimeric combinations of a variety of α- and β-chains that exhibit a remarkable degree of structural conservation across a broad phylogenetic spectrum (1). Specificity is imparted to integrin–ECM interactions by the combination of particular α and β components that recognize single or several ECM proteins (1,3,10). Integrin subunits span the plasma membrane and, in general, have relatively short and highly conserved cytoplasmic domains. Information initiated by the interactions of integrins with ECM proteins may modulate cell behavior through a variety of signaling pathways that are currently the focus of avid exploration (11–14).

## FOCAL ADHESIONS: STRUCTURE AND FUNCTION

In cultured cells, transmembrane integrin receptors establish bridges between ECM proteins and the cytoskeleton at sites known as *focal adhesions*. These are areas of tight (10–15 nm) juxtaposition between the cell surface and the ECM (15,16). Direct interactions between integrins and focal adhesion proteins, including talin and α-actinin (17–19), may facilitate structural continuity and signaling between ECM proteins and the cytoskeleton. Focal adhesions may be involved in diverse cellular responses, including spreading and locomotion on ECM (16,20–22).

Signaling proteins that have been located in focal adhesions include protein kinase C (23–25) and tyrosine kinases (26–29). The description and localization of the nonreceptor focal adhesion kinase (pp125$^{FAK}$; FAK) in the focal adhesions of normal fibroblasts (28,29) has provided a focus for investigation into mechanisms of integrin signaling (30–38).

Recent studies indicate that the tyrosine phosphorylation of focal adhesion proteins may also be linked to other signaling events, including changes in intracellular calcium (30,39). We present data here that implicate integrin-mediated tyrosine phosphorylation as an important component of the cellular response to ECM adhesion.

## DATA

### Phosphotyrosine in Focal Adhesions

Human umbilical vein endothelial cells (HUVECs) and rat embryo fibroblasts (REF52) were obtained and prepared for use in these studies as previously described (32,38,40). Phosphotyrosine localization was studied in actively spreading REF52 cells plated on fibronectin for 75 min in the absence of serum and growth factors (Fig. 1). Cells were prepared with immunofluorescence staining as previously described (32). Anti-phosphotyrosine (Fig. 1A) and anti-talin (Fig. 1B) staining were compared in double-labeled cells. Tyrosine-phosphorylated proteins were mostly located in focal adhesions at the cells' periphery. This observation is consistent with the idea that tyrosine phosphorylation accompanies the development of new focal adhesions at the leading edges of spreading cells (20,41).

We also examined the pattern of tyrosine phosphorylation in migrating cells (Fig. 2). Confluent HUVEC monolayers scratched with a pipette tip were used in an in vitro wound healing model of cell migration. This technique is described elsewhere (38). Quiescent portions of the monolayer distant from the wound edge were remarkable for very faint phosphotyrosine staining. HUVECs actively migrating into the wound breach revealed larger focal adhesions that stained brightly for phosphotyrosine (Figs. 2B,C). Once

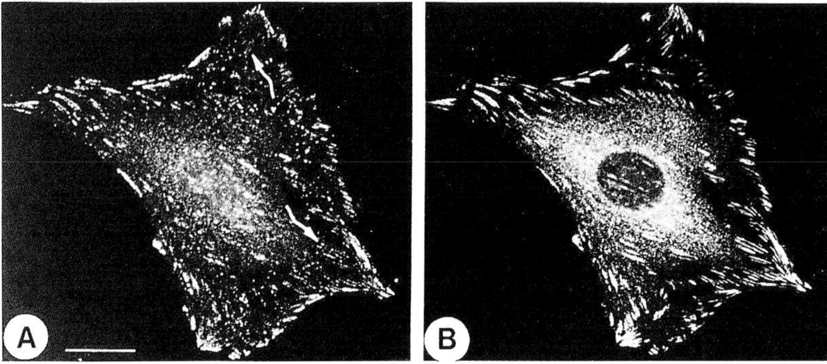


**FIG. 1.** Phosphotyrosine distribution in freshly plated REF52. Cells were stained for immunofluorescence with anti-phosphotyrosine **(A)** or anti-talin **(B)** antibodies 75 min after plating on fibronectin-coated coverslips in serum-free medium. The most marked phosphotyrosine localization is in peripheral focal adhesions. Centrally located focal adhesions (arrows) exhibit little phosphotyrosine staining, although talin staining is prominent. Bar = 20 μm. (From ref. 32, with permission.)

the wound surface area was fully occupied by relocated HUVECs staining for tyrosine phosphorylation was reduced (Fig. 2D).

Western blots of whole-cell lysates of freshly plated cells revealed increased tyrosine phosphorylation in several bands (see 32 and 33 for methods and data). Among the most prominent of these bands were those at 110–130 kDa and 68–70 kDa. We therefore decided to study adhesion-associated tyrosine phosphorylation in the focal adhesion proteins FAK (125 kDa), (28,29) and paxillin (68 kDa) (42,43). FAK and paxillin were immu-

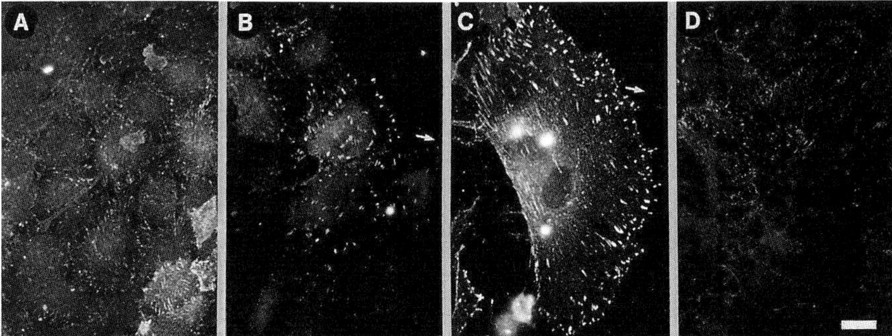


**FIG. 2.** Increased focal adhesion phosphotyrosine in migrating HUVECs. Confluent HUVEC monolayers were wounded with a pipette tip and prepared for immunofluorescence with anti-phosphotyrosine antibody staining during the wound healing process. Phosphotyrosine staining is prominent in new focal adhesions of cells migrating from the wound edge **(B)** and spreading in the middle of the wound breach **(C).** Cells distant from the newly made wound **(A)** or in the midst of the wound after migration has been completed **(D)** have decreased focal adhesion phosphotyrosine staining. Bar = 10 μm. (From ref. 80, with permission.)

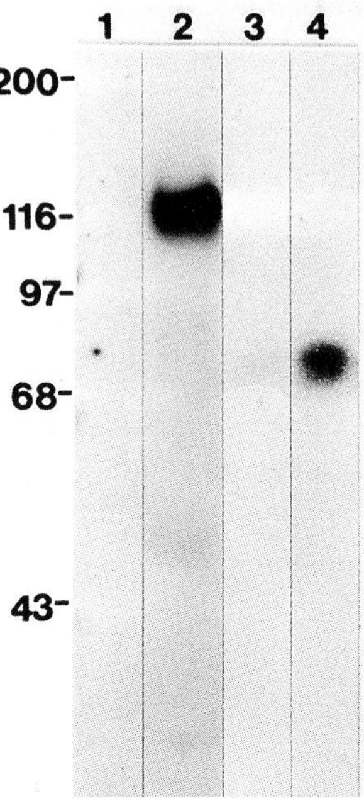


**FIG. 3.** Adhesion-associated tyrosine phosphorylation of FAK and paxillin. A composite of $^{125}$I-labeled anti-phosphotyrosine Western blots is shown. FAK **(lanes 1 and 2)** or paxillin **(lanes 3 and 4)** were immunoprecipitated from lysates of REF52 kept in suspension (lanes 1 and 3) or plated on fibronectin (lanes 2 and 4) for 60 min. Increased tyrosine phosphorylation is seen on each protein in lysates from adhered and spreading cells. Molecular masses are shown to the left of the autoradiograph. (From ref. 33, with permission.)

noprecipitated from REF52 held in suspension (Fig. 3, lanes 1 and 3) or plated for 1 h on fibronectin (lanes 2 and 4) by techniques detailed previously (33). Anti-phosphotyrosine Western blotting of these immunoprecipitates revealed adhesion-associated increases in phosphotyrosine on both proteins.

We then examined freshly plated cells (fibroblasts and HUVECs) and migrating HUVECs for the distribution of FAK (Figs. 4A–C). Immunofluorescence studies revealed that FAK was almost exclusively found in focal adhesions under these two conditions. Focal adhesions staining positively for FAK were frequently oriented along the axis of cell migration (Fig. 4C). The localization and adhesion-associated phosphorylation changes of FAK were therefore consistent with a role in focal adhesion formation.

### FAK Activity and New Focal Adhesion Assembly

As described above and elsewhere, we have noted increased tyrosine phosphorylation of FAK in REF52 and HUVECs adhering to fibronectin

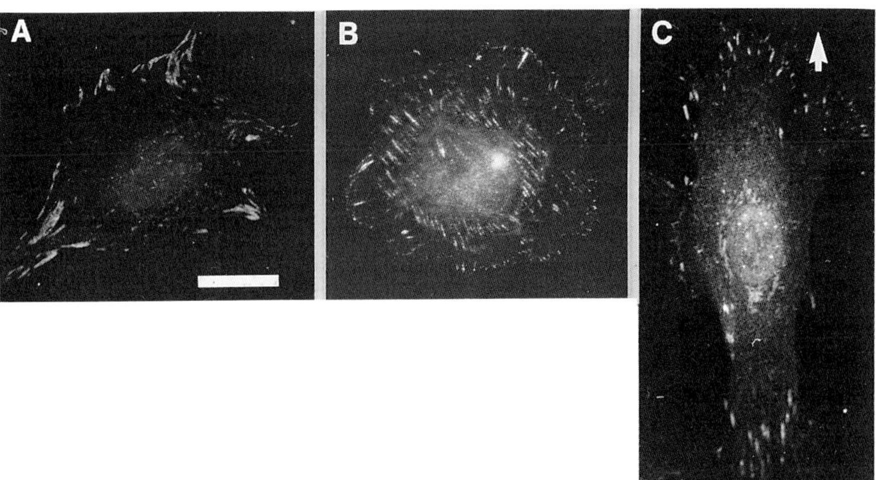


**FIG. 4.** The localization of FAK in freshly plated and migrating cells. Immunofluorescence localization of FAK in CEF and HUVEC **(A,B)** is shown 1 h after plating on fibronectin in serum-free medium. Focal adhesion staining is also seen with anti-FAK antibodies in migrating HUVECs **(C).** The direction of cell migration is shown. Bar = 20 μm. (From ref. 80, with permission.)

(see Fig. 3) (32,38), and elevated FAK activity has previously been noted during fibroblast and platelet adhesion (31,35). We reasoned that increases in FAK tyrosine phosphorylation and activity might also be associated with new focal adhesion assembly in migrating cells.

A nylon comb was used for the study of HUVEC migration in an in vitro wound healing assay involving multiple wounds (38). Lysates were made from intact and multiply wounded monolayers in the presence of protease inhibitors and sodium orthovanadate as described elsewhere, and were normalized for protein content using Coomassie blue-staining (32). Lysates were subjected to SDS-PAGE under reducing conditions, transferred to nitrocellulose, and probed using either $^{125}$I-labeled py20 and autoradiography (32) or anti-FAK antibody and enhanced chemiluminescence (38,44). The results are shown in Fig. 5. Increased tyrosine phosphorylation was seen in lysates from multiply wounded HUVEC monolayers compared to lysates from intact monolayers (Fig. 5A). The predominant signal corresponded to 110–130 kDa. FAK levels, however, were equivalent under the two conditions tested (Fig. 4B). These data are consistent with an increase in phosphotyrosine, but not in FAK expression, during the formation of new focal adhesions in migrating cells.

FAK activity was then compared in intact and multiply wounded HUVEC monolayers (Fig. 5C). Immune complex kinase assays with $^{32}$P-labeled γATP were used to quantify autophosphorylation. This technique is described in our earlier work (38). Briefly, FAK immunoprecipitates were

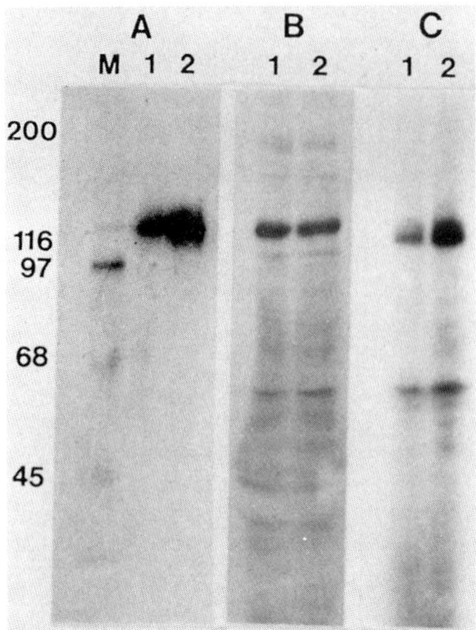


**FIG. 5.** FAK kinase activity is increased in migrating endothelial cells. Intact (lanes 1) and multiply wounded (lanes 2) HUVEC monolayers were cultured on fibronectin and analyzed for phosphotyrosine **(A)**, FAK **(B)**, and FAK activity **(C)**. Protein loads had been normalized using Coomassie blue-stained gels (data not shown). Lysates of the two cell groups were electrophoresed on polyacrylamide gels, transferred to nitrocellulose, and blotted with $^{125}$I-labeled py20 (A) or with anti-FAK antibody (B). Marker proteins ($^{125}$I-labeled in left panel) are shown in the lane labeled "M." The molecular masses of marker proteins are indicated in kDa to the left of panel A. Phosphotyrosine is increased in the lysate from the wounded monolayer in a cluster of dominant bands at 120–130 kDa, although equal amounts of FAK were found in the two cell populations **(B)**. FAK immunoprecipitates from wounded or intact HUVEC monolayers were then used for [$^{32}$P]-γATP–labeled assays of kinase activity (C). FAK autophosphorylation was increased in multiply wounded monolayers compared with the intact controls. (From ref. 38, with permission.)

incubated with $^{32}$P-labeled γATP in the presence or absence of various inhibitors, and autophosphorylation was used as a measure of FAK activity (28,35,38). FAK activity was noted to be higher in migrating HUVECs than it was in their quiescent and sessile counterparts. This observation further supports a role for FAK in locomoting HUVECs.

## Tyrphostin Inhibition of FAK Activity, Focal Adhesion Assembly, and Migration

The roles of FAK in cytoskeletal organization and cell motility were investigated using tyrphostins, specific competitive tyrosine kinase inhibitors (45). Tyrphostins are composed of phenolic rings with a variety of side chains. The high specificity and low cytotoxicity of these compounds is due to their structural similarity to tyrosine residues in a peptide chain (46,47).

Studies were done to test the efficacy of tyrphostins for direct FAK inhibition in immune complex kinase assays (38). As shown in Fig. 6, FAK immunoprecipitates from HUVECs adhering to fibronectin had increased activity compared with FAK from cells held in suspension (Fig. 6, lanes 1 and 2). This increase is consistent with FAK activation by tyrosine phosphorylation during cell–ECM adhesion (29–33,35,38,48). Tyrosine kinase inhibitors were then tested. FAK activity was almost obliterated by addition

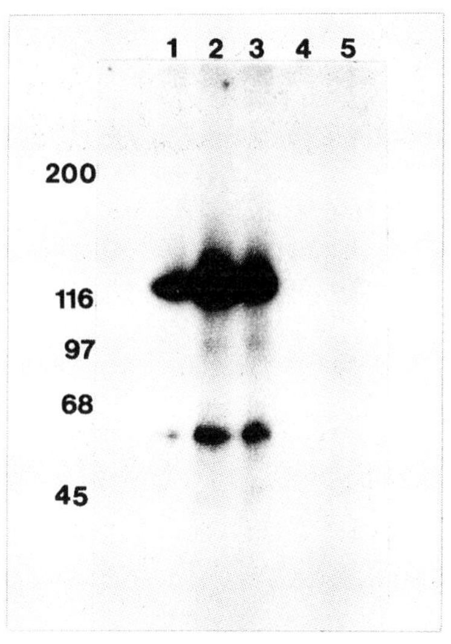


**FIG. 6.** Tyrphostins decrease FAK activity. [$^{32}$P]-$\gamma$ATP–labeled autophosphorylation was measured in FAK immunoprecipitates from HUVECs that were either kept in suspension for 1 h **(lane 1)** or plated for 1 h on fibronectin in serum-free medium **(lanes 2–5).** FAK immunoprecipitates from plated HUVECs were untreated (lane 2), or treated with herbimycin A (875 n*M*, lane 3), or the tyrphostins AG 213 (100 $\mu$*M*, lane 4), or AG 808 (100 $\mu$*M*, lane 5). Tyrphostins substantially reduced FAK activity. Molecular masses of marker proteins are indicated at the left of the autoradiograph. (From ref. 38, with permission.)

of tyrphostins AG 213 and AG 808 to the immunoprecipitates (Fig. 6, lanes 4 and 5, respectively), whereas the addition of herbimycin A had a less pronounced effect (Fig. 6, lane 3).

The role of FAK and other tyrosine kinases in adhesion-induced cytoskeletal organization was assessed by pretreating HUVECs with tyrphostin AG 213 and plating the cells in the presence of this inhibitor (Fig. 7). Focal adhesions were visualized with an anti-talin antibody (Figs. 7A,B), and filamentous actin was studied using rhodamine-conjugated phalloidin (Figs. 7C,D). In comparison with untreated controls (Figs. 7A,C), tyrphostin-treated cells demonstrated a lack of focal adhesions and a sparse and disorganized stress fiber pattern (Figs. 7B,D). This observation is consistent with the effects of other inhibitors on focal adhesion formation and stress fiber assembly in fibroblasts (32,33,49). These data suggest that tyrosine phosphorylation is an essential component of the events leading to cytoskeletal organization after cell adhesion to ECM.

Tyrphostin AG 213 was also used to examine the role of tyrosine kinases in the dynamic assembly of new focal adhesions in HUVECs migrating into in vitro wounds made by pipette scratches in confluent monolayers (Fig. 8; methods detailed in 38). Cells were visualized by epifluorescence microscopy after staining with anti-phosphotyrosine antibody. Untreated control HUVECs formed well-defined, phosphotyrosine-staining focal adhesions as they closed the wound breach by 48 h after wounding (Fig. 8A). The wound path was thoroughly covered by a confluence of the control cells at 96 h, and focal adhesion staining for phosphotyrosine was less prominent (Fig. 8C). Tyrphostin-treated HUVECs barely migrated within the microscopic field at 48 h (Fig. 8B). By 96 h, recovery from tyrphostin treatment was evident,

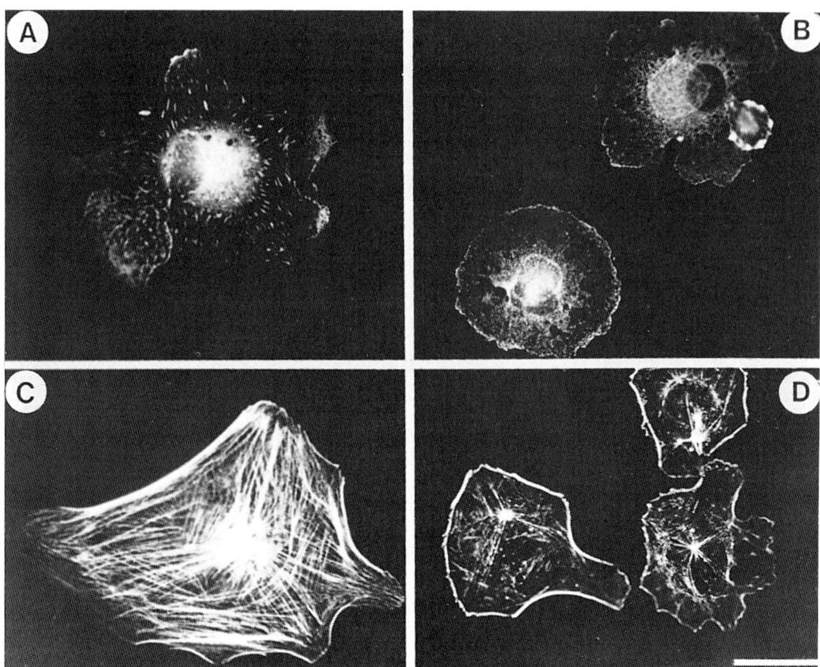


**FIG. 7.** Cytoskeletal organization is decreased in tyrphostin-treated HUVECs. HUVECs were stained for immunofluorescence with anti-talin antibodies **(A,B),** or with rhodamine-conjugated phalloidin **(C,D).** Untreated HUVECs (A,C) and tyrphostin-treated HUVECs (100 μ*M* AG 213; B,D) were compared 1 h after plating in serum-free medium on fibronectin. Bar = 10 μm. (From ref. 38, with permission.)

with improved cell migration and focal adhesion staining for phosphotyrosine (Fig. 8D). These findings support the idea that new focal adhesion formation during cell movement is dependent on tyrosine phosphorylation of focal adhesion proteins.

## Cytosolic Free Calcium Changes in Cell Signaling

Recently, integrin-mediated cell adhesion has been linked to changes in intracellular calcium during neutrophil adhesion (50), migration (51), and transmigration across endothelial monolayers (a process mediated by $\beta_2$ integrins) (52). Studies of intracellular calcium mobilization during $\alpha_v\beta_3$ integrin-mediated adhesion in rat osteoclasts demonstrated a rise in nuclear free calcium and, hence, perhaps a direct link to changes in gene expression (53). Rises in intracellular calcium were also detected during adhesion to fibrinogen in human 293 cells transfected with cDNA for the integrin $\alpha_{IIb}\beta_3$ (30) and during adhesion of human umbilical vein endothelial cells to fibronectin and vitronectin (39,54,55).

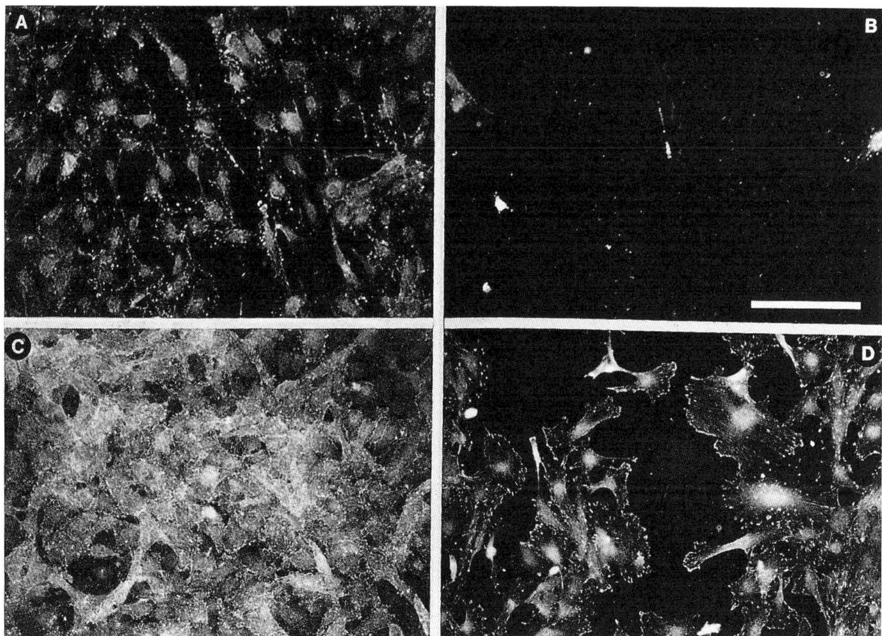


**FIG. 8.** Migration is inhibited in tyrphostin-treated HUVECs. HUVECs were stained for immunofluorescence with anti-phosphotyrosine antibody. Various stages of wound healing are seen after cells were scraped away from a linear wound with a 1-mm pipette tip. All panels were photographed in the center of the wound breach. Untreated HUVEC monolayers **(A,C)** almost closed the wound by 48 h (A) and obliterated all traces of the wound by 96 h (C). Tyrphostin-treated HUVEC (AG 213, 100 μ*M*; **B,D**) showed decreased migration at both 48 and 96 h. Bar = 150 μm. (From ref. 80, with permission.)

We have observed oscillations in cytosolic free calcium levels in HUVECs during adhesion and spreading on fibronectin in the absence of serum and growth factors (data not shown; see 39). These studies have been performed using digitized video microscopy and fura-2 AM-loaded HUVECs. The oscillatory pattern was most marked during the period of rapid cell spreading (20–60 min after plating), had amplitudes of 30 to 150 n*M*, and a frequency of 1–3 $min^{-1}$. These oscillations were also observed in cycloheximide-pretreated HUVECs plated on other ECM proteins, including collagen, laminin, and vitronectin, but not on poly-L-lysine. These adhesion-associated calcium oscillations also occurred in cells plated on the activating anti-$\beta_1$ monoclonal antibody TS-2/16 and on the cell-binding domain of fibronectin (data not shown).

The work of others has implicated the 50-kDa integrin associated protein (IAP) as an important calcium channel that is activated in HUVECs during adhesion to some ECM proteins (56). The biologically active anti-IAP antibody B6H12 (gift of Eric Brown, Washington University, St. Louis, MO) does not effect the adhesion-associated calcium oscillations that we observe when used at 20 μg/ml (data not shown). This observation indicates that the oscillatory changes in intracellular free calcium that we observe during the integrin-dependent adhesion of HUVECs are not mediated by IAP. Furthermore, these oscillations appear to be distinct from the calcium changes in adhering HUVECs described by others (54–56).

A role for calcium oscillations in HUVEC spreading is also supported by the distribution of cytosolic free calcium in these cells as documented using the fluorescent calcium indicator fluo-3 and laser scanning confocal microscopy (data not shown; see 39). Higher concentrations of free calcium were localized to the nucleus and to the areas of the cell that were involved in actively spreading lamellipodia. These data suggest that high concentrations of free calcium may be involved in nuclear events and in local cytoskeletal extension during cell spreading.

## DISCUSSION

### Tyrosine Phosphorylation and Focal Adhesion Assembly

The data summarized here indicate that tyrosine phosphorylation contributes to the process of focal adhesion assembly during cell adhesion. Phosphotyrosine staining was most prominent in the new focal adhesions in the extending lamellipodia of actively spreading (Fig. 1; see 32) and migrating cells (Fig. 2; see 38). In addition, specific focal adhesion proteins demonstrated an increase in tyrosine phosphorylation in association with cell adhesion (Fig. 3; see 33). The tyrosine kinase inhibitors herbimycin A and tyrphostins inhibited focal adhesion formation (Fig. 7; see 32,38), and focal adhesion-associated endothelial cell movement (Fig. 8; see 38). Therefore, it appears that the tyrosine phosphorylation of focal adhesion components serves a role in their orderly assembly.

The adhesion-associated activation of FAK (Figs. 5 and 6; see 29–33,35, 38,48) may provide the means by which some focal adhesion protein constituents are tyrosine-phosphorylated during cell adhesion to ECM. The focal adhesion kinase is a unique cytoplasmic, nonreceptor tyrosine kinase exhibiting a centrally located catalytic domain and a lack of Src homology units (28,29). FAK localizes to the focal adhesions of actively spreading and migrating cells in culture (Fig. 4; see 28,29,38). A great deal of cross-species homology has been demonstrated in the cDNA sequences of FAK cloned from chicken and murine fibroblasts and human lymphocytes (28,29,48,57, 58). This conservation of sequence information is consistent with the involvement of FAK in the basic and essential process of cell–ECM adhesion.

Recent insights into the binding specificities imparted by two portions of the FAK amino-acid sequence support the concept that FAK is important in the initial formation of focal adhesions. A 159-amino-acid sequence at the carboxy terminus directs the incorporation of FAK into focal adhesions (59). Segments of the amino terminus of FAK appear to mediate a direct binding relationship with $\beta_1$ and $\beta_3$ integrins (60). Furthermore, the focal adhesion proteins paxillin and tensin may associate with and be substrates for FAK (42,48). Paxillin and tensin are both tyrosine-phosphorylated during integrin-mediated cell adhesion to ECM (32,61). In addition, tensin contains an SH2 (Src homology 2) domain (62), which may mediate protein binding to phosphorylated tyrosine residues on adjacent proteins (63). This SH2 domain may interact with phosphotyrosines in either paxillin or FAK. Taken together with data about FAK binding to $\beta_1$ integrin and targeting to focal adhesions, these observations suggest that FAK has a significant role in focal adhesion assembly. FAK may facilitate the initial juxtaposition of integrin $\beta$-chains with some cytoskeletal elements. The subsequent phosphorylation of focal adhesion protein constituents by FAK may allow the recruitment, sorting, and orderly incorporation of these proteins into developing focal adhesions.

Tyrosine kinases, including FAK, may be only partly responsible for adhesion-associated increases in the tyrosine phosphorylation of focal adhesion proteins. Cytoskeletal organization during cell adhesion may also be accompanied by changes in the activation states of pertinent phosphatases, although none has yet been localized to focal adhesions. Indeed, vanadyl hydroperoxide inhibition of tyrosine phosphatase activity has been noted to induce cytoskeletal changes and a motile morphology in neutrophils (64). In platelets, $\alpha_{IIb}\beta_3$ integrin-mediated adhesion to fibrinogen has been correlated with changes in the activation state of the tyrosine phosphatase PTP-1B (65). In both of these models, phosphatase activity appears to be linked to changing levels of phosphotyrosine in cytoskeletal proteins during cell adhesion. Focal adhesion disassembly has been correlated with upregulation of phosphatase activity and protein tyrosine dephosphorylation in trypsinized fibroblasts (66). Therefore, the orderly assembly and remodeling of focal adhesions in adherent cells may depend on complex regulatory mechanisms that alter phosphotyrosine levels in constituent proteins. It is intriguing to consider that moving cells may contain regional differences in the local activities of tyrosine kinases and phosphatases. These enzymes may be under the control of regulatory switches that enable cells to form concentrations of heavily tyrosine-phosphorylated focal adhesion proteins at the leading edge while dephosphorylating focal adhesion constituents during retraction of the trailing tail.

## Signaling During Integrin-Mediated Cell Adhesion

The regulation of the phosphotyrosine content of focal adhesion proteins does not appear to be an isolated signaling process during cell adhesion to

ECM. Complex interactions have been observed with several other pathways. Data linking tyrosine kinase-mediated events to the actin cytoskeleton include the inhibition of stress fiber formation by herbimycin A and tyrphostins (Fig. 7; see 32,33,38). This effect was observed under the same conditions that prevented focal adhesion assembly. Furthermore, prevention of actin polymerization with cytochalasin D inhibits both FAK activation during platelet adhesion to fibrinogen (35) and the tyrosine phosphorylation of tensin during fibroblast adhesion (61). FAK signaling has also been linked with MAP kinase activation after integrin-associated cell adhesion (67), and with the activation state of small GTP-binding proteins such as rhoA (68–70).

Connections among these diverse pathways are under intense investigation and may provide insights into important questions. Two issues in integrin-associated signaling that have attracted wide interest are signal specificity and the transmission of signals to the nucleus (1,13). FAK activation appears to be a relatively nonspecific signal that is triggered by diverse integrins and by neuropeptides (48,71). Parallel or downstream signaling events may impart specificity to the integrin-mediated adhesion that activates FAK during ECM adhesion. These signaling cascades have yet to be defined. Nuclear events triggered by cell adhesion to ECM include the activation of immediate early genes (13), anchorage-dependent responsiveness to growth factors (72,73), and the suppression of programmed cell death (74). Signaling pathways that link FAK to these events must interact directly with the nucleus. Therefore, ionic changes such as pH that may be transmitted from cytoplasm to nucleus have been proposed as potential messengers (72,73). Signaling proteins that translocate from cytoplasm to nucleus on activation, such as the MAP kinases (67), are also attractive candidates for this role.

Cytosolic free calcium changes represent another potential pathway by which adhesion-associated tyrosine phosphorylation may be transduced to nuclear signaling. Recent data link tyrosine kinase activity and cytoplasmic calcium changes in fibroblasts and vascular smooth-muscle cells (75,76). Cytosolic free calcium is one determinant of nuclear free calcium (77), and nuclear free calcium levels have been implicated in DNA replication, transcription, and repair (77–79). Therefore, cytosolic free calcium changes may provide a link between adhesion-associated tyrosine kinase activation and gene expression. We have begun to examine the association between FAK and the oscillatory calcium changes that accompany integrin-mediated endothelial cell adhesion (see above).

### Extracellular Matrix Adhesion in the Vascular Endothelium

In the studies presented here, cell adhesion to ECM has been examined in two model systems of new focal adhesion formation: freshly plated actively spreading cells and cells migrating into wounds in monolayers. We have used vascular endothelial cells in several of these studies. The growth behavior and morphology of these cells are particularly well suited to these investigations. Adaptation to the ECM environment is a well-developed feature of endothelial cells in culture, as evidenced by marked changes in growth behavior associated with changes in matrix components. The apparent association between focal adhesion formation and endothelial cell movement (38) further supports the concept that ECM is an important determinant of structure and function in vascular endothelium. Endothelial cells in culture have a flat, well-spread morphology with a relatively short dorsal–ventral axis, and these features are conducive to imaging focal adhesion and cytoskeletal proteins. These cells also achieve quiescence in contact-inhibited monolayer culture. This phenomenon has allowed the comparison of tyrosine phosphorylation and calcium signaling in quiescent cells with those of cells that actively reorder the cytoskeleton in response to wound healing or cytokine activation.

The coordination of signaling pathways is central to the function of vascular endothelial cells. These cells must interact with other vascular cells, elaborate humoral messengers, and maintain a selective permeability barrier. Each of these endothelial functions is dependent on appropriate adaptation of cell shape and function to the extracellular environment in response to signaling data transmitted by adhesion receptors. Further studies are needed to define processes that connect cell surface–cytoskeleton signaling and the regulation of gene expression, protein synthesis, and cell shape. Understanding these processes will be essential to the appreciation of the complex gatekeeper functions of vascular endothelial cells.

## SUMMARY

Focal adhesions are transmembrane bridges between the cytoskeleton and the extracellular matrix. They function both as structural adhesive links and as signaling pathways. We have investigated signaling functions in focal adhesions.

New focal adhesion assembly during cell adhesion, spreading, and migration is associated with increased phosphotyrosine in focal adhesions. The focal adhesion proteins FAK and paxillin undergo increased tyrosine phosphorylation during integrin-mediated cell adhesion to ECM. FAK is localized to focal adhesions in spreading and migrating cells and exhibits increased kinase activity during the new focal adhesion formation associated with cell migration.

The tyrphostin family of specific tyrosine kinase inhibitors was used to explore the functional significance of tyrosine phosphorylation and increased FAK activity in the endothelial cells during adhesion and migration.

Tyrphostins AG 213 and AG 808 inhibited the kinase activity of FAK in immune complex kinase assays. Endothelial cell focal adhesion and stress fiber formation were inhibited by the same concentration of tyrphostin AG 213. Tyrphostins also inhibited endothelial cell migration in an in vitro wound healing model.

Changes in intracellular calcium have also been noted during integrin-mediated adhesion. Cytosolic free calcium is localized to extending lamellipodia in spreading endothelial cells. Endothelial cell levels of cytosolic free calcium oscillate in response to integrin-mediated adhesion to ECM proteins. These oscillations are associated with the period of rapid cell spreading in these cells.

These data suggest that FAK-mediated tyrosine phosphorylation events and cytosolic free calcium oscillations may have a vital role in the complex and multifactorial process of endothelial cell wound healing. Therapeutic strategies that modulate these signaling processes may have an important impact on the treatment of vascular disease and injury.

## ACKNOWLEDGMENT

Gratitude is extended to Keith Burridge for his magnanimous mentorship, collaboration, and support, and to Natalie V. McLean for her superlative and untiring technical contributions. Thanks also to colleagues Ammasi Periasamy and Brian Herman at the University of North Carolina for their collaboration on the calcium studies. Generous gifts of reagents are appreciated from Michael Schaller and J. Thomas Parsons (University of Virginia at Charlottesville), Alexander Levitski (Hebrew University, Jerusalem, Israel), Harold Erickson (Duke University, Durham, North Carolina), and Eric Brown (Washington University, St. Louis, Missouri). This work was supported by University Research Council and Junior Faculty Development Awards from the University of North Carolina at Chapel Hill and by HL 45100 and GM 29860.

## REFERENCES

1. Hynes RO, Integrins: versatility, modulation, and signaling in cell adhesion. *Cell* 1992;69:11–25.
2. Takeichi M. The cadherins: cell-cell adhesion molecules controlling animal morphogenesis. *Development* 1988;102:639–55.
3. Albelda SM, Buck CA. Integrins and other cell adhesion molecules. *FASEB J* 1990;4:2868–71.
4. Edelman GM. Cell adhesion molecules in the regulation of animal form and tissue pattern. *Annu Rev Cell Biol* 1986;2:81–116.
5. Buck CA, Horwitz AF. Cell surface receptors for extracellular matrix molecules. *Annu Rev Cell Biol* 1987;3:179–205.
6. Takeichi M. Cadherin cell adhesion receptors as a morphogenetic regulator. *Science* 1991;251:1451–5.

7. Kubota Y, Kleinman HK, Martin GR, Lawley TJ. Role of laminin and basement membrane in the morphological differentiation of human endothelial cells. *J Cell Biol* 1988;107:1589–98.
8. Madri JA, Pratt BM, Yannariello-Brown J. Matrix-driven cell size change modulates aortic endothelial cell proliferation and migration. *Am J Pathol* 1988;132:18–27.
9. Yost JC, Herman IM. Substratum-induced stress fiber assembly in vascular endothelial cells during spreading *in vitro*. *J Cell Sci* 1990;95:507–20.
10. Hemler M. VLA proteins in the integrin family: structure, functions and their role on leukocytes. *Annu Rev Immunol* 1990;8:365–400.
11. Guan J-L, Trevithick, JE, Hynes RO. Fibronectin/integrin interaction induces tyrosine phosphorylation of a 120 kDa protein. *Cell Regul* 1991;2:951–64.
12. Kornberg LJ, Earp HS, Turner CE, Prockop C, Juliano RL. Signal transduction by integrins: increased protein tyrosine phosphorylation caused by clustering $\beta_1$ integrins. *Proc Natl Acad Sci USA* 1991;88:8392–6.
13. Juliano RL, Haskill S. Signal transduction from ECM. *J Cell Biol* 1993;120:577–85.
14. Burridge K, Petch L, Romer L. Signals from focal adhesions. *Curr Biol* 1992;2:537–9.
15. Izzard CS, Lochner LR. Cell-to-substrate contacts in living fibroblasts: an interference reflection study with an evaluation of the technique. *J Cell Sci* 1976;21:129–59.
16. Burridge K, Fath KT, Kelly T, Nuckolls G, Turner C. Focal adhesions: transmembrane junction between the extracellular matrix and the cytoskeleton. *Annu Rev Cell Biol* 1988; 4:487–525.
17. Horwitz A, Duggan K, Buck C, Beckerle MC, Burridge K. Interaction of plasma membrane fibronectin receptor with talin—a transmembrane linkage. *Nature* 1986;320:531–3.
18. Otey CA, Pavalko FM, Burridge K. An interaction between α-actinin and the $\beta_1$ integrin subunit in vitro. *J Cell Biol* 1990;111:721–9.
19. Otey CA, Vasquez GB, Burridge K, Erickson BW. Mapping of the α-actinin binding site within the $\beta_1$ integrin cytoplasmic domain. *J Biol Chem* 1993;268:21193–7.
20. DePasquale JA, Izzard CS. Evidence for an actin-containing cytoplasmic precursor of the focal contact and the timing of incorporation of vinculin at the focal contact. *J Cell Biol* 1987;105:2803–9.
21. Regen CM, Horwitz AF. Dynamics of β1 integrin-mediated adhesive contacts in motile fibroblasts. *J Cell Biol* 1992;119:1347–59.
22. Nuckolls GH, Romer LH, Burridge K. Microinjection of antibodies against talin inhibits the spreading and migration of fibroblasts. *J Cell Sci* 1992;102:753–62.
23. Jaken S, Leach K, Klauck T. Association of type 3 protein kinase C with focal contacts in rat embryo fibroblasts. *J Cell Biol* 1989;109:697–704.
24. Woods A, Couchman JR, Protein kinase C involvement in focal adhesion formation. *J Cell Sci* 1992;101:277–90.
25. Woods A, McCarthy J, Furcht L, Couchman JR. Synthetic peptide from the heparin-binding domain of fibronectin promotes focal adhesions. *Mol Biol Cell* 1993;4:605–13.
26. Rohrschneider LR. Adhesion plaques of Rous sarcoma virus-transformed cells contain the *src* gene product. *Proc Natl Acad Sci USA* 1980;77:3514–8.
27. Maher PA, Pasquale EB, Wang JYJ, Singer SJ. Phosphotyrosine-containing proteins are concentrated in focal adhesions and intercellular junctions in normal cells. *Proc Natl Acad Sci USA* 1985;82:6576–80.
28. Schaller MD, Borgman CA, Cobb BS, Vines RR, Reynolds AB, Parsons JT. $pp125^{FAK}$, a structurally unique protein tyrosine kinase associated with focal adhesions. *Proc Natl Acad Sci USA* 1992;89:5192–6.
29. Hanks SK, Calalb MB, Patel SK. Focal adhesion protein-tyrosine kinase phosphorylated in response to cell attachment. *Proc Natl Acad Sci USA*. 1992;89:8487–91.
30. Pelletier AJ, Bodary SC, Levinson AD. Signal transduction by the platelet integrin $\alpha_{IIb}\beta_3$: Induction of calcium oscillations required for protein-tyrosine phosphorylation and ligand-induced spreading of stably transfected cells. Mol Biol Cell 1992;3:989–98.
31. Guan J-L, Shalloway D. Regulation of focal adhesion-associated protein tyrosine kinase by cellular adhesion and transformation. *Nature* 1992;358:690–2.
32. Burridge K, Turner C, Romer L. Tyrosine phosphorylation of paxillin and $pp125^{FAK}$ accompanies cell adhesion to extracellular matrix. *J Cell Biol* 1992;119:893–903.
33. Romer LH, Burridge K, Turner CE. Signaling between the extracellular matrix and the

cytoskeleton: tyrosine phosphorylation and focal adhesion assembly. In: *The cell surface. Cold Spring Harbor Laboratory Symposium on Quantitative Biology, Vol. 57*. Plainview, NY: Cold Spring Harbor Laboratory Press, 1992;193–202.

34. Kornberg L, Earp HS, Parsons JT, Schaller M, Juliano RL. Cell adhesion or integrin clustering increases phosphorylation of a focal adhesion-associated tyrosine kinase. *J Biol Chem* 1992;267:23439–42.
35. Lipfert L, Haimovich B, Schaller MD, Cobb BS, Parsons JT, Brugge JS. Integrin-dependent phosphorylation and activation of the protein tyrosine kinase $pp125^{FAK}$ in platelets. *J Cell Biol* 1992;119:905–12.
36. Fuortes J, Jin W-W, Nathan C. Adhesion-dependent protein tyrosine phosphorylation in neutrophils. *J Cell Biol* 1993;120:777–84.
37. Weiner TM, Liu ET, Craven RJ, Cance WG. Expression of focal adhesion kinase gene and invasive cancer. *Lancet* 342:1024–5.
38. Romer LH, McLean NV, Turner CE, Burridge K. Tyrphostins inhibit focal adhesion kinase activation, cytoskeletal organization, and motility in human vascular endothelial cells. *Mol Biol Cell* 1994;5:349–61.
39. Romer LH, Periasamy A, Herman B, Burridge K. Oscillations in cytosolic free calcium in human vascular endothelial cells: association with cell spreading [Abstract]. *Mol Biol Cell* 1993;4:S365a.
40. Gimbrone MA Jr, Cotran RS, Folkman J. Human vascular endothelial cells in culture. *J Cell Biol* 1984;60:673–81.
41. DePasquale JA, Izzard CS. Accumulation of talin in nodes at the edge of the lamellipodium and separate incorporation into adhesion plaques of focal contacts in fibroblasts. *J Cell Biol* 1991;113:1351–9.
42. Turner CE. Paxillin is a major phosphotyrosine-containing protein during embryonic development. *J Cell Biol* 1991;115:201–7.
43. Turner CE, Glenney JR, Burridge K. Paxillin: a new vinculin-binding protein present in focal adhesions. *J Cell Biol* 1990;111:1059–68.
44. Bockholt SM, Otey CA, Glenney JR Jr, Burridge K. Localization of a 215-kDa tyrosine-phosphorylated protein that cross-reacts with tensin antibodies. *Exp Cell Res* 1992;203:39–46.
45. Levitski A. Tyrphostins—novel molecular tools. *Biochem Pharmacol* 1990;40:913–18.
46. Dvir A, Milner Y, Chomsky O, Gilon C, Gazit A, Levitski, A. The inhibition of EGF-dependent proliferation of keratinocytes by tyrphostin tyrosine kinase blockers. *J Cell Biol* 1991;113:857–65.
47. Lyall RM, Zilberstein A, Gazit A, Gilon C, Levitski A, Schlessinger J. Tyrphostins inhibit epidermal growth factor (EGF)-receptor tyrosine kinase activity in living cells and EGF-stimulated cell proliferation. *J Biol Chem* 1989;264:14503–9.
48. Schaller MD, Parsons JT. Focal adhesion kinase. *Trends Cell Biol* 1993;3:258–62.
49. Chrzanowska-Wodnicka M, Burridge K. The focal adhesion kinase is involved in reorganization of the actin cytoskeleton in response to serum or LPA stimulation. *J Cell Sci* 1994;107:3643–54.
50. Jaconi MEE, Theler JM, Schlegel W, Appel RD, Wright SD, Lew PD. Multiple elevations of cytosolic-free $Ca^{2+}$ in human neutrophils: initiation by adherence receptors of the integrin family. *J Cell Biol* 1991;112:1249–57.
51. Marks PW, Hendey B, Maxfield FR. Attachment to fibronectin or vitronectin makes human neutrophil migration sensitive to alterations in cytosolic free calcium concentration. *J Cell Biol* 1991;112:149–58.
52. Huang AJ, Manning JE, Bandak TM, Ratau MC, Hanser KR, Silverstein SC. Endothelial cell cytosolic free calcium regulates neutrophil migration across monolayers of endothelial cells. *J Cell Biol* 1993;120:1371–80.
53. Shankar G, Davison I, Mason WT, Horton MA. Integrin receptor-mediated mobilisation of intranuclear calcium in rat osteoclasts. *J Cell Sci* 1993;105:61–8.
54. Schwartz MA. Spreading of human endothelial cells on fibronectin or vitronectin triggers elevation of intracellular free calcium. *J Cell Biol* 1993;120:1003–10.
55. Leavesley DI, Schwartz MA, Cheresh DA. Integrin β1- and β3-mediated endothelial cell migration triggered by distinct mechanisms. *J Cell Biol* 1993;121:163–70.
56. Schwartz MA, Brown EJ, Fazeli B. A 50-kDa integrin-associated protein is required for integrin regulated calcium entry in endothelial cells. *J Biol Chem* 1993;268:19931–3.

57. Whitney GS, Chan P-Y, Blake J, Aruffo A, Kanner SB. Human T and B lymphocytes express a conserved focal adhesion kinase. *DNA Cell Biol* 1993;12:823–30.
58. Devor BB, Zhang X, Patel SK, Polte TR, Hanks SK. Chicken and mouse focal adhesion kinases are similar in structure at their amino termini. *Biochem Biophys Res Commun* 1993;190:1084–9.
59. Hildebrand JD, Schaller MD, Parsons JT. Identification of sequences required for localization of focal adhesion kinase to focal adhesion. *J Cell Biol* 1993;123:993–1005.
60. Otey CA, Schaller M, Parsons JT. Focal adhesion kinase binds directly to the cytoplasmic domain of $\beta_1$ integrin in vitro. *Mol Biol Cell* 1993;4:S347a(abst).
61. Bockholt SB, Burridge K. Cell spreading on extracellular matrix proteins induces tyrosine phosphorylation of tensin. *J Biol Chem* 1993;268:14565–9.
62. Davis S, Lu ML, Lin S, et al. Presence of an SH2 domain in the actin-binding protein tensin. *Science* 1991;252:712–5.
63. Koch D, Anderson D, Moran MF, Ellis C, Pawson T. SH2 and SH3 domains: elements that control interaction of cytoplasmic signaling proteins. *Science* 1991;252:668–74.
64. Bennett PA, Dixon RJ, Kellie S. The phosphotyrosine phosphatase inhibitor vanadyl hydroperoxide induces morphological alterations, cytoskeletal rearrangements and increased adhesiveness in rat neutrophil leucocytes. *J Cell Sci* 1993;106:891–901.
65. Frangioni J, Oda A, Smith M, Salzman EW, Neel BG. Calpain-catalyzed cleavage and subcellular relocation of protein phosphotyrosine phosphatase 1B in human platelets. *EMBO J* 1993;12:4843–56.
66. Maher P. Activation of phosphotyrosine phosphotase activity by reduction of cell-substrate adhesion. *Proc Natl Acad Sci USA* 1993;90:11177–81.
67. Chen Q, Kinch M, Lin T, Burridge K, Juliano RL. Integrin-mediated cell adhesion activates MAP kinases. *J Biol Chem* 1994;269:26602–5.
68. Kumagai N, Morii N, Fujisawa K, Nemoto Y, Narumiya S. ADP-ribosylation of *rho* p21 inhibits lysophosphatidic acid-induced protein tyrosine phosphorylation in cultured Swiss 3T3 cells. *J Biol Chem* 1993;268:24535–8.
69. Seufferlein T, Rozengurt E. Lysophosphatidic acid stimulates tyrosine phosphorylation of focal adhesion kinase, paxillin, and p130. *J Biol Chem* 1994;269:9345–51.
70. Kumagai N, Morii N, Fujisawa K, Yoshimasa T, Nakao K, Narumiya S. Lysophosphatidic acid induces tyrosine phosphorylation and activation of MAP-kinase and focal adhesion kinase in cultured Swiss 3T3 cells. *FEBS Lett* 1993;329:273–6.
71. Zachary I, Sinnett-Smith J, Rozengurt E. Bombesin, vasopressin, and endothelin stimulation of tyrosine phosphorylation. *J Biol Chem* 1992;267:19031–4.
72. Schwartz MA, Rupp E, Frangioni JV, Lechene CP. Cytoplasmic pH and anchorage-independent growth induced by V-Ki-ras, V-src, and polyoma middle T. *Oncogene* 1990; 5:55–88.
73. Schwartz MA, Ingber DE, Lawrence M, Springer TA, Lechene CP. Multiple integrins share the ability to induce elevation of intracellular pH. *Exp Cell Res* 1991;195:533–5.
74. Meredith JE Jr, Fazeli B, Schwartz MA. The extracellular matrix as a cell survival factor. *Mol Biol Cell* 1993;4:953–61.
75. Lee K-M, Toscas K, Villereal ML. Inhibition of bradykinin- and thapsigargin-induced $Ca^{2+}$ entry by tyrosine kinase inhibitors. *J Biol Chem* 1993;268:9945–8.
76. Wijetunge S, Aalkjaer C, Hughes AD. Tyrosine kinase inhibitors block calcium channels in vascular smooth muscle cells. *Biochem Biophys Res Commun* 1992;189:1620–23.
77. Bachs O, Agell N, Carafoli E. Calcium and calmodulin in cell nucleus. *Biochem Biophys Acta* 1992;1113:259–70.
78. Rodland KD, Muldoon LL, Lenormand P, Magun BE. Modulation of RNA expression by intracellular calcium. *J Biol Chem* 1990;265:11000–7.
79. Ryazanov A, Spirin A. Phosphorylation of elongation factor 2. *New Biologist* 1990;2:843–50.
80. Romer RH. Endothelial cell signalling during wound healing. In: Lemasters J, Oliver J, eds. *Cell biology of trauma*. Boca Raton, FL: CRC Press, 1994:262–87.

*Topics in Molecular Medicine, Volume 1,*
edited by Wolfgang Siess, Reinhard Lorenz,
and Peter C. Weber. Raven Press, Ltd.,
New York © 1995.

# 5

# Mechanisms Involved in Adherence-Dependent Activation of Monocyte Gene Expression

Alan K. Lofquist, Laura K. Price, *Krishna Mondal, and †J. Stephen Haskill

*UNC Lineberger Comprehensive Cancer Center, *Department of Microbiology and Immunology and *†Department of Obstetrics and Gynecology, University of North Carolina at Chapel Hill, Chapel Hill, North Carolina 27599-7295*

At least three families of cellular adhesion molecules (CAMs) play distinct and important roles in the extravasation of leukocytes into sites of injury, infection, and wound repair (see 1 for a recent review). Although the particular steps involved for monocyte localization are not as well characterized as those for neutrophils, the general characteristics are likely to be similar. The initial tethering of leukocytes in regions of endothelial cell activation is dependent on the selectin family of adhesion receptors. Attachment to vascular endothelium is then secured by the leukocyte CD18 ($\beta_2$ integrin) recognition of ICAM-1, ICAM-2, and VCAM-1 on activated endothelial cells (2,3). The association is further strengthened by interactions between members of the VLA/$\beta_1$ family of integrins and endothelial cell VCAM-1. Interactions between the leukocyte integrins and extracellular matrix proteins augment the adhesive interaction and facilitate leukocyte invasion of extracellular spaces.

During extravasation, both neutrophils and monocytes synthesize and release a number of inflammatory mediator cytokines, such as TNF$\alpha$, IL-8, and IL-1$\beta$. This suggests that adhesive interactions with the extracellular matrix can have profound effects on gene expression. Indeed, several previous studies have provided evidence for adherence- and matrix-dependent regulation of gene expression (4–8). Because transient adhesive interactions are clearly important for monocyte diapedesis, we have been investigating the mechanisms by which adhesion stimulates gene expression in primary human monocytes. We have specifically focused on the role(s) played by monocyte integrins in this process.

## MONOCYTE ADHESION STIMULATES ACCUMULATION OF INFLAMMATORY MEDIATOR RNAs

Isolation of peripheral monocytes from the blood of healthy normal donors requires extensive density or elutriation procedures to remove contaminating lymphocytes. Alternatively, monocytes can be purified by adherence, but compared with monocytes isolated by Ficoll (9) and Percoll (10) density centrifugation methods, monocytes purified by adherence to tissue culture plastic appear highly activated, as demonstrated by an increase in steady-state levels for a number of immediate early transcripts such as IL-1β, TNFα, and c-fos (11). Our initial screening of a cDNA library constructed from 30-min adhered monocytes (12) identified many unique cDNA clones, indicating that adhesion is probably a global activator of immediate early monocyte genes associated with the early stages of inflammation. We subsequently demonstrated that although adhesion to surfaces other than plastic also induced expression of the immediate early genes, the induction patterns differed in an apparently substrate-dependent manner (12,13). For

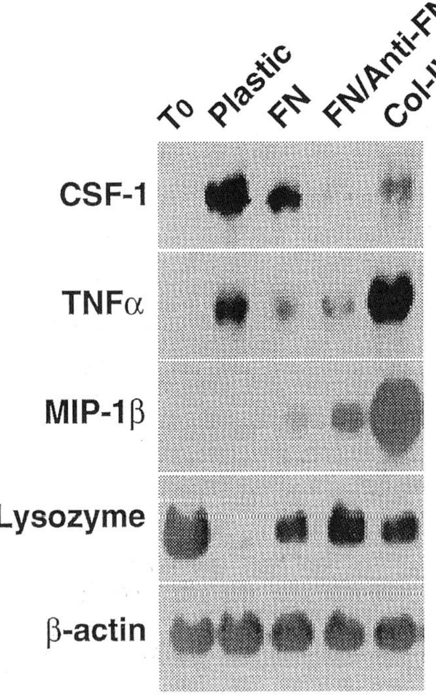


**FIG. 1.** Monocyte adhesion stimulates accumulation of inflammatory mediator RNAs. CSF-1, TNFα, MIP-1β, and lysozyme are differentially expressed during monocyte adherence to biologic substrates. Total RNA was isolated from freshly isolated monocytes ($T_0$) and from monocytes that were adhered for 4 h to tissue culture dishes (plastic), fibronectin-coated dishes (FN), fibronectin/anti-fibronectin antibody-coated dishes (FN/Anti-FN), and to collagen type IV (Col-IV)-coated dishes as indicated. The mRNA levels for CSF-1, TNFα, MIP-1β, lysozyme, and β-actin were determined by Northern analysis as previously described (13).

**TABLE 1.** *Secretion of TNFα and CSF-1 from adhered monocytes requires a second signal*

| Time | Units TNFα/ml | |
|---|---|---|
| | −LPS | +LPS |
| 20 min | 0 | 0 |
| 40 min | 4 | 120 |
| 90 min | 12 | 210 |
| 2 h | 12 | 175 |
| | Units CSF-1/ml | |
| 0 | 0 | 0 |
| 4.5 h | 22 | 22 |
| 20 h | 18 | 103 |

example, induction of CSF-1 mRNA was greater on both plastic- and fibronectin-coated surfaces than on collagen type IV (Fig. 1). TNFα message was only marginally induced by adherence to fibronectin but rapidly accumulated after adherence to either plastic or collagen. MIP-1β was induced primarily by adherence to collagen (Fig. 1). In contrast to those genes for which mRNA levels increase during adherence, steady-state levels for genes constitutively expressed in monocytes, such as lysozyme, were downregulated by adhesion, albeit to different levels on different substrates. Immediate early genes were not induced when purified monocytes were prevented from adhering by continuous rolling in polypropylene tubes.

Although the above data clearly indicated that adherence was sufficient to induce expression of immediate early response genes, both translation (data not shown) and subsequent secretion, if warranted, of the induced messages required a signal in addition to that provided by adhesion. As shown in Table 1, we failed to detect secretion of TNFα or CSF-1 in monocyte culture supernatants unless a second signal was supplied [e.g., bacterial endotoxin (11) or cytomegalovirus (14)]. From these data, we concluded that monocyte adhesion was sufficient to prime monocytes for production of key mediators of inflammation but that a second activation signal was required for translation and secretion. Presumably, this activation signal would confer specificity of response to particular pathogens.

## MONOCYTE ENDOTHELIAL CELL INTERACTIONS RESULT IN BIDIRECTIONAL GENE ACTIVATION

Our observation that adherence-induced expression of immediate early genes led us to examine whether the interaction of monocytes with normal or activated endothelial cells would also lead to immediate early gene induction. We compared the gene expression profiles of monocytes that had been adhered to (and subsequently removed from) control or LPS-activated

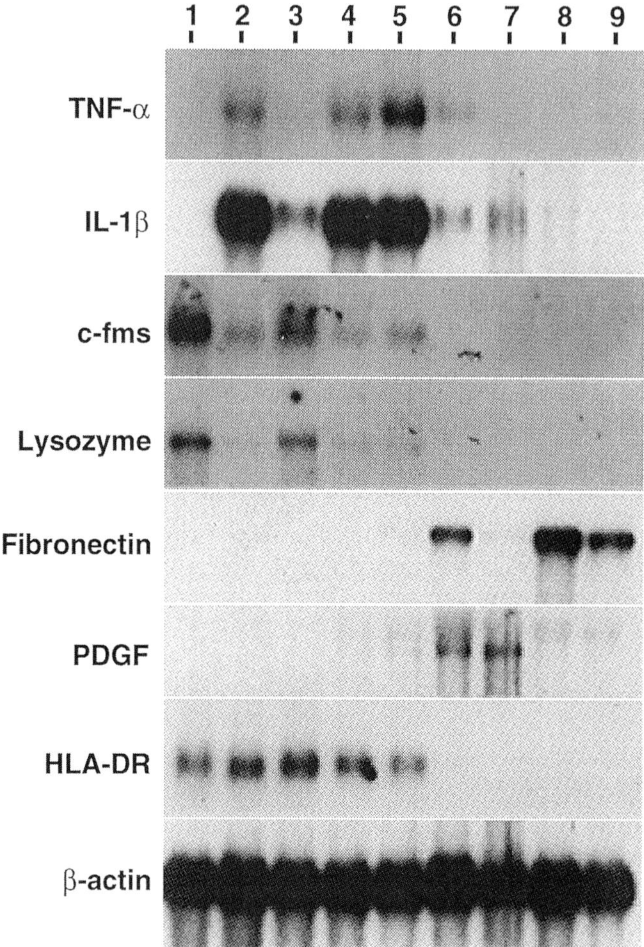


**FIG. 2.** Monocyte–endothelial cell interactions result in bidirectional gene activation. Total RNA was prepared from monocytes isolated by nonadherent, gradient methods and either harvested immediately **(lane 1)** or cultured for 4 h adherently **(lane 2)** or nonadherently **(lane 3).** EA926 endothelial cells were pretreated with medium **(lanes 4, 6, 8)** or medium containing 1 μg/ml LPS **(lanes 5, 7, 9)** for 4 h and then washed twice with medium containing 10 μg/ml polymyxin B (to inactivate residual LPS). Monocytes were adhered to the EA926 cells for 4 h and subsequently recovered from the monolayers before RNA isolation. Lanes 4 and 5, monocytes removed from endothelial cell monolayers; Lanes 6 and 7, residual EA926 cells; Lanes 8 and 9, control EA926 endothelial cells to which monocytes were never adhered. Steady-state mRNA levels for the genes indicated as determined by Northern analysis are presented.

endothelial cells with those of both nonadherent and plastic-adhered monocytes (15). Monocyte adherence to control and LPS-activated endothelial cell layers induced expression of both TNFα and IL-1β mRNA (Fig. 2). The level of message induction in monocytes adhered to endothelium was similar to that seen during adherence to plastic (Fig. 2, lanes 2, 4, and 5). Both c-fms and lysozyme were downregulated by monocyte adherence, regardless of substrate (Fig. 2, lanes 1–5). Not surprisingly, monocytes are not unilaterally affected by adherence to the endothelial cells. Although expression of fibronectin message was confined to the endothelial cell fractions, endothelial cells to which monocytes had previously adhered accumulated less fibronectin than cells that had not interacted with monocytes (Fig. 2, lanes 6–9). Perhaps more dramatic was the induction of PDGF transcripts in only those endothelial cells that had previously been adhered to by monocytes (Fig. 2, lanes 6–9). The absence of MHC class II transcripts in those endothelial cells to which monocyte had been adhered demonstrates that the modulation of endothelial gene expression we observed was not merely a consequence of residual monocyte adherence (compare HLA-DR expression in Fig. 2, lanes 4–7). Although these data clearly indicate that monocyte–endothelial cell interactions activate gene expression in both cell types, our attempts to reproducibly demonstrate cytokine secretion from monocyte–endothelial cell cultures have proven unsuccessful. We suggest, therefore, that as in the case of monocyte adherence to isolated matrix components, a second activation signal is required for translation of the induced messages.

## $\beta_1$ AND NOT $\beta_2$ INTEGRINS REGULATE IMMEDIATE EARLY GENE EXPRESSION

It seemed likely that signaling through one or more cell adhesion molecules was responsible for initiating the gene expression patterns observed during monocyte adherence. To address this, we investigated whether direct activation of either $\beta_1$ or $\beta_2$ integrin receptors would be a sufficient signal for gene activation. Nonadherent incubation of monocytes with primary antibodies directed against the common CD29 ($\beta_1$) chain of the VLA integrins had a potent stimulatory influence on expression of all the immediate early genes induced by monocyte adherence to plastic. The anti-$\beta_1$ antibody TS2/16 stimulated expression of IL-1β, IL-1ra, and MAD-6/A20 (Fig. 3) (16). Cross-linking integrins with secondary antibodies had no influence on gene induction (data not shown). Nonadherent incubation of monocytes with primary monoclonal antibodies (MAbs) directed against the common CD18 ($\beta_2$) chain or against distinct CD11a, b, or c chains (unpublished data) failed to stimulate immediate early gene expression (Fig. 3) (16). Although the anti-$\beta_2$ MAb 60.3 did not directly stimulate gene expression, engagement of $\beta_2$ integrin receptors did modulate $\beta_1$-mediated gene activation events. Pretreatment with the anti-$\beta_2$ MAb inhibited the anti-$\beta_1$–induced expression of

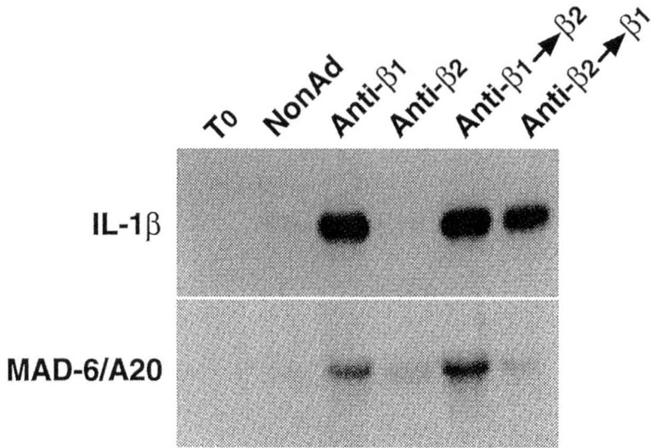


**FIG. 3.** Monoclonal antibodies (MAbs) against $\beta_1$ but not $\beta_2$ integrins regulate immediate early gene expression. Total RNA was isolated from freshly isolated human monocytes ($T_0$) or from monocytes cultured for 60 min nonadherently (NonAd), nonadherently in the presence of the anti-$\beta_1$ integrin MAb TS2/16 (Anti-$\beta_1$), nonadherently in the presence of the anti-$\beta_2$ integrin MAb 60.3 (Anti-$\beta_2$), from monocytes preincubated with MAb TS/2/16 before culturing nonadherently for 60 min in the presence of MAb 60.3 (Anti-$\beta_1 \rightarrow \beta_2$), or preincubated with MAb 60.3 before culturing nonadherently for 60 min in the presence of MAb TS2/16 (Anti-$\beta_2 \rightarrow \beta_1$). Antibodies were used at a final concentration of 1 μg/ml. Message levels for IL-1β and MAD-6/A20 were determined by Northern analysis. [Portions of this figure have previously been published (16).]

MAD-6/A20 (Fig. 3) and post-treatment with the anti-$\beta_2$ MAb superinduced the anti-$\beta_1$–induced expression of IL-1ra (16). Control studies using $F(ab')_2$ and F(ab) fragments in place of intact antibodies indicated that dimerization of the β-chains of $\beta_1$ integrins was required for induction and that Fc receptors did not mediate immediate early gene induction.

The selective expression of cytokine genes observed during monocyte adhesion to various matrix components suggested that nonadherent engagement of VLA-4 (the integrin receptor specific for fibronectin and VCAM-1) might induce expression of immediate early genes normally characteristic of monocyte adherence to fibronectin. Although direct engagement of monocyte VLA-4 chains with three different MAbs induced immediate early gene expression, the pattern of expression showed no apparent selectivity (16). We conclude from these data that engagement of the $\beta_1$ family of integrins may induce immediate early gene transcripts. However, the observed selectivity of expression must be associated with additional events triggered when monocytes actually adhere to the appropriate ligand and reorganize their cytoskeleton.

## TRANSCRIPTIONAL EVENTS ASSOCIATED WITH INTEGRIN ENGAGEMENT

The transient, integrin-mediated contacts by which extravasating monocytes interact with endothelial cells and with components of the extracellular matrix (ECM) trigger phosphorylation cascades and rapid changes in both intracellular pH and calcium (17–20). Our observations that adhesion of primary human monocytes to ECM components (or to tissue culture plastic) dramatically alters the population of certain mRNA species suggested that this signaling through integrin receptors may ultimately influence gene expression at the transcriptional level. We recently have demonstrated that both adhesion and $\beta_1$ integrin engagement lead to cytoplasmic degradation of IκBα protein, a critical inhibitor of the transcription factor NF-κB (21). Degradation of IκBα closely parallels the increases in nuclear NF-κB binding activity after adherence and integrin stimulation (Fig. 4). Interestingly, there seems to be a general activation of transcription factor activities in the

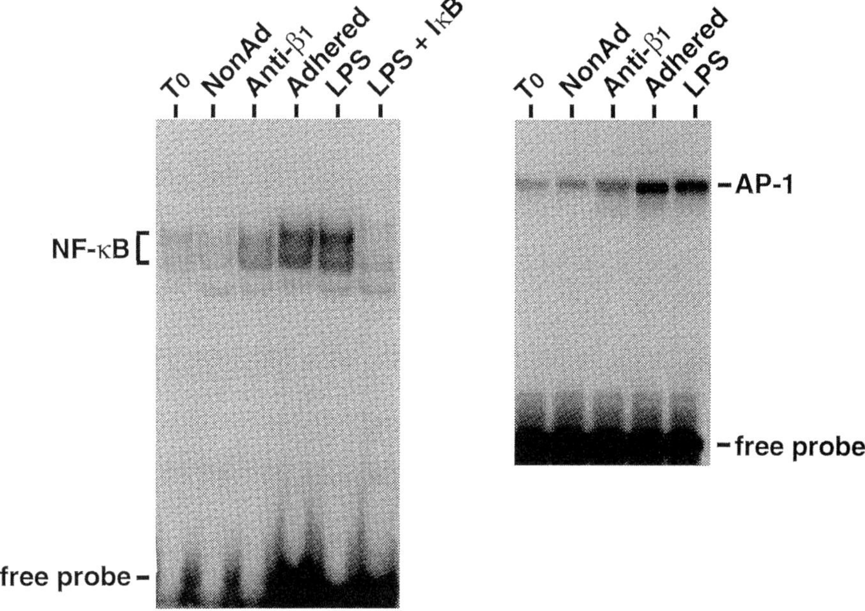


**FIG. 4.** Monocyte adherence and $\beta_1$ integrin engagement result in activation of transcription factor-binding activities. Freshly isolated human monocytes were stimulated either by adherence to plastic (Adhered) or by nonadherent incubation with either the anti-$\beta_1$ integrin MAb TS2/16 (Anti-$\beta_1$), or LPS (1 μg/ml) for 1 h. Nuclear extracts were prepared and electrophoretic mobility shift assays (EMSAs) were performed as previously described (30,31). As controls, EMSAs were also performed on extracts prepared from freshly isolated cells ($T_0$) and from unstimulated nonadherent cells (NonAd). **A:** The mobility of the class I MHC enhancer probe (32) in the above extracts suggest the presence of both p65/50 (upper) and p50/50 (lower) NF-κB/DNA complexes. To confirm the identity of p65/50 NF-κB complexes, exogenous IκBα was added to nuclear extract from LPS-treated nonadherent cells (LPS + IκBα). Individual NF-κB complexes were identified by supershifting antibodies raised against NF-κB subunits (K. Mondal, data not shown). **B:** Changes in mobility of the human collagenase (coll I) AP-1 enhancer probe (33) in the nuclear extracts described above indicates enhancement of AP-1 binding activity in adherence, anti-$\beta_1$ integrin- and LPS-stimulated monocytes. The identity of AP-1 complexes was confirmed by competition using unlabeled AP-1–binding site probe and by supershifting anti-fos and anti-jun antibodies (L. K. Price, data not shown).

nuclei of adhered monocytes. Increased DNA-binding activity after monocyte adherence or after engagement of monocyte $\beta_1$ integrins was observed for AP-1 (Fig. 4) and other DNA-binding activities required for transcription of many immediate early genes (i.e., NF-IL6, NF$\beta_1$, NF$\beta_2$, NF$\beta$A, and Oct-1), (L. K. Price and K. Mondal, unpublished data). The rate at which DNA-binding activities increase (within 5 min of adherence or integrin engagement) suggests this is not a result of new protein synthesis but rather is a result of modification of pre-existing factors.

The rapid increase in DNA-binding activities for various transcription factors suggested that adhesion-dependent transcriptional activation could account for increased mRNA levels observed for the immediate early genes. Such a response would be analogous to reports that bacterial toxins transcriptionally activate the IL-1$\beta$ and TNF$\alpha$ genes in adherence-purified monocytes (22,23). We have recently demonstrated, by nuclear run-on analysis, that for certain immediate early genes this is indeed the case. For example, transcriptional rates for both IL-8, TNF$\alpha$ (Fig. 5), and other immediate early cytokine genes (data not shown) were markedly increased 15 to 30 min after either adhesion or engagement of the $\beta_1$-chains of integrin adhesion receptors. In agreement with our previous studies, which demonstrated that dimerization of $\beta_1$-chains was sufficient for accumulation of

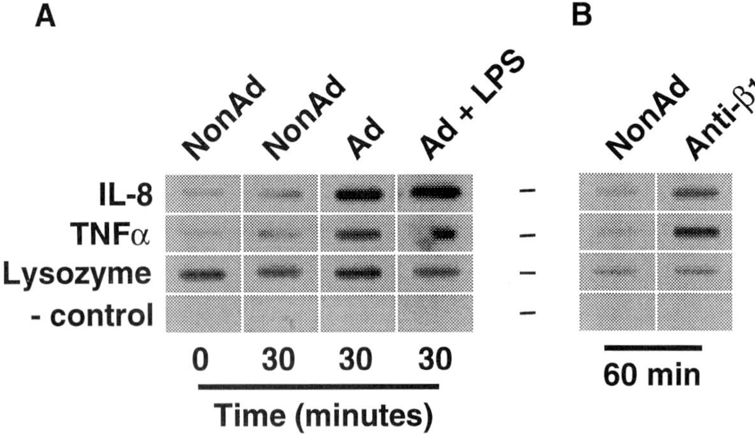


**FIG. 5.** Transcription of IL-8 and TNF$\alpha$ in monocytes is activated by both adherence and $\beta_1$ integrin engagement. **A:** Nuclei were prepared from monocytes immediately after isolation and after 30 min of nonadherent culture (NonAd), after 30 min of adherent culture on tissue culture plastic (Ad), or after 30 min of adherent culture in the presence of 10 ng/ml LPS (Ad + LPS). $^{32}$P-labeled nuclear run-on RNA was prepared from each sample as previously described (22) and hybridized to slot-blots of control DNA (no insert) and cDNAs for IL-8, TNF$\alpha$, and lysozyme as indicated. **B:** Nuclei were prepared from monocytes after 60 min of nonadherent culture (NonAd) stimulated or after 60 min of nonadherent incubation with the anti-$\beta_1$ integrin MAb TS2/16 (Anti-$\beta_1$). $^{32}$P-labeled run-on RNA from each sample was hybridized to slot-blots of control DNA and cDNAs for IL-8, TNF$\alpha$, or lysozyme as indicated.

immediate early messages (Fig. 3) (16), $\beta_1$ integrin-mediated transcriptional activation requires only bivalent cross-linking of integrin receptors. In contrast, other studies have indicated that the signaling cascades induced by integrin engagement require clustering of the receptors (18,19). This apparent divergence between the stimulatory requirements for currently identified integrin-dependent signaling pathways and those required for integrin-mediated transcriptional activation appears to indicate that an important gap exists in our understanding of the adhesion-dependent signals required for increased nuclear activities.

## STABILIZATION MAY BE REGULATED BY CYTOSKELETON-DEPENDENT EVENTS

Previous reports have indicated that RNAs containing AU-rich sequence elements (AREs) in their 3′-untranslated regions are more rapidly degraded than those which lack AREs (24,25). Because the 3′-untranslated regions of many adherence-induced monocyte transcripts contain ARE-like sequences, it seemed likely that their accumulation could occur, in part, by activation of one or more message stabilization mechanisms. To address this, we compared mRNA stabilities in monocytes cultured nonadherently with those cultured adherently or cultured nonadherently in the presence of an anti-$\beta_1$ integrin MAb. After stimulation, actinomycin D was added to the cultures and decay rates for various transcripts were assessed by Northern analysis (Fig. 6). Compared with nonadherent monocytes, the stability of immediate early cytokine genes such as IL-8 (Fig. 6) and IL-1β (26) was markedly enhanced during monocyte adhesion. For example, the calculated stability of IL-1β message in adhered monocytes (approximately 6 h) is more than twice that reported for monocytes cultured in the presence of LPS (22) and almost 20-fold that in monocytes cultured nonadherently. From these data, we conclude that adhesion signaling in monocytes involves one or more message stabilization mechanisms. Furthermore, because adhesion had no effect on message stability for other immediate early genes such as c-fos (Fig. 6) and IκBα (26), the stabilization appears to be a selective rather than a general phenomenon. Bivalent engagement of $\beta_1$ integrins under nonadherent conditions did not affect the stability of adherence-induced immediate early messages (Fig. 6). We propose that the different message stabilities observed in adhesion and anti-$\beta_1$ integrin-stimulated monocytes indicate differences in signaling pathways induced by adhesion and integrin receptor engagement. Monocytes undergo dramatic morphologic changes during adherence to immobile substrates, suggesting the induction of signaling cascades in addition to those triggered by receptor cross-linking of integrin adhesion receptors. Cytoskeletal involvement in regulation of gene ex-

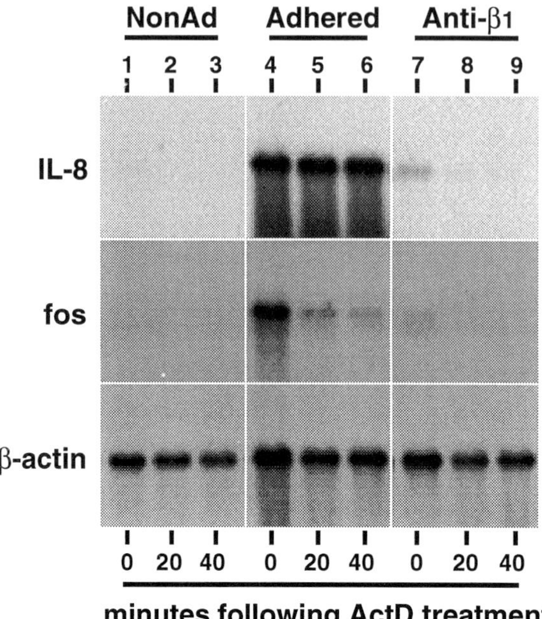


**FIG. 6.** Monocyte adherence results in selective mRNA stabilization events. Freshly isolated human monocytes were cultured nonadherently with no stimulus (NonAd), adherently on plastic (Adhered), or nonadherently with the anti-$\beta_1$ integrin MAb TS2/16 (Anti-$\beta_1$). After 30 min, the monocyte cultures were treated with 5 μg/ml actinomycin D (ActD) for the times indicated before collection of the cells and isolation of the RNA for Northern analysis. The kinetics of IL-8, c-fos, and β-actin mRNA degradation are presented for a single representative donor.

pression has previously been documented at both the transcriptional (27,28) and the post-transcriptional level (29). It does not seem improbable, therefore, that the initial signals generated by activating monocyte integrin receptors could be modulated by additional signals generated during adherence-induced cytoskeletal reorganization.

## SUMMARY AND FUTURE CONSIDERATIONS

The selectin- and integrin-mediated interactions between vascular endothelium and circulating lymphocytes are critical first steps during an inflammatory response to infection or injury. These adhesive interactions trigger a variety of signal transduction pathways during the early stages of monocyte extravasation (1,2) (Fig. 7). Collectively, our own studies have shown that monocytes ultimately respond to adhesion by selectively altering gene expression profiles. Adhesion-dependent regulation of gene expression in monocytes occurs by direct modification of transcriptional rates and by

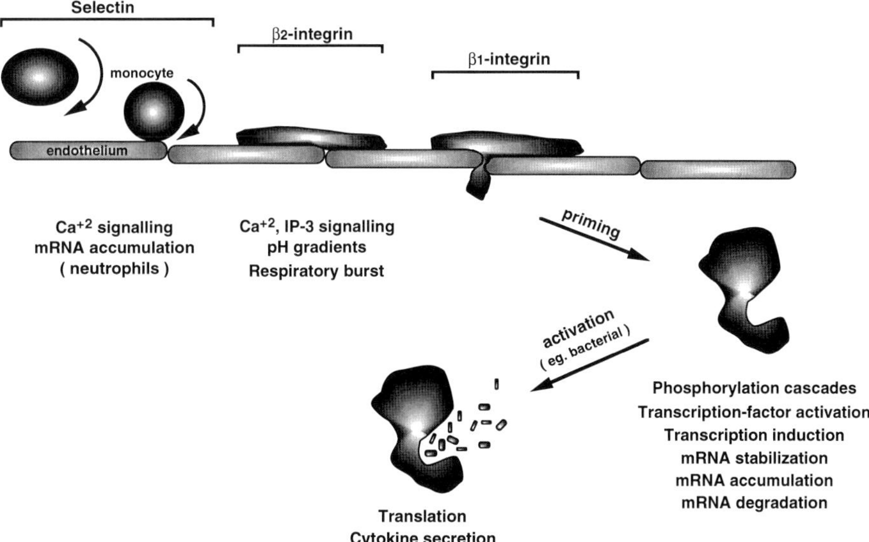


**FIG. 7.** Stages in adhesion-dependent monocyte gene expression. Selectin- and integrin-mediated adherent interactions stimulate a wide variety of signal transduction pathways (summarized below each step). Although adhesion ultimately influences gene expression at both transcriptional and post-transcriptional levels, whether any of the currently identified signaling cascades induced by adhesion are directly affecting gene expression remains to be determined. Translation and subsequent secretion of adherence-induced cytokine messages requires an additional activation signal, suggesting an additional level of complexity (see text).

modulation of one or more message stabilization mechanisms. Interestingly, although a number of genes associated with inflammation and wound repair are induced during monocyte adherence, thereby priming the cells for protein synthesis, no translation of the induced messages occurs unless a second signal is provided to activate protein synthesis. We have suggested that this signal may be pathogen (or injury)-specific, but we have not yet fully examined this issue. Our current studies are focused on determining which, if any, of the presently identified adhesion-induced signaling pathway(s) are important for regulating transcriptional events that occur during the early (0- to 30-min) stages of monocyte adhesion. In addition, we have recently initiated a study addressing the mechanism(s) by which monocyte adhesion induces cytokine message stabilization, with particular emphasis on the importance of cytoskeletal reorganization.

## ACKNOWLEDGMENT

We thank Francisco Sánchez-Madrid (Universidad Autónoma de Madrid) for his gift of the anti-$\beta_1$ integrin monoclonal antibody TS2/16 and John

Harlin (University of Washington) for his gift of the anti-$\beta_2$ integrin monoclonal antibody 60.3. We also thank past and present members of the Haskell Laboratory for assistance at various stages of this work. This research was supported by National Institutes of Health grant AI 26774.

## REFERENCES

1. Albelda SM, Smith CW, Ward PA. Adhesion molecules and inflammatory injury. *FASEB J* 1994;8:504–12.
2. Beekhuizen H, van Furth R. Monocyte adherence to vascular endothelium. *J Leukocyte Biol* 1993;54:363–78.
3. Springer TA. Traffic signals for lymphocyte recirculation and leukocyte emigration: the multistep paradigm. *Cell* 1994;76:301–14.
4. Werb Z. Tremble PM, Behrendtsen O, Crowley E, Damsky CH. Signal transduction through the fibronectin receptor induces collagenase and stromelysin gene expression. *J Cell Biol* 1989;109:877–89.
5. Streuli CH, Bissell MJ. Expression of extracellular matrix components is regulated by substratum. *J Cell Biol* 1990;110:1405–15.
6. Schmidhauser C, Bissell MJ, Myers CA, Casperson GF. Extracellular matrix and hormones transcriptionally regulate bovine β-casein 5′ sequences in stably transfected mouse mammary cells. *Proc Natl Acad Sci USA* 1990;87:9118–22.
7. Dhawan J, Farmer SR. Regulation of alpha 1 (I)-collagen gene expression in response to cell adhesion in Swiss 3T3 fibroblasts. *J Biol Chem* 1990;265:9015–21.
8. Dhawan J, Lichtler AC, Rowe DW, Farmer SR. Cell adhesion regulates pro-alpha 1 (I) collagen mRNA stability and transcription in mouse fibroblasts. *J Biol Chem* 1991;266: 8470–5.
9. Böyum A. Isolation of mononuclear cells and granulocytes from human blood. *Scand J Clin Lab Invest* 1968;21(suppl 97):77–89.
10. Ulmer AJ, Flad H-D. Discontinuous density gradient separation of human mononuclear leukocytes using Percoll as gradient medium. *J Immunol Methods* 1979;30:1–10.
11. Haskill S, Johnson C, Eierman D, Becker S, Warren K. Adherence induces selective mRNA expression of monocyte mediators and proto-oncogenes. *J Immunol* 1988;140: 1690–4.
12. Sporn SA, Eierman DF, Johnson CE, et al. Monocyte adherence results in selective induction of novel genes sharing homology with mediators of inflammation and tissue repair. *J Immunol* 1990;144:4434–41.
13. Eierman DF, Johnson CE, Haskill JS. Human monocyte inflammatory mediator gene expression is selectively regulated by adherence substrates. *J Immunol* 1989;142:1970–6.
14. Dudding L, Haskill S, Clark BD, Auron PH, Sporn S, Huang E-S. Cytomegalovirus infection stimulates expression of monocyte-associated mediator genes. *J Immunol* 1989;143:3343–52.
15. Johnson CE. *Characterization of inflammatory gene expression in monocytes and in vascular and connective tissue cells*. PhD Dissertation, The University of North Carolina at Chapel Hill, 1989.
16. Yurochko AD, Liu DY, Eierman D, Haskill S. Integrins as a primary signal transduction molecule regulating monocyte immediate-early gene induction. *Proc Natl Acad Sci USA* 1992;89:9034–8.
17. Ferrell JE, Martin GS. Tyrosine-specific protein phosphorylation is regulated by glycoprotein IIb-IIIa in platelets. *Proc Natl Acad Sci USA* 1989;86:2234–8.
18. Kornberg LJ, Earp HS, Turner CE, Prokop C, Juliano RL. Signal transduction by integrins: increased protein tyrosine phosphorylation caused by clustering of $\beta_1$ integrins. *Proc Natl Acad Sci USA* 1991;88:8392–6.
19. Schwartz MA, Lechene C, Ingber DE. Insoluble fibronectin activates the Na/H antiporter by clustering and immobilizing integrin $\alpha_5\beta_1$, independent of cell shape. *Proc Natl Acad Sci USA* 1991;88:7849–53.
20. Schwartz MA. Spreading of human endothelial cells on fibronectin or vitronectin triggers elevation of intracellular free calcium. *J Cell Biol* 1993;120:1003–10.

21. Lenardo MJ, Baltimore D. NF-κB: a pleiotropic mediator of inducible and tissue specific gene control. *Cell* 1989;58:227–9.
22. Arend WP, Gordon DF, Wood WM, Janson RW, Joslin FG, Jameel S. IL-1β production in cultured human monocytes is regulated at multiple levels. *J Immunol* 1989;143:118–26.
23. Osipovich OA, Fegeding KV, Misuno NI, et al. Differential action of cycloheximide and activation stimuli on transcription of tumor necrosis factor-a, IL-1b, IL-8, and P53 genes in human monocytes. *J Immunol* 1993;150:4958–65.
24. Aharon T, Schneider RJ. Selective destabilization of short-lived mRNAs with the granulocyte-macrophage colony-stimulating factor AU-rich 3′ noncoding region is mediated by a cotranslational mechanism. *Mol Cell Biol* 1993;13:1971–80.
25. Chen C-YA, Chen T-M, Shyu A-B. Interplay of two functionally and structurally distinct domains of the *c-fos* AU-rich element specifies its mRNA-destabilizing function. *Mol Cell Biol* 1994;14:416–26.
26. Lofquist AK, Mondal K, Morris JS, Haskill JS. Transcription-independent turnover of IκBα during monocyte adherence: implications for a translational component regulating IκBα MAD-3 message levels. *Mol Cell Biol* [*in press*].
27. Zambetti G, Ramsey-Ewing A, Bortell R, Stein G, Stein J. Disruption of the cytoskeleton with cytochalasin D induces c-fos gene expression. *Exp Cell Res* 1991;192:93–101.
28. Ulevitch RJ, Kline L, Schreiber RD, et al. Hyperexpression of interferon-gamma-induced MHC class II genes associated with reorganization of the cytoskeleton. *Am J Pathol* 1991; 139:287–96.
29. Santell L, Morotti K, Bartfeld NS, Baynham P, Levin EG. Disruption of microtubules inhibits the stimulation of tissue plasminogen activator expression and promotes plasminogen activator inhibitor type 1 expression in human endothelial cells. *Exp Cell Res* 1992; 201:358–65.
30. Baldwin AS. Analysis of sequence-specific DNA-binding proteins by the gel mobility shift assay. *DNA Protein Eng Tech* 1990;2:73–6.
31. Haskill S, Beg AA, Tompkins SM, et al. Characterization of an immediate-early gene induced in adherent monocytes that encodes IκB-like activity. *Cell* 1991;65:1281–9.
32. Baldwin AS, Sharp PA. Binding of a nuclear factor to a regulatory sequence of the mouse H-2K$^b$ class I major histocompatibility gene. *Mol Cell Biol* 1987;7:305–13.
33. Stein B, Rahmsdorf HJ, Steffen A, Litfin M, Herrlich P. UV-induced DNA damage is an intermediate step in UV-induced expression of human immunodeficiency virus type I, collagenase, *c-fos*, and metallothionein. *Mol Cell Biol* 1989;9:5169–81.

*Topics in Molecular Medicine, Volume 1,*
edited by Wolfgang Siess, Reinhard Lorenz,
and Peter C. Weber. Raven Press, Ltd.,
New York © 1995.

# 6

# Integrin-Binding Peptides

Erkki Ruoslahti

*Cancer Research Center, La Jolla Cancer Research Foundation,
La Jolla, California 92037*

The integrins have surprising binding specificities in that the large extracellular domain of an integrin recognizes a very small binding site in the equally large ligand proteins; the RGD cell-adhesion sequence is the prototype of such binding sites. Early observations on RGD proteins and conformationally restrained RGD peptides led to the hypothesis that the presentation of the RGD site in a protein or peptide was important in determining integrin specificity and affinity (1). The RGD adhesion site generally appears to be present in a turn or loop, as has been particularly thoroughly documented for fibronectin (2–4).

Regardless of the recently demonstrated contribution to the binding of RGD-dependent integrins to RGD-containing ligands by sites outside the RGD (5–8), it has been possible to show that the correct stereochemistry of an RGD peptide can impart high affinity and receptor specificity. Applying this principle, it has been possible to develop RGD peptides that are specific for the platelet $\alpha_{IIb}\beta_3$ integrin, and peptides that demonstrate some selectivity for the $\alpha_v\beta_3$ and $\alpha_5\beta_1$ integrins have also been obtained (9–14). Moreover, the affinities of the most advanced peptides for the $\alpha_{IIb}\beta_3$ integrin are very high; they inhibit platelet aggregation at nanomolar concentrations, a 20,000-fold improvement over the standard GRGDSP peptide. Our aim has been to accomplish a similar improvement with peptides that bind to the $\alpha_5\beta_1$ integrin, as this integrin appears to play a role in cell growth and tumor metastasis.

We use phage display libraries to identify peptides that bind to $\alpha_5\beta_1$ (7,15). In this method, random peptides are expressed as part of a phage surface protein, which enables one to affinity-select the desired phage from a mixture (16,17). Peptide libraries can also be used to identify conformationally restrained peptides with improved integrin-binding properties (7,18). This is accomplished by introducing a cysteine residue on each side of the random sequence and thereby having the phage display a peptide sequence cyclized by a disulfide bond between the cysteine residues. These libraries can yield improved peptide probes for the study of RGD-directed integrins.

## SCREENING OF PHAGE LIBRARIES FOR PEPTIDES BINDING TO $\alpha_5\beta_1$

### RGD Sequences and RGD Homologues

Our initial experiments with peptide screening, which utilized a library displaying six–amino-acid linear peptides, yielded primarily RGD and related sequences on screening with the $\alpha_5\beta_1$ integrin (15). On one level, these results are a strong confirmation of the importance of RGD in integrin binding. Another significant finding was the emergence of the sequence CRGDC which, when reproduced as a cyclic peptide, was highly active in integrin binding. This structure prompted us to design libraries with the general structure of $CX_nC$ (7). Such libraries potentially present all of the random sequences as cyclic structures. A $CX_7C$ library also yielded a high proportion of RGD peptides. Screening at progressively higher stringencies led to a remarkable convergence of the sequence on the C-terminal side of the RGD. At the highest level of stringency, almost all of the RGD sequences had a glycine residue following the RGD and the next residue was either tryptophan or phenylalanine (Table 1). The residues on the N-terminal side

**TABLE 1.** *RGD-containing sequences from a cyclic (CX7C) phage display library selected by $\alpha_5\beta_1$ binding at increasing stringencies*

| | Integrin coating concentration for plastic | |
|---|---|---|
| 50,000 ng/ml (2nd panning) | 100 ng/ml (3rd panning) | 10 ng/ml (5th panning) |
| L S **R G D** T P | I P **R G D** G W (3) | I P **R G D** G W (2) |
| D R **R G D** G F | Y R **R G D** G H | V A **R G D** G W |
| F T **R G D** A P | E L **R G D** G W | Q T **R G D** G W |
| T S **R G D** M P | V A **R G D** G W | L F **R G D** G W |
| Q L **R G D** G W | T **R G D** G W F (3) | L S **R G D** G W |
| E G **R G D** W H | E **R G D** L R M (2) | F **R G D** G F V |
| T L **R G D** N H | T **R G D** M Q W | **R G D** G F G S |
| H L **R G D** G W | S **R G D** G W I | I P **R G D** G W |
| M L **R G D** S F | L **R G D** G F L | M T **R G D** G F |
| M P **R G D** G F | Y **R G D** H L L | L F **R G D** G W |
| S **R G D** G F S (2) | L **R G D** A R F | **R G D** G F G S |
| F **R G D** H V R | T **R G D** G W P | |
| G **R G D** S V P | G **R G D** R P Q | |
| S **R G D** G F R | K **R G D** G F W | |
| G **R G D** N L P | **R G D** F S Y M | |
| **R G D** L R F N | | |

Details of the methods are described in Koivunen et al. (7). The number of clones encoding the same peptide is shown in parentheses.

of the RGD varied. Peptides of this general type are the most potent RGD peptides available for $\alpha_5\beta_1$ at this time. Surprisingly, potent $\alpha_5\beta_1$ binders with no resemblance to RGD were also found (see below).

Among the clones that deviated from the RGD sequence, many were apparent RGD homologues. One set of such peptides had the motif NGR. These sequences may be mimics of the NGR sequence in the ninth type III repeat of fibronectin. Variation in the residue between the arginine and aspartic acid of RGD was also seen. All of the RGD homologues had low affinities for $\alpha_5\beta_1$.

### Relationships Among Integrin Recognition Specificities

One motif found in these searches that is strikingly different from RGD was IDI/SDV. These sequences are homologous to the LDV sequence in the alternatively spliced IIICS region of fibronectin that binds to the $\alpha_4\beta_1$ integrin (19–22). Indeed, we found that a peptide containing the IDI sequence interacted with $\alpha_4\beta_1$. This result and recent results from others (23) suggest a previously unsuspected relationship between the auxiliary $\alpha_5\beta_1$ binding sites and the binding specificities of other integrins, $\alpha_4\beta_1$ in particular.

The RGD tripeptide sequence is recognized at least by the integrins $\alpha_5\beta_1$, $\alpha_v\beta_1$, $\alpha_{IIb}\beta_3$, $\alpha_v\beta_3$, $\alpha_v\beta_5$ (24), $\alpha_v\beta_6$ (25), and $\alpha_v\beta_8$ (26). Moreover, $\alpha_2\beta_1$ (27) and $\alpha_3\beta_1$ (28) have also been reported to recognize RGD under some conditions. In addition to the LDV sequence that binds to $\alpha_4\beta_1$, other integrin-binding peptide motifs include KQAGDV for $\alpha_{IIb}\beta_3$ (29) and DGEA, which is thought to bind to $\alpha_2\beta_1$ (30). Moreover, $\alpha_1\beta_1$ binds to a configuration of residues formed by arginine and aspartic acid residues on the surface of the type I collagen triple helix (31), and $\alpha_M\beta_2$ recognizes a KRLDGS sequence in fibrinogen (32).

The various integrin-binding sequences have an aspartic acid residue and sometimes both an aspartic acid and arginine residue in common, suggesting that the sites recognized by the integrins evolved from one primordial recognition site. That this is the case is suggested by the discovery of peptides that bind to both $\alpha_5\beta_1$ and $\alpha_4\beta_1$ from peptide libraries and in chemical synthesis (7,23). In fact, it has recently been shown that $\alpha_4\beta_1$ can also bind to the RGD region in fibronectin (33). These results indicate that the sites characterized by the RGD and LDV motifs are somehow related. One possibility is that RGD and LDV are homologues of one another. The KQAGDV, which is an alternative ligand for the RGD-directed binding site of $\alpha_{IIb}\beta_3$ (29,34) but resembles LDV more closely than RGD, could be an intermediate between the two prototype structures. Another possibility, discussed below, is that RGD and LDV represent a main binding site and an auxiliary site for an integrin.

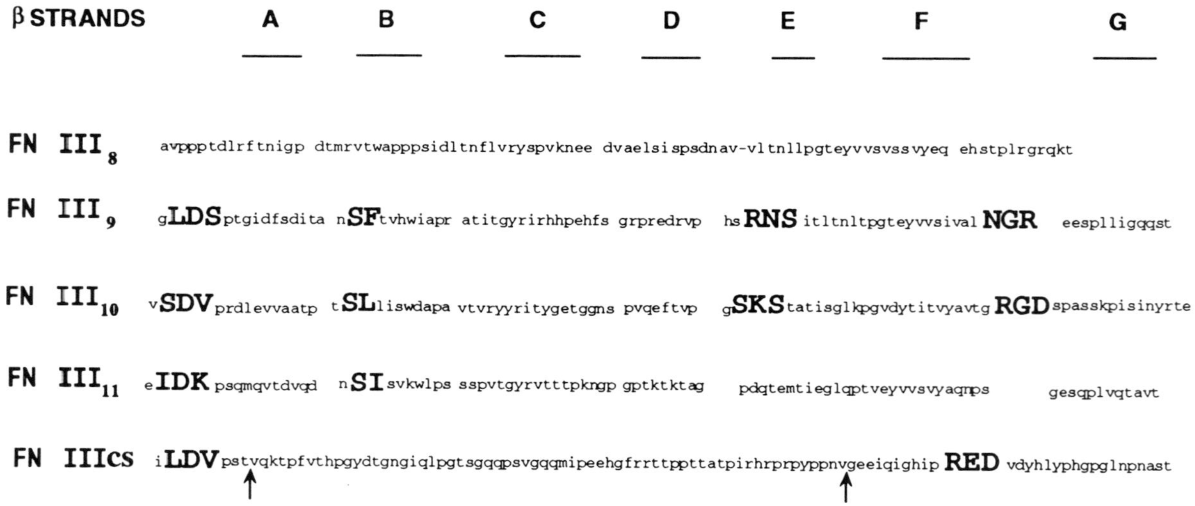

β STRANDS A B C D E F G

FN $III_8$ avppptdlrftnigp dtmrvtwapppsidltnflvryspvknee dvaelsispsdnav-vltnllpgteyvvsvssvyeq ehstplrgrqkt

FN $III_9$ g**LDS**ptgidfsdita n**SF**tvhwiapr atitgyrirhhpehfs grpredrvp hs**RNS**itltnltpgteyvvsival**NGR** eesplligqqst

FN $III_{10}$ v**SDV**prdlevvaatp t**SL**liswdapa vtvryyritygetggns pvqeftvp g**SKS**tatisglkpgvdytitvyavtg**RGD**spasskpisinyrte

FN $III_{11}$ e**IDK**psqmqvtdvqd n**SI**svkwlps sspvtgyrvtttpkngp gptktktag pdqtemtieglqptveyvvsvyaqnps gesqplvqtavt

FN IIIcs i**LDV**pstvqktpfvthpgydtgngiqlpgtsgqqpsvgqqmipeehgfrrttppttatpirhrprpyppnvgeeiqighip**RED** vdyhlyphgpglnpnast

**FIG. 1.** Sequence alignment of the eighth, ninth, tenth, and eleventh type III repeats with the III CS segment of fibronectin. The sequence elements represented among the sequences isolated from peptide libraries by screening on the $\alpha_5\beta_1$ integrin (7) are highlighted.

### Peptides Representing Auxiliary Integrin-Binding Sites

Our results suggest that the peptide libraries can also give information on the size and nature of the ligand-binding site in integrins. Auxiliary integrin binding sites outside the RGD loop have been placed in the eighth and ninth type III repeats of fibronectin, based on the ability of these domains to enhance the activity of the RGD-containing tenth repeat (5,6,35). We have found $\alpha_5\beta_1$-binding peptides that appear to align with segments of the fibronectin cell attachment domain other than the RGD-containing loop of the tenth type III repeat (7). These results suggest that it will be possible to identify auxiliary integrin-binding sites by using peptide libraries, because the libraries can yield peptides representing those sites.

Judging from our results presented above, it appears that the $\alpha_5\beta_1$ integrin binds to RGD as its main recognition sequence in the tenth type III module but is also capable of recognizing sequences that represent various loops in the tenth and ninth type III modules (7). Therefore, it is possible that the integrin contacts fibronectin over an extended molecular surface on which the RGD site contributes the main part of the binding energy. Sequences such as the SDV sequence at the N terminus of the $III_{10}$ module would add incrementally to the energy of the binding.

In the $\alpha_4\beta_1$ integrin, the roles of these binding sites appear to be reversed: the SDV homologue of the IIICS module (LDV) would provide for the bulk of the binding force and the RGD (rat) or RED (human) sequence elsewhere in this module would play a minor role. Interestingly, a minor degree of cell attachment activity has been assigned to the RED site (19,22). Although the folding of the IIICS domain is not known, computer predictions show a structure dominated by β strands as in the type III modules. Moreover, it is interesting to note that the distances in linear sequence between the RGD/RED and SDV/LDV sites in the $III_{10}$ module and the unspliced IIICS module are similar (Fig. 1). Moreover, the $III_9$ module also contains the related sequences (NGR and LDS) at equivalent positions. The RNS site in the $III_9$ overlaps with an auxiliary binding site proposed for $\alpha_{IIb}\beta_3$ (8). $III_{11}$, which does not appear to contribute to the integrin binding, lacks the hydrophobic–aspartic–hydrophobic site and the RGD-like site. However, $III_8$, which is believed to contribute to the $\alpha_5\beta_1$ binding, also does not contain recognizable RGD- or LDV-type sequences. One possible integrin binding sequence in $III_8$ is the RLRGRQKT sequence at its C terminus that resembles the basic Tat protein and vitronectin sequences we have identified as $\alpha_v\beta_5$-binding sequences (36).

### High-Affinity Non-RGD Binder of $\alpha_5\beta_1$

The most prevalent sequence in some high-stringency phage screening experiments was CRRETAWAC (7). As a peptide, this sequence is a potent and specific ligand for $\alpha_5\beta_1$. It blocks the binding of fibronectin to $\alpha_5\beta_1$ and inhibits $\alpha_5\beta_1$-mediated cell attachment to fibronectin. Among the species tested thus far, this peptide binds to human $\alpha_5\beta_1$ but not to $\alpha_5\beta_1$ in CHO cells or mouse 3T3 cells. The CRRETAWAC peptide is essentially specific for $\alpha_5\beta_1$; $\alpha_v\beta_1$, another fibronectin receptor, is inhibited slightly, and other $\alpha_v$ integrins and $\alpha_{IIb}\beta_3$ are unaffected. This peptide therefore is an excellent new probe for $\alpha_5\beta_1$-mediated biologic activities. It may also be a useful probe for the $\alpha_5\beta_1$-binding site because its binding to $\alpha_5\beta_1$ is distinct from RGD binding; CRRETAWAC is a better inhibitor of the binding of CRRETAWAC-displaying phage to $\alpha_5\beta_1$ than CRGDC, whereas CRGDC inhibits better the binding of CRGDC-displaying phage. Therefore, the binding sites for these two types of peptides may be a least partially different on $\alpha_5\beta_1$.

## RGD IN TUMOR TREATMENT

Two integrin-directed manipulations have been shown to suppress tumorigenicity or metastasis. One is elevating the expression of $\alpha_5\beta_1$ integrin by means such as gene transfer. The other is the use of peptides containing the RGD sequence or other integrin-binding sequences. Although the expectation that $\alpha_5$ gene transfer would be an efficacious anti-tumor strategy is based on in vitro manipulation of tumorigenic cells (e.g., 37,38), an extensive literature exists to show that RGD peptides have anti-tumor effects in vivo (39–44). These approaches have not received the attention from cancer researchers and the pharmaceutical industry that they may deserve. One of the reasons is perhaps that it has not been determined how the integrin or peptides mediate these anti-tumor effects.

When peptides containing the RGD sequence are added to a cell culture of almost any type, the cells detach from the substrate. The reason for the detachment is apparently competition by the peptide for the integrins responsible for cell attachment. Interestingly, the RGD-treated cells also stop proliferating (45). Moreover, as has been shown quite recently, epithelial cells detached in this manner not only do not proliferate, they undergo apoptosis (46). On the other hand, the migration and invasion of melanoma, glioma, and osteosarcoma cells through amniotic membrane tissue can be inhibited with synthetic RGD peptides in an in vitro tumor cell invasion assay (47). Peptides that are partially selective for the $\alpha_5\beta_1$ integrin were particularly effective in this regard.

The mechanism by which the RGD peptides inhibit cell proliferation, migration, and tumor invasion in the experimental systems described above could be inhibition of cell adhesion. In this scenario, loss of adhesion would physically deny the cells anchorage and traction for migration. Another possibility relates to the recently discovered role of the $\alpha_v\beta_3$ integrin in angiogenesis (48); RGD peptides could reduce tumor growth by inhibiting angiogenesis.

The peptides could also be delivering a signal into the tumor cells rather than inhibiting adhesion. This is an attractive possibility because it would mean that binding of an RGD peptide to one integrin would render that peptide capable of suppressing invasion, regardless of what other adhesion proteins the tumor cells might encounter and be able to adhere to; the peptide would be acting as a traditional agonist-type drug.

Several pieces of evidence suggest that signaling may indeed underlie the antimetastatic effect of the RGD peptides. First, it is difficult to explain the apparent efficacy of the peptides in inhibiting tumor dissemination in the co-injection experiments by postulating an anti-adhesion effect, because the peptides have a half-life of only minutes in vivo. It is more plausible that the cells receive a signal from the peptide–integrin interactions and that this signal persists longer than the peptide. Second, peptides that are partially selective for the $\alpha_5\beta_1$ fibronectin receptor inhibit invasion even when the barrier used in the assay contains no fibronectin (M. Hendrix and E. Ruoslahti, unpublished results). It has also been shown that treatment of cells with *soluble* RGD peptide can direct the accumulation of the $\alpha_5\beta_1$ into adhesion plaques under the cells (49), indicating that binding of the peptides to the integrin causes integrin redistribution. This redistribution is probably driven from inside the cell, since the peptide is soluble and cannot cross-link the receptors. Finally, it has been suggested, on the basis of cDNA transfer results with the $\alpha_5\beta_1$ integrin, that this integrin, when not ligated to a fibronectin matrix, transmits a negative growth signal to cells (38). One possible mechanism of action for the RGD peptides would be to keep such a signal active by preventing the binding of $\alpha_5\beta_1$ to matrix fibronectin. The availability of integrin-specific peptides from peptide libraries will enable these questions to be answered in the near future and may yield improved anti-tumor compounds.

## ACKNOWLEDGMENT

The author's work is supported by grants CA 28896 and CA 42506, with institutional support from Cancer Center Support Grant CA 30199 awarded by the National Cancer Institute, DHHS.

## REFERENCES

1. Ruoslahti E, Pierschbacher MD. New perspectives in cell adhesion: RGD and integrins. *Science* 1987;238:491–7.
2. Pierschbacher MD, Ruoslahti E. Cell attachment activity of fibronectin can be duplicated by small fragments of the molecule. *Nature* 1994;309:30–3.
3. Main AL, Harvey TS, Baron M, Boyd J, Campbell ID. The three-dimensional structure of the tenth type III module of fibronectin: an insight into RGD-mediated interactions. *Cell* 1992;71:671–78.
4. Dickinson CD, Veerapandian B, Dai X, et al. Crystal structure of the tenth type III cell adhesion module of human fibronectin. *J Mol Biol* 1994;236:1079–92.

5. Obara M, Kang MS, Yamada KM. Site-directed mutagenesis of the cell-binding domain of human fibronectin: separable, synergistic sites mediate adhesive function. *Cell* 1988;53: 649–57.
6. Nagai T, Yamakawa N, Aota S, et al. Monoclonal antibody characterization of two distant sites required for function of the central cell-binding domain of fibronectin in cell adhesion, cell migration, and matrix assembly. *J Cell Biol* 1991;114:1295–1305.
7. Koivunen E, Wang B, Ruoslahti E. Isolation of a highly specific ligand for the $\alpha_5\beta_1$ integrin from a phage display library. *J Cell Biol* 1994;124:373–80.
8. Bowditch RD, Hariharan M, Tominna EF, et al. Identification of a novel integrin binding site in fibronectin. *J Biol Chem* 1994;269:10856–63.
9. Pierschbacher MD, Ruoslahti E. Influence of stereochemistry of the sequence Arg-Gly-Asp-Xaa on binding specificity in cell adhesion. *J Biol Chem* 1987;262:17294–8.
10. Kirchhofer D, Gailit J, Ruoslahti E, Grzesiak J, Pierschbacher MD. Cation-dependent modulation of GPIIb/IIIa ligand specificity. *J Biol Chem* 1990;265:18525–30.
11. Samanen J, Ali F, Romoff T, et al. Development of a small RGD peptide fibrinogen receptor antagonist with potent antiaggregatory activity in vitro. *J Med Chem* 1991;34:3114–25.
12. Gurrath M, Müller G, Kessler H, Aumailley M, Timpl R. Conformation/activity studies of rationally designed potent anti-adhesive RGD peptides. *Eur J Biochem* 1992;210:911–21.
13. Scarborough RM, Naughton MA, Teng W, et al. Design of potent and specific integrin antagonists. *J Biol Chem* 1993;268:1066–73.
14. Tschopp JF, Driscoll EM, Mu D-X, Black SC, Pierschbacher MD, Lucchesi BR. Inhibition of coronary artery reocclusion after thrombolysis with an RGD-containing peptide with no significant effect on bleeding time. *Coronary Artery Dis* 1993;4:1–9.
15. Koivunen E, Gay DA, Ruoslahti E. Selection of peptides binding to the $\alpha_5\beta_1$ integrin from phage display library. *J Biol Chem* 1993;268:20205–10.
16. Scott JK, Smith GP. Searching for peptide ligands with an epitope library. *Science* 1990; 249:386–90.
17. Smith GP, Scott JK. Libraries of peptides and proteins displayed on filamentous phage. *Method Enzymol* 1993;217:228–57.
18. O'Neil KT, Hoess RH, Jackson SA, Ramachandran NS, Mousa SA, DeGrado WF. Identification of novel peptide antagonists for GPIIb/IIIa from a conformationally constrained phage peptide library. *Peptides* 1992;14:509–15.
19. Humphries MJ, Komoriya A, Akiyama SK, Olden K, Yamada KM. Identification of two distinct regions of the type III connecting segment of human plasma fibronectin that promote cell type-specific adhesion. *J Biol Chem* 1987;262:6886–92.
20. Guan J-L, Hynes RO. Lymphoid cells recognize an alternatively spliced segment of fibronectin via the integrin receptor $\alpha_4\beta_1$. *Cell* 1990;60:53–61.
21. Garcia-Pardo A, Wayner EA, Carter WG, Ferreira OC Jr. Human B lymphocytes define an alternative mechanism of adhesion to fibronectin. *J Immunol* 1990;90:3361–6.
22. Mould AP, Komoriya A, Yamada KM, Humphries MJ. The CS5 peptide is a second site in the IIICS region of fibronectin recognized by the integrin $\alpha_4\beta_1$. Inhibition of $\alpha_4\beta_1$ function by RGD peptide homologues. *J Biol Chem* 1991;266:3579–85.
23. Nowlin DM, Gorcsan F, Moscinski M, Chiang S-L, Lobl TJ, Cardarelli PM. A novel cyclic pentapeptide inhibits $\alpha_4\beta_1$ and $\alpha_5\beta_1$ integrin-mediated cell adhesion. *J Biol Chem* 1993;268: 20352–9.
24. Ruoslahti E. Integrins. *J Clin Invest* 1991;87:1–5.
25. Busk M, Pytela R, Sheppard D. Characterization of the integrin $\alpha_v\beta_6$ as a fibronectin-binding protein. *J Biol Chem* 1992;267:5790–6.
26. Nishimura SL, Pytela R. Characterization of the integrin $\alpha_1\beta_8$ [Abstract]. *Mol Biol Cell* 1993;4(suppl):285.
27. Cardarelli PM, Yamagata S, Taguchi I, Gorscan F, Chiang SL, Lobl T. The collagen receptor $\alpha_2\beta_1$ from MG-63 and HT1080 cells interacts with a cyclic RGD peptide. *J Biol Chem* 1992;267:23159–64.
28. Elices MJ, Urry LA, Hemler ME. Receptor functions for the integrin VLA-3: fibronectin, collagen, and laminin binding are differentially influenced by ARG-GLY-ASP peptide and by divalent cations. *J Cell Biol* 1991;112:169–81.
29. Kloczewiak M, Timmons S, Hawiger J. Recognition site for the platelet receptor is present

on the 15-residue carboxy-terminal fragment of the gamma chain of human fibrinogen and is not involved in the fibrin polymerization reaction. *Thromb Res* 1983;29:249–55.
30. Santoro SA, Lawing WJ Jr. Modulation of an RGDS binding site on the platelet membrane glycoprotein IIb-IIIa complex. *Biochem Biophys Res Commun* 1989;160:189–95.
31. Eble JA, Golbik R, Mann K, Kühn K. The $\alpha_1\beta_1$ integrin recognition site of the basement membrane collagen molecule $[\alpha1(IV)]_2\alpha2(IV)$. *EMBO J* 1993;12:4795–802.
32. Altieri DC, Plescia J, Plow EF. The structural motif glycine 190-valine 202 of the fibrinogen $\gamma$ chain interacts with CD11b/CD18 integrin ($\alpha_M\beta_2$, Mac-1) and promotes leukocyte adhesion. *J Biol Chem* 1993;268:1847–53.
33. Sánchez-Aparicio P, Dominguez-Giménez C, Garcia-Pardo A. Activation of the $\alpha_4\beta_1$ integrin through the $\beta_1$ subunit induces recognition of the RGDS sequence in fibronectin. *J Cell Biol* 1994;126:271–9.
34. Plow EF, Srouji AH, Meyer D, Marguerie G, Ginsberg MH. Evidence that three adhesive proteins interact with a common recognition site on activated platelets. *J Biol Chem* 1984; 259:5388–91.
35. Bowditch RD, Halloran CE, Aota S, et al. Integrin $\alpha_{IIb}\beta_3$ (platelet GPIIb-IIIa) recognizes multiple sites in fibronectin. *J Biol Chem* 1991;266:23323–28.
36. Vogel BE, Lee S-J, Hildebrand A, et al. A novel integrin specificity exemplified by binding of the $\alpha_v\beta_5$ integrin to the basic domain of the HIV Tat protein and vitronectin. *J Cell Biol* 1993;121:461–8.
37. Giancotti FG, Ruoslahti E. Elevated levels of the $\alpha_5\beta_1$ fibronectin receptor suppress the transformed phenotype of Chinese hamster ovary cells. *Cell* 1990;60:849–59.
38. Juliano RL, Varner JA. Adhesion molecules in cancer: the role of integrins. *Curr Opin Cell Biol* 1993;5:812–8.
39. Humphries MJ, Olden K, Yamada KM. A synthetic peptide from fibronectin inhibits experimental metastasis of murine melanoma cells. *Science* 1986;233:467–70.
40. Tressler RJ, Belloni PN, Nicolson GL. Correlation of inhibition of adhesion of large cell lymphoma and hepatic sinusoidal endothelial cells by RGD-containing peptide polymers with metastatic potential: role of integrin-dependent and -independent adhesion mechanisms. *Cancer Commun* 1989;1:55–63.
41. Saiki I, Iida J, Murata J, et al. Inhibition of the metastasis of murine malignant melanoma by synthetic polymeric peptides containing core sequences of cell-adhesive molecules. *Cancer Res* 1989;49:3815–22.
42. Ugen KE, Mahalingam M, Klein PA, Kao K-J. Inhibition of tumor cell-induced platelet aggregation and experimental tumor metastasis by the synthetic Gly-Arg-Gly-Asp-Ser peptide. *J Natl Cancer Inst* 1988;80:1461–6.
43. Komazawa H, Saiki I, Igarashi Y, et al. The conjugation of RGDS peptide with CM-chitin augments the peptide-mediated inhibition of tumor metastasis. *Carbohydr Polymers* 1993; 21:299–307.
44. Hardan I, Weiss L, Hershkoviz R, et al. Inhibition of metastatic cell colonization in murine lungs and tumor-induced morbidity by non-peptidic arg-gly-asp mimetics. *Int J Cancer* 1993;55:1023–8.
45. Hayman EG, Pierschbacher MD, Ruoslahti E. Detachment of cells from culture substrate by soluble fibronectin peptides. *J Cell Biol* 1985;100:1948–54.
46. Frisch SM, Francis H. Disruption of epithelial cell-matrix interactions induces apoptosis. *J Cell Biol* 1994;124:619–26.
47. Gehlsen KR, Argraves WS, Pierschbacher MD, Ruoslahti E. Inhibition of *in vitro* tumor cell invasion by Arg-Gly-Asp-containing synthetic peptides. *J Cell Biol* 1988;106:925–30.
48. Brooks PC, Clark RAF, Cheresh DA. Requirement of vascular integrin avb3 for angiogenesis. *Science* 1994;264:569–71.
49. LaFlamme SE, Akiyama SK, Yamada KM. Regulation of fibronectin receptor distribution. *J Cell Biol* 1992;117:437–47.

*Topics in Molecular Medicine, Volume 1,*
edited by Wolfgang Siess, Reinhard Lorenz,
and Peter C. Weber. Raven Press, Ltd.,
New York © 1995.

# 7

# Rapid Recruitment of Leukocytes to the Vascular Wall: A Juxtacrine System Using Selectins and Phospholipid Autacoids

Stephen M. Prescott, Guy A. Zimmerman, and
Thomas M. McIntyre

*The Eccles Program in Human Molecular Biology and Genetics, The Nora Eccles Harrison Cardiovascular Research & Training Institute, and the Departments of Internal Medicine and Biochemistry, University of Utah, Salt Lake City, Utah 84112*

Binding of leukocytes to endothelium and their subsequent emigration and activation are essential responses to maintain homeostasis of the organism, a process we refer to as *physiologic inflammation.* For example, these steps are crucial in the early response to bacterial infection. In contrast to this beneficial effect, adhesion and activation of leukocytes can be disastrous if they occur at the wrong time or place—*pathologic inflammation.* Intravascular aggregation and activation of neutrophils are responsible for some or all of the damage in various types of vascular injury, including ischemia–reperfusion and acute lung injury, and may contribute to atherogenesis. Therefore, it is crucial that the interactions between endothelial cells and leukocytes be strictly controlled. We have studied both physiologic inflammation and its pathologic counterpart in vitro to dissect the molecular events that regulate adhesion and activation.

## POLYMORPHONUCLEAR LEUKOCYTES AND MONOCYTES ADHERE TO P-SELECTIN BUT ARE NOT ACTIVATED IN THE PROCESS

Our previous investigations emphasized the endothelial cell responses that result in a surface that mediates the binding of neutrophilic polymorphonuclear leukocytes (PMNs). In more recent experiments, we have studied human monocytes in the same systems and have found that similar mechanisms operate. We found that endothelial cells stimulated with thrombin, histamine, bradykinin, or $LTC_4/D_4$ acquire a surface that is adhesive for PMNs (1–3). This response begins within seconds, is maximal at 5 to 15 min, and returns to baseline within an hour after addition of the agonist. The complete response requires that the endothelial cell express an adhesion protein that mediates the initial binding, a process we have termed *tethering,* and that the neutrophil subsequently activates one of its adhesion proteins. In this rapid form of adhesion, the initial binding is to the endothelial cell's P-selectin, a lectin that is stored in the Weibel–Palade bodies and which is transferred to the plasma membrane in response to agonists. It is able to tether leukocytes by binding to a counter-receptor on their surfaces.

One possibility was that the binding to P-selectin would be sufficient to explain all of the leukocyte responses. That is, PMNs adherent to activated endothelial cells are polarized, have upregulated their $\beta_2$ integrins, have an increased cytosolic calcium concentration, and are primed for secretion of the oxidative burst and enzyme secretion. We asked whether adhesion to P-selectin could account for all of the changes. To clarify this problem, we carried out experiments in which we measured the adhesion, and potentially other response(s), of PMNs incubated with purified immobilized P-selectin or CHO cells transfected with a cDNA for P-selectin. We found that the tethering step does not activate the PMNs (4). This indicated that there must be another signal for the responses of the leukocytes.

## THE SIGNAL THAT ACTIVATES LEUKOCYTES UNDER THESE CONDITIONS IS PLATELET-ACTIVATING FACTOR

We had found previously that, in response to the same stimuli, endothelial cells synthesize platelet-activating factor (PAF), which appeared to have some role (5). PAF (1-O-alkyl-2-acetyl-*sn*-glycero-3-phosphocholine) is a potent agonist for many types of cells (6,7). The common name is inaccurate, however, it refers to only one of PAF's many effects. For example, PAF is likely to be important in inflammation because it is a potent activator of PMNs, monocytes, and macrophages, and it induces increased vascular permeability. It acts through a receptor on the surface of target cells, and is the first receptor described, at least at a molecular level, for a phospholipid. The PAF receptor has been well characterized pharmacologically, and many potent antagonists have been developed. Most of those described are competitive antagonists, although they usually share no structural homology with PAF. Each of the structural elements of PAF is important for recognition by the receptor. Modification of the ether-linkage at the *sn*-1 position, the acetate at *sn*-2, or the phosphocholine head group usually results in reduced potency, although we have found that some oxidized phospholipids with 1-acyl groups and longer *sn*-2 groups are similar to PAF in potency (unpublished observations). The oxidized phospholipids and the other PAF analogues may have substantial bioactivity and may mediate responses in vivo.

We found that the PAF is expressed on the endothelial cell surface, but its physical orientation there is unknown (8,9). The P-selectin and PAF are both necessary for maximal adhesion and activation of PMNs to activated endothelial cells, and the effect of the adhesion protein and the autacoid lipid is synergistic (3). The endothelium-derived PAF activates the PMNs, a component of which is the upregulation of their adhesion proteins, the $\beta_2$ integrins (CD11b/CD18). This sequential process, with different mediators on the surface of endothelium responsible for the initial adhesion and then for the leukocyte activation, has several potential advantages. First, the messages to the leukocyte come from the surface of the endothelial cell rather than from free solution, therefore the signals are spatially discrete, which enables the leukocytes to bind precisely where they are needed. Second, the tethering in the absence of activation allows an editing step by the PMN: P-selectin captures the leukocyte and holds it close to the surface of the endothelial cell, and the PMN (or monocyte) then scans the surface for an activator signal. If an agonist such as PAF is present, the leukocyte undergoes the responses described above, a key feature of which is upregulation of its integrins, which strengthen the adhesion and cause the change in shape. In contrast, if no activator is encountered, the leukocyte can release from the tether and continue to circulate in an unactivated form. Third, the tethering of leukocytes by P-selectin provides specificity, because PMNs and monocytes have receptors for P-selectin, whereas other blood cells, such as platelets, do not. This sequential binding and activation response represents a form of juxtacrine activation, which was originally described as a mechanism by which membrane-anchored growth factors activated their target cells. The response described by us in endothelial cells that have undergone a rapid response may apply to the adhesion of leukocytes to endothelial cells in other circumstances.

## ENDOTHELIAL CELLS EXPOSED TO OXIDANTS EXPRESS P-SELECTIN AND RELEASE BIOACTIVE OXIDIZED PHOSPHOLIPIDS

The studies described thus far were focused on the events involved in physiologic inflammation. In other experiments, we have compared these responses with those that occur in response to stimuli likely to be involved in pathologic inflammation. One such circumstance is oxidative stress, which we mimicked by exposing endothelial cells to oxidants. We found that these conditions provoked changes similar to those described above. In particular, the oxidant-treated endothelial cells expressed a surface that is adhesive for PMNs. These effects were reversible, and the "injury" to the endothelial cells was not lethal. Adhesion of leukocytes to the oxidant-treated endothelium was blocked by antibodies to P-selectin, which was abundantly expressed on the endothelial surface (10). However, unlike the response to receptor-mediated agonists, the response was markedly prolonged, and P-selectin could be detected on the endothelial cell surface for hours. In these experiments we also observed that exposure to oxidants resulted in a dramatic change in the appearance of the endothelial cells. Most remarkably, the endothelial cells developed large protrusions of the plasma membrane (''blebs'') and subsequently shed many vesicles into the culture medium (11). These vesicles from oxidant-treated endothelial cells activated PMNs through the receptor for PAF (11,12). This process may be relevant to reperfusion injury and other strongly inflammatory situations in which potent oxidants are generated. Indeed, other investigators have found accumulation of PAF bioactivity during reperfusion of ischemic tissues in animal models, and were able to block the neutrophil accumulation with antagonists to the PAF receptor. In addition, several groups have reported blebs in histologic specimens from animal models or postmortem samples from patients in conditions such as renal ischemia and the adult respiratory distress syndrome.

We next examined the basis for generation of the blebs and the PAF-like lipids when endothelial cells were exposed to oxidants. We excluded the possibility that authentic PAF was synthesized, but we found that a structurally similar phospholipid accumulated. The unsaturated fatty acids of phospholipids in the membranes are targets for oxidation, and evidence for lipid oxidation has been found in many human diseases that have a component of strong inflammation. We found that exposure of synthetic phospholipids to a strongly oxidizing environment fragmented the unsaturated fatty acid at the *sn*-2 position and generated a structural analogue(s) of PAF. These compounds act through the PAF receptor to reproduce its effects (12). We found that similar compounds are produced when endothelial cells are exposed to oxidants, and that they are released as a component of the vesicles (blebs). This process is in marked contrast to the synthesis of PAF, as the latter is tightly regulated. For example, both of the enzymatic steps in PAF synthesis have a phosphorylation–dephosphorylation cycle that controls activity (13,14), whereas the PAF-like oxidized phospholipids result from unregulated chemical reactions. Therefore, they could be produced in much larger amounts than PAF, and at inappropriate times and places. In such circumstances, degradation of the bioactive phospholipids might be crucial in protecting the tissues from injury.

## THE PAF ACETYLHYDROLASE MAY BE A KEY REGULATORY STEP BY REMOVING THE PAF AND/OR OXIDIZED PHOSPHOLIPIDS

In either the physiologic or the pathologic response, the ''off'' signal is as important as the ''on'' signal. Because PAF or closely related lipids act as a crucial signal for leukocyte activation, their removal is probably a crucial step in suppressing the inflammation. We have shown that the enzyme that catalyzes the degradation of PAF or oxidatively fragmented phospholipids is the PAF acetylhydrolase (15,16). This enzyme has several remarkable features: it is specific for substrates with short acyl groups (such as the acetyl moiety in PAF); it is calcium-independent; and the plasma form is bound to lipoproteins. There are at least two forms of this enzyme, a mammalian plasma form (15) and an intracellular form (17). They catalyze the same reaction and have the same substrate specificity, but are different proteins as judged by several criteria. Degradation of PAF and related phospholipids by these isoenzymes may be a crucial step that regulates inflammation.

For the plasma enzyme, this trait is essential if the enzyme is to circulate in an active form, because if typical phospholipids were substrates the enzyme would continuously hydrolyze the phospholipids of lipoproteins and cell membranes. Macrophages also may play a role in the local regulation of PAF levels, as the precursor monocytes do not produce the enzyme. However, on differentiation to macrophages they begin to produce and secrete PAF acetylhydrolase (18).

## CONCLUSION

A key early step in physiologic inflammation is recruitment of leukocytes to the appropriate site. This is a complicated task because the leukocytes are flowing past the endothelium overlying the site where they are needed. Therefore, the response must be rapid, spatially specific, and of appropriate affinity to remove the cells from the bloodstream. Moreover, it would be dangerous to activate the leukocytes while they are still in flowing blood, because this could lead to damage of the endothelium through the release of oxygen radicals and proteases. We have shown that endothelial cells exposed to agonists such as thrombin, histamine, and bradykinin rapidly become adhesive for leukocytes. The initial step is to tether the unactivated leukocytes to the endothelial wall by binding of P-selectin to its counterligand on the leukocyte surface. Subsequently, the isolated leukocyte is activated by PAF on the surface of endothelial cells. This causes functional upregulation of the leukocyte integrins, which completes the adhesion. A similar, but unregulated and potentially injurious, set of responses occurs when endothelium is exposed to oxidants. P-Selectin is expressed on the cell surface, and membrane phospholipids are chemically converted into analogues of PAF. In both the physiologic and pathologic state, the degradation of PAF and/or oxidized phospholipids by PAF acetylhydrolase may be essential in limiting the extent of the inflammatory response.

## ACKNOWLEDGMENT

Our work has been supported by the Nora Eccles Treadwell and the George and Dolores Doré Eccles Foundations, and by grants from the National Institutes of Health and the American Heart Association. We thank the dedicated and skilled students, postdoctoral fellows, and technical assistants who have worked with us through the years and on whom this study depended.

## REFERENCES

1. Zimmerman GA, McIntyre TM, Prescott SM. Thrombin stimulates the adherence of neutrophils to human endothelial cells in vitro. *J Clin Invest* 1985;76:2235–46.
2. Geng J-G, Bevilacqua MP, Moore KL, et al. GMP-140 mediates rapid neutrophil adhesion to activated endothelium. *Nature* 1990;343:757–60.
3. Lorant DE, Patel KD, Prescott SM, McEver RP, McIntyre TM, Zimmerman GA. Coexpression of GMP-140 and PAF by endothelium stimulated by histamine or thrombin: a juxtacrine system for adhesion and activation of neutrophils. *J Cell Biol* 1991;113:223–34.
4. Lorant DE, Topham MK, Whatley RE, et al. Inflammatory roles of P-selectin. *J Clin Invest* 1993;92:559–70.
5. Zimmerman GA, McIntyre TM, Mehra M, Prescott SM. Endothelial cell-associated platelet-activating factor: a novel mechanism for signaling intercellular adhesion. *J Cell Biol* 1990;110:529–40.
6. Prescott SM, Zimmerman GA, McIntyre TM. Platelet-activating factor. *J Biol Chem* 1990; 265:17381–4.
7. Venable MA, Zimmerman GA, McIntyre TM, Prescott SM. Platelet-activating factor: a phospholipid autacoid with diverse actions. *J Lipid Res* 1993;34:691–702.
8. Prescott SM, Zimmerman GA, McIntyre TM. Human endothelial cells in culture produce platelet-activating factor (1-alkyl-2-acetyl-*sn*-glycero-3-phosphocholine) when stimulated with thrombin. *Proc Natl Acad Sci USA* 1984;81:3534–8.
9. McIntyre TM, Zimmerman GA, Satoh K, Prescott SM. Cultured endothelial cells synthesize both platelet-activating factor and prostacyclin in response to histamine, bradykinin, and adenosine triphosphate. *J Clin Invest* 1985;76:271–80.
10. Patel KD, Zimmerman GA, Prescott SM, McEver RP, McIntyre TM. Oxygen radicals induce human endothelial cells to express GMP-140 and bind neutrophils. *J Cell Biol* 1991;112:749–59.
11. Patel KD, Zimmerman GA, Prescott SM, McIntyre TM. Novel leukocyte agonists are released by endothelial cells exposed to peroxide. *J Biol Chem* 1992;267:15168–75.
12. Smiley PL, Stremler KE, Prescott SM, Zimmerman GA, McIntyre TM. Oxidatively-fragmented phosphatidylcholines activate human neutrophils through the receptor for platelet-activating factor. *J Biol Chem* 1991;266:11104–10.
13. Whatley RE, Fennell DF, Kurrus JA, Zimmerman GA, McIntyre TM, Prescott SM. Synthesis of platelet-activating factor by endothelial cells: the role of G proteins. *J Biol Chem* 1990;265:15550–9.
14. Holland MR, Venable ME, Whatley RE, Zimmerman GA, McIntyre TM, Prescott SM. Activation of the acetyl-coenzyme A: lysoplatelet-activating factor acetyltransferase regulates platelet-activating factor synthesis in human endothelial cells. *J Biol Chem* 1992;267: 22883–90.
15. Stafforini DM, Prescott SM, McIntyre TM. Human plasma platelet-activating factor acetylhydrolase: purification and properties. *J Biol Chem* 1987;262:4223–30.
16. Stremler KE, Stafforini DM, Prescott SM, McIntyre TM. Human plasma PAF acetylhydrolase: oxidatively-fragmented phospholipids as substrates. *J Biol Chem* 1991;266:11095–103.
17. Stafforini DM, Rollins EN, Prescott SM, McIntyre TM. The platelet-activating factor acetylhydrolase from human erythrocytes: purification and properties. *J Biol Chem* 1993; 268:3857–65.
18. Stafforini DM, Elstad MR, Zimmerman GA, McIntyre TM, Prescott SM. Human macrophages secrete platelet-activating factor acetylhydrolase. *J Biol Chem* 1990;265:9682–7.

*Topics in Molecular Medicine, Volume 1,*
edited by Wolfgang Siess, Reinhard Lorenz,
and Peter C. Weber. Raven Press, Ltd.,
New York © 1995.

# 8

# Adhesive Function and Regulation of Expression of P-Selectin

Rodger P. McEver

*W. K. Warren Medical Research Institute, Departments of Medicine and Biochemistry, University of Oklahoma Health Sciences Center, and Cardiovascular Biology Research Program, Oklahoma Medical Research Foundation, Oklahoma City, Oklahoma 73104*

The selectins are a group of three related membrane proteins that mediate the initial rolling of leukocytes on the blood vessel wall in response to tissue injury or infection. L-Selectin is constitutively expressed on leukocytes and binds to ligands on high endothelial venules of lymph nodes and to inducible ligands on activated endothelium at sites of inflammation. E-Selectin is expressed by cytokine-stimulated endothelium, where it mediates adhesion of leukocytes. P-Selectin is expressed by activated platelets and endothelial cells, where it also mediates leukocyte adhesion. The transient adhesion mediated by the selectins allows rolling leukocytes to be activated by locally expressed signaling molecules. Leukocyte activation increases the avidity of integrins for their immunoglobulin-like counter-receptors on the endothelium, thereby strengthening adhesion. Several general reviews on the selectins have appeared (1–4). This report is focused on recent studies of P-selectin.

## LIGANDS FOR P-SELECTIN

Each of the selectins contains an N-terminal carbohydrate-recognition domain like those in $Ca^{2+}$-dependent animal lectins, followed by an epidermal growth factor-like motif, a series of consensus repeats related to those in complement-binding proteins, a transmembrane domain, and a short cytoplasmic tail. The selectins mediate cell adhesion through $Ca^{2+}$-dependent interactions of the carbohydrate recognition domain with cell surface oligosaccharides. The nature of the carbohydrate ligands for the selectins has elicited intense interest. All three selectins can bind with low affinity to sialylated, fucosylated tetrasaccharides such as sialyl Lewis x (sLe$^x$; Siaα2-3Galβ1-4[Fucβ1-3]GlcNAc-R). Both the sialic acid and fucose moieties are required for recognition. Because sLe$^x$ is found on many glycolipids and glycoproteins of myeloid cells, a frequent interpretation of early studies was that this tetrasaccharide represented the physiologic ligand for E- and P-selectin. However, accumulating data from several laboratories indicate that the selectins bind to specific glycoprotein ligands with affinities that are much higher than those to simple oligosaccharides, such as sLe$^x$.

P-Selectin isolated from human platelet membranes binds to a limited number of sites on human neutrophils and HL-60 cells; these sites are protease-sensitive, suggesting that they are glycoproteins rather than glycolipids (5). Recombinant forms of P-selectin lacking the transmembrane domain are monomeric, whereas the membrane protein is oligomeric in detergent levels below the critical micellar concentration (6). In studies of equilibrium binding to myeloid cells, both the monomers and oligomers appear to interact with the same limited number of sites, as assessed by competitive binding analysis. Although the oligomers bind with higher avidity, even the monomers bind well to human neutrophils and HL-60 cells. These sites can be contrasted with much lower-affinity binding sites for P-selectin on Chinese hamster ovary cells transfected with a fucosyltransferase, which express higher levels of cell surface sLe$^x$ than do myeloid cells (7). These data suggest that mere expression of sLe$^x$ is not sufficient to confer preferential binding to P-selectin.

We have identified a high-affinity glycoprotein ligand for P-selectin in extracts of human neutrophils and HL-60 cells (8,9). The ligand, a minor component of total plasma membrane glycoproteins, is an extensively sialylated dimer consisting of two disulfide-linked subunits of $M_r$ 120,000. It carries α2,3-linked sialic acids and the sLe$^x$ antigen. Almost all of the sialic acids are unmodified *N*-acetylneuraminic acid. The protein contains a limited number of *N*-linked glycans that are not required for binding to P-selectin. In contrast, it contains a large number of sialylated, *O*-linked oligosaccharides that are released by β-elimination. The ligand is cleaved by the enzyme *O*-sialoglycoprotease from *Pasteurella hemolytica,* which recognizes only proteins containing clustered, sialylated *O*-linked glycans. Treatment of HL-60 cells with this enzyme eliminates the high-affinity binding sites for P-selectin and abolishes adhesion to immobilized P-selectin, without affecting total surface expression of sLe$^x$. These data suggest that the ligand accounts for the high-affinity binding sites on intact myeloid cells and may play an important role in mediating cell adhesion to P-selectin.

The P-selectin ligand has been purified from human neutrophils and radioiodinated to facilitate additional structural and functional characterization (10). Experiments with glycosidases and plant lectins indicate that the molecule contains poly-*N*-acetyllactosamine, most of which is on the *O*-linked glycans. Most of the Le$^x$ and sLe$^x$ on the ligand are also on the

*O*-linked chains, and at least a portion of both moieties are on poly-*N*-acetyllactosamine. These data suggest that at least some of the *O*-linked oligosaccharides on the ligand have relatively large, complex structures. This interpretation is consistent with the observation that very few of the *O*-linked glycans are removed by sequential treatment with sialidase and endo-α-*N*-galactosaminidase; the latter enzyme can cleave only simple, core 1 Galβ1–3GalNAc disaccharides linked to serine or threonine residues.

The structural basis for high-affinity binding of the ligand to P-selectin remains to be determined. The *O*-linked oligosaccharides may contain features not found on $sLe^x$ or related known sialylated, fucosylated lactosaminoglycans. Closely packed glycans might also form a rigid "clustered saccharide patch" in which components of multiple oligosaccharides create a unique recognition structure (9). Structural characterization of the individual oligosaccharides and determination of the sites of attachment may help to explain why this ligand interacts so well with P-selectin. It is noteworthy that another mucin-like protein, leukosialin (CD43), is also expressed by myeloid cells but is not a ligand for P-selectin (8). Therefore, in a given cell type the polypeptide chain may play a critical role in determining the structure of the *O*-linked oligosaccharides that are attached during passage through the endoplasmic reticulum and the Golgi complex.

Recently, a cDNA encoding the protein component of the P-selectin ligand was isolated by expression cloning in cells co-transfected with fucosyltransferase III (11). The ligand, termed P-selectin glycoprotein ligand 1 (PSGL-1), is a type I membrane protein that includes 15 decameric repeats rich in threonine residues. The repeats have many potential sites for addition of clustered *O*-linked oligosaccharides that would render the protein sensitive to *O*-sialoglycoprotease. The expressed recombinant protein, like PSGL-1 from human neutrophils, has both *N*- and *O*-linked carbohydrate and binds to myeloid cells. How closely the carbohydrate modifications of the recombinant protein match those of the native protein in myeloid cells is unknown, a relevant issue in determining the degree to which the function of the recombinant molecule is equivalent to that of the native glycoprotein. Significantly, however, many of the *O*-linked oligosaccharides on recombinant PSGL-1 are removed by sialidase and endo-α-*N*-galactosaminidase. This finding suggests that recombinant PSGL-1, unlike the native glycoprotein from human neutrophils, contains many simple core 1 Galβ1–3GalNAc *O*-linked disaccharides after enzymatic removal of terminal sialic acids.

Because PSGL-1 from human neutrophils expresses $sLe^x$, it could conceivably also serve as a ligand for E-selectin. In cell adhesion experiments, recombinant PSGL-1 does interact with E-selectin as well as P-selectin (11). Furthermore, a glycoprotein that appears to correspond to PSGL-1 in mouse and human neutrophils is precipitated by both E-selectin and P-selectin Ig chimeras coupled to protein A beads (12). However, in an equilibrium competitive binding assay, monomeric E-selectin binds to PSGL-1 from human neutrophils with at least 50-fold weaker affinity than does monomeric P-selectin (10). Therefore, under equilibrium conditions, PSGL-1 has a distinct preference for binding to P-selectin. Whether the different affinities of PSGL-1 for E- and P-selectin are physiologically relevant is unknown. Indeed, it has not yet been determined whether PSGL-1 actually mediates leukocyte adhesion to either P- or E-selectin under shear forces.

A 160-kDa glycoprotein ligand for P-selectin, but not for E-selectin, has also been described in mouse and human myeloid cells. This molecule may contain functional *N*-linked and *O*-linked oligosaccharides (12). However, we have been unable to isolate this ligand by P-selectin affinity chromatography of lysates from human neutrophils or metabolically labeled HL-60 cells (8,9). High-affinity glycoprotein ligands for both L- and E-selectin have also been identified (13–16). At least some of these ligands contain many sialylated *O*-linked oligosaccharides. Although these glycoprotein ligands for selectins are attractive candidates for mediating selectin-dependent cell adhesion, further studies are required to confirm this hypothesis. It remains possible that larger numbers of low-affinity interactions also contribute to cell adhesion. The mechanisms by which selectins mediate cell adhesion under shear forces are even less well understood. It is likely that high on- and off-rates of bond formation are required, but the structural features that favor such kinetics are unknown. Many factors, such as receptor and ligand length, mobility, and clustering, may contribute to bond formation, and hence to adhesion, under shear conditions. P-selectin itself is a rigid, asymmetric protein that should project the carbohydrate recognition domain approximately 40 nm from the cell surface, where it might rapidly contact a ligand on a circulating leukocyte (6).

## MEMBRANE TRAFFICKING OF P-SELECTIN

Regulated expression of the selectins plays a critical role in controlling the duration of leukocyte adhesion to the blood vessel wall. The control of expression of P-selectin is particularly complex. The protein is constitutively synthesized by megakaryocytes and venular endothelial cells, where it is delivered to secretory granules ($\alpha$ granules in platelets and Weibel–Palade bodies in endothelial cells) by a sorting signal in the cytoplasmic domain (17). On stimulation of these cells by agonists such as thrombin or histamine, P-selectin is redistributed to the cell surface as granule membranes fuse with the plasma membrane. On activated endothelial cells, P-selectin is then rapidly internalized, a mechanism that limits the adhesive phenotype of the vessel wall (18). Internalization occurs in clathrin-coated pits (H. Setiadi and R. P. McEver, unpublished observations) and requires a signal in the cytoplasmic domain (19; and H. Setiadi and R. P. McEver, unpublished observations). Although antibody-binding studies have suggested that P-selectin is efficiently recycled into Weibel–Palade bodies after endocytosis (19), P-selectin antigen disappears in dermal venules after intradermal injection of agents that induce degranulation, suggesting rapid degradation after endocytosis (20,21). The latter observations may reflect a signal in the cytoplasmic domain that mediates rapid movement of P-selectin from endosomes to lysosomes (22). Rapid delivery of P-selectin from endosomes to lysosomes is an effective means for limiting recycling to the plasma membrane and, hence, for controlling the duration in which the endothelium remains adhesive for leukocytes. Therefore, the movements of P-selectin within cells are controlled by information in the cytoplasmic domain that directs sorting into secretory granules, endocytosis, and targeting to lysosomes. The structural features in the cytoplasmic domain required for each of these signals, and the degree to which the signals overlap, are unknown. The single cysteine in the cytoplasmic domain is acylated (23). The cytoplasmic domain is also phosphorylated, although the sites of phosphorylation are controversial (24,25). Whether these post-translational modifications affect the trafficking of P-selectin is unclear.

## REGULATION OF SYNTHESIS OF P-SELECTIN

Inflammatory cytokines increase synthesis of P-selectin in at least some tissues, providing an additional level of regulation (26–29). Transcripts in both platelets and endothelial cells have been identified that encode an alternatively spliced soluble form of P-selectin lacking the transmembrane domain (30,31). The physiologic significance of soluble P-selectin is unknown. Because the sorting signal no longer faces the cytoplasm, this protein is constitutively secreted rather than sorted into secretory granules (17). Very low levels of P-selectin antigen are found in human plasma, and the levels may increase slightly in some thrombotic or inflammatory disorders (6,32–34).

The 5′-flanking region of the human P-selectin gene has been cloned and sequenced (35). Transcription of the gene is initiated from multiple sites, consistent with the lack of a canonical TATA box that would facilitate transcription initiation at a single site. A relatively short segment of the 5′-flanking region directs specific expression of a reporter gene in cultured endothelial cells but not in several other cells tested, suggesting that it contains at least some of the elements required for cell type-specific expression. The region contains a number of candidate regulatory elements, including a GATA element that was demonstrated to be functional. A consensus site for binding of members of the NFκB/rel family of transcription factors is also present. Whether or not this site plays a role in the cytokine-induced increase in P-selectin synthesis observed in some tissues is unknown (26–28).

Dysregulated expression of selectins is a likely mechanism for contributing to the pathogenesis of some inflammatory and thrombotic disorders. In vitro, oxygen radicals cause P-selectin to remain on the surface of endothelial cells for several hours, probably by impairing endocytosis (36). In vivo, monoclonal antibodies to P-selectin reduce tissue injury in animal models of lung damage (37) and ischemia–reperfusion injury (38,39). Prolonged expression of P-selectin on the endothelial cell surface at sites of acute tissue damage (39) and overlying atherosclerotic plaques (40) has been documented, consistent with a role for the protein in mediating pathologic leukocyte adhesion. The antibodies presumably inhibit the initial rolling of neutrophils on venules where P-selectin has been inappropriately expressed. Indeed, other studies confirm a role for P-selectin in mediating leukocyte rolling on venules of tissues subjected to mild injury after exteriorization (41–43). Additional in vivo studies, coupled with further biochemical analysis, should enhance our understanding of the role of P-selectin and the other selectins in physiologic inflammation and hemostasis, and may suggest approaches to inhibit selectin function during pathologic inflammation, thrombosis, and tumor metastasis.

## ACKNOWLEDGMENT

Work in the author's laboratory was supported by National Institutes of Health grants HL 34363 and HL 45510.

## REFERENCES

1. McEver RP. Leukocyte-endothelial cell interactions. *Curr Opin Cell Biol* 1992;4:840–9.
2. Lasky LA. Selectins: interpreters of cell-specific carbohydrate information during inflammation. *Science* 1992;258:964–9.
3. Bevilacqua MP, Nelson RM. Selectins. *J Clin Invest* 1993;91:379–87.
4. McEver RP. Selectins. *Curr Opin Immunol* 1994;6:75–84.
5. Moore KL, Varki A, McEver RP. GMP-140 binds to a glycoprotein receptor on human neutrophils: evidence for a lectin-like interaction. *J Cell Biol* 1991;112:491 9.
6. Ushiyama S, Laue TM, Moore KL, Erickson HP, McEver RP. Structural and functional characterization of monomeric soluble P-selectin and comparison with membrane P-selectin. *J Biol Chem* 1993;268:15229–37.
7. Zhou Q, Moore KL, Smith DF, Varki A, McEver RP, Cummings RD. The selectin GMP-140 binds to sialylated, fucosylated lactosaminoglycans on both myeloid and nonmyeloid cells. *J Cell Biol* 1991;115:557–64.
8. Moore KL, Stults NL, Diaz S, et al. Identification of a specific glycoprotein ligand for P-selectin (CD62) on myeloid cells. *J Cell Biol* 1992;118:445–56.
9. Norgard KE, Moore KL, Diaz S, et al. Characterization of a specific ligand for P-selectin on myeloid cells. A minor glycoprotein with sialylated O-linked oligosaccharides. *J Biol Chem* 1993;268:12764–74.
10. Moore KL, Eaton SF, Lyons DE, Lichenstein HS, Cummings RD, McEver RP. The P-selectin glycoprotein ligand from human neutrophils displays sialylated, fucosylated, *O*-linked poly-*N*-acetyllactosamine. *J Biol Chem* 1994;269:23318–27.
11. Sako D, Chang X-J, Barone KM, et al. Expression cloning of a functional glycoprotein ligand for P-selectin. *Cell* 1993;75:1179–86.

12. Lenter M, Levinovitz A, Isenmann S, Vestweber D. Monospecific and common glycoprotein ligands for E- and P-selectin on myeloid cells. *J Cell Biol* 1994;125:471–81.
13. Imai Y, Singer MS, Fennie C, Lasky LA, Rosen SD. Identification of a carbohydrate-based endothelial ligand for a lymphocyte homing receptor. *J Cell Biol* 1991;113:1213–22.
14. Lasky LA, Singer MS, Dowbenko D, et al. An endothelial ligand for L-selectin is a novel mucin-like molecule. *Cell* 1992;69:927–38.
15. Levinovitz A, Mühloff J, Isenmann S, Vestweber D. Identification of a glycoprotein ligand for E-selectin on mouse myeloid cells. *J Cell Biol* 1993;121:449–59.
16. Baumhueter S, Singer MS, Henzel W, et al. Binding of L-selectin to the vascular sialomucin CD34. *Science* 1993;262:436–8.
17. Disdier M, Morrissey JH, Fugate RD, Bainton DF, McEver RP. Cytoplasmic domain of P-selectin (CD62) contains the signal for sorting into the regulated secretory pathway. *Mol Biol Cell* 1992;3:309–21.
18. Hattori R, Hamilton KK, Fugate RD, McEver RP, Sims PJ. Stimulated secretion of endothelial von Willebrand factor is accompanied by rapid redistribution of the cell surface of the intracellular granule membrane protein GMP-140. *J Biol Chem* 1989;264:7768–71.
19. Subramaniam M, Koedam JA, Wagner DD. Divergent fates of P- and E-selectins after their expression on the plasma membrane. *Mol Biol Cell* 1993;4:791–801.
20. Smith CH, Barker JNWN, Morris RW, MacDonald DM, Lee TH. Neuropeptides induce rapid expression of endothelial cell adhesion molecules and elicit granulocytic infiltration in human skin. *J Immunol* 1993;151:3274–82.
21. Silber A, Newman W, Reimann KA, Hendricks E, Walsh D, Ringler DJ. Kinetic expression of endothelial adhesion molecules and relationship to leukocyte recruitment in two cutaneous models of inflammation. *Lab Invest* 1994;70:163–75.
22. Green SA, Setiadi H, McEver RP, Kelly RB. The cytoplasmic domain of P-selectin contains a sorting determinant that mediates rapid degradation in lysosomes. *J Cell Biol* 1994; 124:435–48.
23. Fujimoto T, Stroud E, Whatley RE, et al. P-selectin is acylated with palmitic acid and stearic acid at cysteine 766 through a thioester linkage. *J Biol Chem* 1993;268:11394–400.
24. Fujimoto T, McEver RP. The cytoplasmic domain of P-selectin is phosphorylated on serine and threonine residues. *Blood* 1993;82:1758–66.
25. Crovello CS, Furie BC, Furie B. Rapid phosphorylation and selective dephosphorylation of P-selectin accompanies platelet activation. *J Biol Chem* 1993;268:14590–3.
26. Weller A, Isenmann S, Vestweber D. Cloning of the mouse endothelial selectins. Expression of both E- and P-selectin is inducible by tumor necrosis factor. *J Biol Chem* 1992;267: 15176–83.
27. Sanders WE, Wilson RW, Ballantyne CM, Beaudet AL. Molecular cloning and analysis of in vivo expression of murine P-selectin. *Blood* 1992;80:795–800.
28. Hahne M, Jäger U, Isenmann S, Hallmann R, Vestweber D. Five tumor necrosis factor-inducible cell adhesion mechanisms on the surface of mouse endothelioma cells mediate the binding of leukocytes. *J Cell Biol* 1993;121:655–64.
29. Gotsch U, Jager U, Dominis M, Vestweber D. Expression of P-selectin on endothelial cells is upregulated by LPS and TNF-α in vivo. *Cell Adhes Commun* 1994;2:7–14.
30. Johnston GI, Cook RG, McEver RP. Cloning of GMP-140, a granule membrane protein of platelets and endothelium: sequence similarity to proteins involved in cell adhesion and inflammation. *Cell* 1989;56:1033–44.
31. Johnston GI, Bliss GA, Newman PJ, McEver RP. Structure of the human gene encoding GMP-140, a member of the selectin family of adhesion receptors for leukocytes. *J Biol Chem* 1990;265:21381–5.
32. Dunlop LC, Skinner MP, Bendall LJ, et al. Characterization of GMP-140 (P-selectin) as a circulating plasma protein. *J Exp Med* 1992;175:1147–50.
33. Katayama M, Handa M, Araki Y, et al. Soluble P-selectin is present in normal circulation and its plasma level is elevated in patients with thrombotic thrombocytopenic purpura and haemolytic uraemic syndrome. *Br J Haematol* 1993;84:702–10.
34. Wu G, Li F, Li P, Ruan C. Detection of plasma alpha-granule membrane protein GMP-140 using radiolabeled monoclonal antibodies in thrombotic diseases. *Haemostasis* 1993;23: 121–8.

35. Pan J, McEver RP. Characterization of the promoter for the human P-selectin gene. *J Biol Chem* 1993;268:22600–8.
36. Patel KD, Zimmerman GA, Prescott SM, McEver RP, McIntyre TM. Oxygen radicals induce human endothelial cells to express GMP-140 and bind neutrophils. *J Cell Biol* 1991;112:749–59.
37. Mulligan MS, Polley MJ, Bayer RJ, Nunn MF, Paulson JC, Ward PA. Neutrophil-dependent acute lung injury. Requirement for P-selectin (GMP-140). *J Clin Invest* 1992;90: 1600–7.
38. Weyrich AS, Ma X, Lefer DJ, Albertine KH, Lefer AM. *In vivo* neutralization of P-selectin protects feline heart and endothelium in myocardial ischemia and reperfusion injury. *J Clin Invest* 1993;91:2620–9.
39. Winn RK, Liggitt D, Vedder NB, Paulson JC, Harlan JM. Anti-P-selectin monoclonal antibody attenuates reperfusion injury to the rabbit ear. *J Clin Invest* 1993;92:2042–7.
40. Johnson-Tidey RR, McGregor JL, Taylor PR, Poston RN. Increase in the adhesion molecule P-selectin in endothelium overlying atherosclerotic plaques. Coexpression with intercellular adhesion molecule-1. *Am J Pathol* 1994;144:952–61.
41. Bienvenu K, Granger DN. Molecular determinants of shear rate-dependent leukocyte adhesion in postcapillary venules. *Am J Physiol* 1993;264:H1504–8.
42. Dore M, Korthuis RJ, Granger DN, Entman ML, Smith CW. P-selectin mediates spontaneous leukocyte rolling in vivo. *Blood* 1993;82:1308–16.
43. Mayadas TN, Johnson RC, Rayburn H, Hynes RO, Wagner DD. Leukocyte rolling and extravasation are severely compromised in P selectin-deficient mice. *Cell* 1993;74:541–54.

*Topics in Molecular Medicine, Volume 1,*
edited by Wolfgang Siess, Reinhard Lorenz,
and Peter C. Weber. Raven Press, Ltd.,
New York © 1995.

# 9

# P-Selectin: Its Role in Cell Adhesion and Stimulus–Response Coupling of Effector Function

Bruce Furie and Barbara C. Furie

*Center for Hemostasis and Thrombosis Research and the Division of Hematology–Oncology, New England Medical Center, and Departments of Medicine and Biochemistry, Tufts University School of Medicine, Boston, Massachusetts 02111*

P-Selectin is a cell adhesion molecule that resides in the α-granules of resting platelets and the Weibel–Palade bodies of endothelial cells. On stimulation of these cells the protein is translocated to the plasma membrane, where it functions as a leukocyte receptor. To date, four families of adhesion molecules have been defined (1,2): members of the Ig superfamily; the integrin family; the mucin-like ligands of the selectins; and the selectins (P-selectin, L-selectin, and E-selectin). The selectins contain a common domain structure, including a lectin domain, an EGF domain, a variable number of consensus repeats, a transmembrane domain, and a cytoplasmic tail (3–6). P-Selectin is stored in the storage granules of platelets (7,8) and endothelial cells (9,10). On stimulation, these cells undergo degranulation, with rapid translocation of P-selectin to the plasma membrane (7,10,11). After expression on the plasma membrane of platelets and endothelial cells, this protein is a receptor for neutrophils and monocytes (12).

The selectins are important receptors in cell–cell interactions that involve circulating neutrophils and monocytes in the blood and their association with endothelium and activated platelets in the sites of tissue injury. After tissue injury, leukocytes roll along the vessel wall. Lawrence and Springer (13) demonstrated that P-selectin mediates this rolling phenomenon; recent work on the P-selectin null mouse (14) confirms these findings, in that leukocytes become associated (roll) on the stimulated vasculature of the normal mouse but not the P-selectin null mouse. In summary, the selectins play a role in inflammation (2) and P-selectin, in particular, plays a role in thrombosis (15).

Critical questions remain. First, what are the structural elements on P-se-

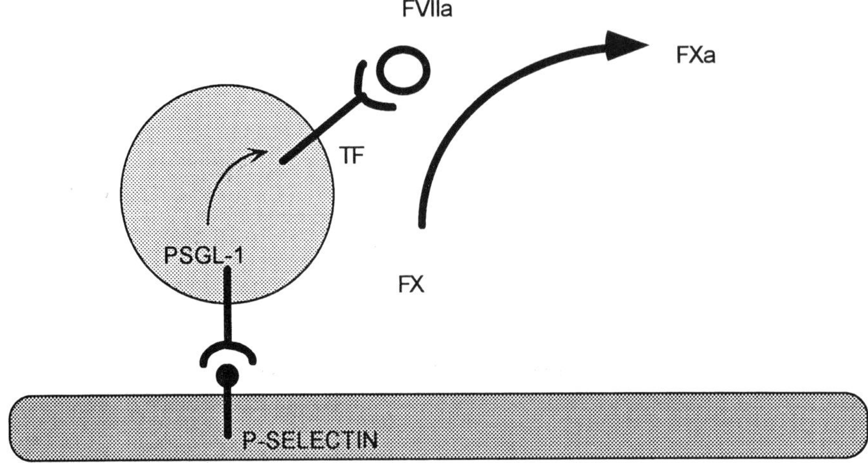


**FIG. 1.** Cell adhesion and signal transduction mediated by P-selectin. P-Selectin is expressed on the surface of stimulated platelets and endothelial cells. Through P-selectin, platelets and endothelial cells interact with monocytes by binding to the P-selectin ligand PSGL-1. This interaction primes the monocyte to initiate de novo synthesis of tissue factor (TF). Tissue factor, in complex with Factor VIIa, activates Factor X to initiate blood coagulation and fibrin formation.

lectin that are involved in leukocyte recognition, and what is the role of phosphorylation of P-selectin during cell stimulation? Second, what are the chemical features on the P-selectin ligand that allow recognition by P-selectin, and what is the physiologic role of the P-selectin ligand? Third, does P-selectin merely perform a cell–cell docking function, or is this interaction associated with cell signaling? If the latter case is true, how are signals transmitted by P-selectin binding to the P-selectin ligand (Fig. 1)?

## STRUCTURE–FUNCTION RELATIONSHIPS IN P-SELECTIN

Although considerable attention has been focused on the lectin domain in cell recognition, we have attempted to establish whether or not the EGF domain also plays a direct or indirect role. The P-selectin EGF domain, consisting of 40 amino acids and three disulfide bonds, was prepared by solid-phase synthesis. By oxidation, the EGF domain was refolded. After the EGF domain was purified by HPLC, it was sequenced to confirm its primary structure and the disulfide bond structure was determined by sequence analyses of a thermolysin digest of the EGF domain. In collaboration with Dr. David Sanford, we have solved the three-dimensional structure of the P-selectin EGF domain by two-dimensional NMR spectroscopy. The EGF structure is homologous to the murine epidermal growth factor (EGF) and to other EGF domains whose structure have been solved. Comparison to the first EGF domain of factor IX reveals general similarity in the folding pattern, but several differences, particularly near the N terminus.

In collaboration with Dr. Tom Tedder, we have prepared chimeras of P-selectin and L-selectin to determine the specific functions of the domains of these selectins. Using the cDNA of P-selectin and the cDNA of L-selectin, we expressed the following chimeras: PLL, containing the P-selectin lectin domain on an L-selectin background; PPL, containing the lectin and EGF domain of P-selectin on an L-selectin background; and PPP, containing the lectin, EGF, and consensus repeats of P-selectin on an L-selectin background. These chimeras were expressed in CHO cells in parallel with the wild-type P-selectin and L-selectin. Stable cell lines were cloned and these clones were amplified to similar selectin surface density with methotrexate. Clones expressing the various chimeras *at similar protein density* (150 molecules/$\mu m^2$) were compared for their ability to bind to HL60 cells. P-selectin, but not L-selectin, bound to HL60 cells. The selectin chimeras PPL and PPP bound well to HL60 cells, whereas the PLL chimera bound HL60 cells much less efficiently. Our results indicate that although the lectin domain is sufficient to support cell adhesion, the lectin–EGF domain represents the cell recognition unit on P-selectin.

To determine the relationship of P-selectin density on the membrane surface to the cell adhesion properties of a cell, we performed quantitative experiments to better understand the contradictions that appear in the P-selectin literature regarding potential ligands. We believe that the basis for these differences has to do with the density of P-selectin and the P-selectin ligand on membrane surfaces of the cells studied. Purified P-selectin was incorporated into lipospheres, uniform microglass beads encapsulated with phospholipid-containing P-selectin (16). The known density of P-selectin in the membrane was varied, and the binding of lipospheres to HL60 cells was analyzed. A critical P-selectin density of about 50 molecules/$\mu m^2$ is required to sustain cell adhesion; maximal cell adhesion is observed at 100–150 molecules/$\mu m^2$. A similar analysis of the surface density of P-selectin on CHO cell clones indicates optimal HL60 cell adhesion at about 150–200 molecules/$\mu m^2$.

We expressed the lectin domain in *E. coli* with the pT7-7–derived expression plasmid. The DNA insert coding for the lectin domain or the lectin-EGF domain was derived from a human P-selectin cDNA template by PCR. Cells were stimulated with IPTG and bacteria were grown and harvested. On SDS gel electrophoresis of extract from bacteria transfected with the cDNA encoding the lectin domain, a new band of $M_r$ 16,000 could be identified. The N-terminal sequence of the first 25 residues of this band confirmed the presence of the P-selectin lectin domain sequence. Furthermore, Western blot analysis of the bacterial lysate using polyclonal anti-P-selectin antibodies revealed reactivity with this $M_r$ 16,000 species. The lectin domain was solubilized in 6 *M* guanidine HCl and allowed to oxidize as the material was dialyzed against buffer containing 10 m*M* $CaCl_2$. The oxidized lectin domain, purified by HPLC, yielded a single band on an SDS gel. The role of individual isolated P-selectin domains in cell adhesion was evaluated in a cell adhesion assay based on the binding of CHO–P-selectin to HL60 cells. The isolated, folded P-selectin lectin domain expressed in bacteria was a potent inhibitor of cell adhesion; half-maximal inhibition was observed at 200 n*M* (17). In contrast, the P-selectin EGF domain showed no inhibition even at concentrations up to 2 m*M*. In combination with our results that the chimera of L-selectin containing the P-selectin EGF domain is capable of binding HL60 cells (18), unlike L-selectin, and that the recognition unit on P-selectin is the lectin–EGF domain, these data suggest that the P-selectin EGF domain does not make direct contact with the P-selectin ligand but rather influences the structure of the adjacent lectin domain (be it from P-selectin or L-selectin) to enable it to bind to the P-selectin ligand.

## Phosphorylation of P-Selectin

To determine whether phosphorylation of P-selectin might accompany platelet activation, P-selectins from resting platelets and from thrombin-stimulated platelets were compared for the presence of phosphorylated amino acids (19). Purified platelets were incubated with [$^{32}$P]orthophosphoric acid, then stimulated with α-thrombin. Platelets were lysed with detergent and P-selectin was immunoprecipitated with the monoclonal antibody AC1.2 directed against P-selectin. SDS gel electrophoresis of the immunoprecipitate indicated about 10- to 20-fold higher levels of $^{32}$P were incorporated into P-selectin from thrombin-activated platelets than into P-selectin from resting platelets (Fig. 2). As a lower limit, the molar ratio of phosphate to P-selectin in activated platelets is 6:10 at 15 s, thus indicating the very significant rapid phosphorylation of P-selectin. Other platelet agonists, including the thrombin receptor peptide (SFLLR), epinephrine, ADP, and collagen, similarly stimulated phosphorylation of P-selectin. The kinetics of P-selectin phosphorylation after thrombin stimulation indicate rapid modification (detectable at 5 s), with maximal incorporation observed at 15 to 30 s. Phosphoamino acid analysis of the phosphorylated P-selectin revealed the presence of phosphoserine, phosphothreonine, and phosphotyrosine, but phosphotyrosine and phosphothreonine rapidly disappeared 60 s after platelet activation. The C-terminal peptide consisting of the cytoplasmic domain contained the phosphoamino acids. We have recently extended these studies to show that P-selectin in thrombin-activated, unstirred platelets undergoes phosphorylation but not dephosphorylation. The rapid phosphorylation and selective dephosphorylation of tyrosine and threonine on the cytoplasmic domain of P-selectin after platelet activation may be important for P-selectin function (inside-out signaling) and signal transduction within platelets.

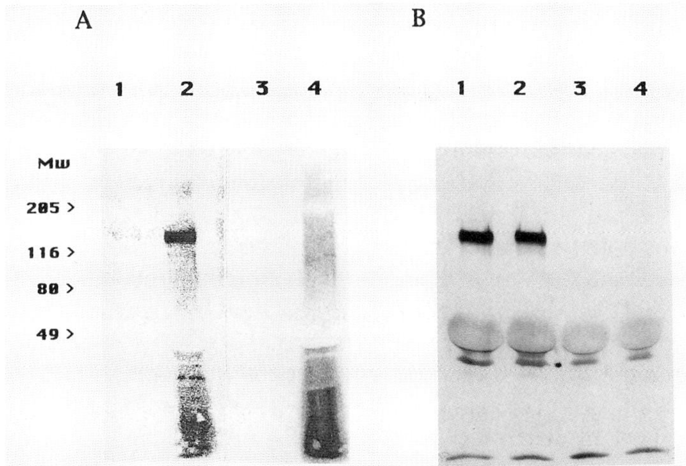


**FIG. 2.** Phosphorylation of P-selectin in platelets during thrombin stimulation. Purified platelets were activated with thrombin, lysed, and the lysate treated with MOPC-21–Sepharose, then AC1.2–Sepharose to immunoprecipitate P-selectin. The immunoprecipitates were subjected to SDS-PAGE and electrophoretic transfer to Immobilon P. $^{32}P$ was detected by autoradiography and P-selectin was detected by immunoblot with polyclonal anti–P-selectin antibodies. Lane 1, resting platelets; lane 2, thrombin-activated platelets; lane 3, resting platelets; lane 4, thrombin-activated platelets. Lanes 1 and 2 contain proteins immunoprecipitated with AC1.2, an anti–P-selectin antibody; lanes 3 and 4 contain proteins immunoprecipitated with MOPC-21, an irrelevant isotype-matched antibody. Molecular weight markers are shown to the left. **A:** Autoradiograph detecting $^{32}P$. **B:** Western blot detecting P-selectin protein antigen. (From ref. 19, with permission.)

Stimulation of endothelial cells with thrombin similarly leads to rapid phosphorylation of P-selectin. Primary cultures of human umbilical vein endothelial cells (first passage) were used. With the use of techniques similar to those described above, P-selectin initially undergoes phosphorylation and dephosphorylation, with peak phosphorylation at approximately 1 to 2 min. A second phosphorylation event, centered at approximately 8 to 10 min, includes rapid phosphorylation followed by dephosphorylation.

## P-SELECTIN LIGAND

The P-selectin ligand was first described in our laboratory as being distributed on neutrophils and monocytes (12). It now appears that a small subpopulation of T lymphocytes express this ligand (20) and that certain malignant cells, including HL60 cells, U937 cells, small-cell lung cancer, and neuroblastoma cells (21) express a P-selectin ligand. We first described the importance of Lewis x ($Le^x$) as a component of the P-selectin ligand (22), and Corral et al. (23) emphasized that sialic acid was also a component. Although $sLe^x$ may be required for P-selectin interaction on cell surfaces, it is not sufficient for high-affinity binding, and a protein component is also important (24,25). A putative P-selectin ligand of $M_r$ 120,000 had been identified from myeloid cells by blotting techniques (26). Others have claimed that sulfatides are also P-selectin ligands (27) and that L-selectin can serve as the P-selectin ligand (28). With Larsen and colleagues at Genetics Institute, we have isolated a cDNA that encodes the protein component of the P-selectin ligand; this protein, termed P-selectin glycoprotein ligand-1 (PSGL-1), has significant potential for *O*-linked glycosylation and is heavily glycosylated (29). PSGL-1 is a mucin-like protein with a monomeric molecular weight of 110,000 that behaves as a homodimer. It is an integral membrane protein containing an extracellular domain composed of a series of 15 repeating units that are rich in threonine, a single transmembrane domain, and a cytoplasmic domain. This protein, when expressed in COS cells co-transfected with specific fucosyltransferases, yields a biologically active ligand that binds to P-selectin. Polyclonal antibodies to PSGL-1 inhibit the binding of P-selectin to HL60 cells.

To study the function of PSGL-1 in the mouse, a murine homologue of human PSGL-1 was cloned using the human cDNA sequences as probes of a cDNA library prepared from the WEHI-3 cell line and a genomic DNA library prepared from 129 SVJ mouse tissue. The deduced amino-acid sequence of the murine PSGL-1 has an open reading frame encoding a protein of 397 amino acids, five residues shorter than human PSGL-1 (30). The murine and human PSGL-1 show an overall homology of 67% and identity of 50%. They contain similar domain structures, including a signal peptide, propeptide, homologous repeat units, transmembrane region, and cytoplasmic tail. A much higher degree of homology (88%) is found in the transmembrane domain and the cytoplasmic tail. The mucin-like domain is less conserved at the primary sequence level. There are 15 threonine-rich decameric repeats in the human PSGL-1, but only 10 such repeats in murine PSGL-1. Instead of threonines as potential *O*-linked glycosylation sites in the repeat region as in human PSGL-1, the murine PSGL-1 repeat sequence contains threonine and serine residues. Compared to the three potential *N*-glycosylation sites in human PSGL-1, the murine PSGL-1 carries only one conserved site and an additional unique site. The murine PSGL-1 functions as a ligand for human P-selectin in a cell-binding assay. Furthermore, we have established by Northern analysis that WEHI-3 cells, a murine myeloid leukemia cell line, synthesize the PSGL-1 mRNA and that these cells express the P-selectin ligand in a functional form that interacts with human P-selectin. Another cell line, PU-5, also contains the PSGL-1 mRNA but does not bind P-selectin; the defect in this PSGL-1 function may be due to the absence of a co-receptor or incomplete post-translational processing.

Other mucins have been described on the surface of leukocytes, and their relationship to selectin ligands remains unknown. CD68 is an $M_r$ 110,000 transmembrane protein expressed on monocytes and macrophages (31). The cDNA for human CD68 predicts a heavily glycosylated extracellular domain with many potential *N*-linked and *O*-linked glycosylation sites; it is distinct from PSGL-1. The function of this protein is unknown. Similarly, leukosialin/CD43 is a major mucin on hematopoietic cells (32). To test the hypothesis that CD68 may be an E-selectin or P-selectin ligand, the CD68 cDNA was cloned to explore its interaction with E-selectin and P-selectin. We co-transfected COS cells with cDNA for a 1/3,4 fucosyltransferase and CD68 and determined their ability to bind to CHO cells expressing either E-selectin or P-selectin (33). COS cells expressing CD68 alone or coexpressing CD68 and $sLe^x$ did not acquire an enhanced ability to bind to CHO cells expressing P-selectin or E-selectin compared with COS cells expressing $sLe^x$ only. Co-transfection of COS cells with PSGL-1 and fucosyltransferase, in contrast, enhanced COS cell binding to CHO P-selectin. These experiments suggest that CD68 is not a ligand for either E-selectin or P-selectin.

## LEUKOCYTE EFFECTOR FUNCTION INDUCED BY P-SELECTIN

P-Selectin is believed to play a physiologic role in inflammation and thrombosis by mediating the interaction of leukocytes with platelets bound in the region of tissue injury and with stimulated endothelium. To evaluate the role of P-selectin in platelet–leukocyte adhesion in vivo, we employed an arteriovenous shunt model in baboons. The accumulation of leukocytes within an experimental thrombus was explored in a primate model to yield insight into the physiologic role of P-selectin. We determined the effect of anti–P-selectin antibodies that inhibit P-selectin–mediated platelet–leuko-

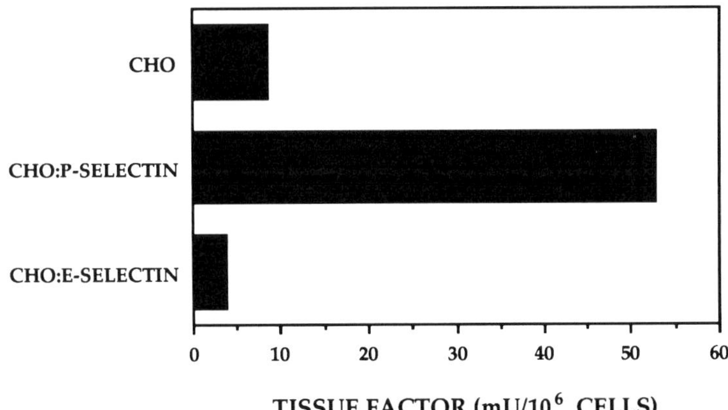


**FIG. 3.** CHO P-selectin induces expression of tissue factor in monocytes. Chinese hamster ovary cells, either naive or expressing P-selectin or E-selectin, were cultured to confluency. Monocytes were then added and incubated for 6 h before assay for tissue factor activity. (From ref. 35, with permission.)

cyte interaction on the uptake of leukocytes and the deposition of fibrin in an experimental thrombus. A Dacron graft implanted within an arteriovenous shunt is thrombogenic, accumulating platelets and fibrin within its lumen. These bound platelets express P-selectin. Using anti–P-selectin antibodies and their Fab$'_2$ fragments, we observed inhibition of leukocyte binding to P-selectin expressed on platelets immobilized on the graft. This antibody blocks leukocyte accumulation and inhibits the deposition of fibrin within the

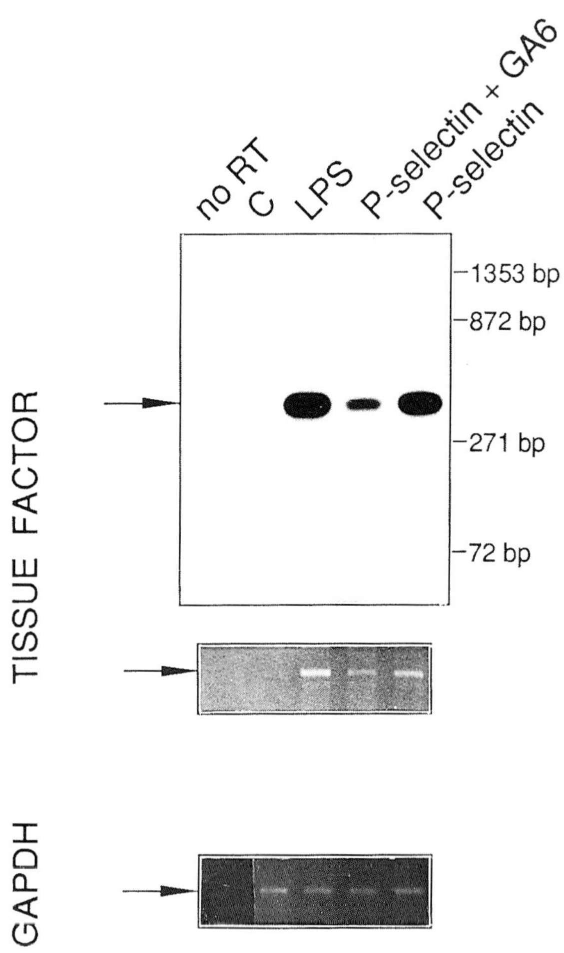


**FIG. 4.** Effect of P-selectin stimulation on tissue factor mRNA levels. Reverse transcriptase/PCR analysis of tissue factor and GAPDH mRNA expression in monocytes untreated (control, C) or exposed to LPS, P-selectin plus anti–P-selectin antibody, or P-selectin alone for 6 h at 37°C. **Upper panel:** Analysis of tissue factor mRNA by Southern blot after PCR. The specific tissue factor cDNA product (317 bp) is indicated by the arrow. **Middle panel:** Ethidium bromide staining of same PCR products. **Lower panel:** Ethidium bromide staining of reaction products after PCR. (From ref. 35, with permission.)

thrombus. These results indicate that P-selectin is an important adhesion molecule on platelets which mediates platelet-leukocyte binding in vivo, that the presence of leukocytes in thrombi is mediated by P-selectin, and that these leukocytes promote fibrin deposition.

To determine whether P-selectin plays only a cell-docking function or whether it might induce biologic activities in leukocytes bound to activated platelets, we examined the effect of P-selectin on the expression of tissue factor activity in monocytes. Platelets induced procoagulant activity on mononuclear cells, as previously demonstrated by Niemetz and Marcus (34). The activity was inhibited with antibodies directed against tissue factor or blocking antibodies against P-selectin (35). Purified P-selectin stimulated tissue factor expression in a dose-dependent manner. Heat denaturation of P-selectin markedly reduced the level of tissue factor stimulation, demonstrating that P-selectin rather than contaminating endotoxin was responsible for this effect. To explore this observation systematically, we isolated monocytes from mononuclear cells (35). These populations were shown to be almost homogeneous by morphologic criteria and expression of CD13 and CD33 antigen. These monocytes could be isolated in their unstimulated form. CHO cells expressing P-selectin stimulated tissue factor expression in monocytes, whereas untransfected CHO cells and CHO cells expressing E-selectin did not (Fig. 3). By reverse transcriptase/PCR, we demonstrated that unstimulated monocytes had no detectable tissue factor mRNA, whereas monocytes stimulated with CHO P-selectin contained tissue factor mRNA (Fig. 4). These results indicate that P-selectin upregulates the expression of tissue factor on monocytes. Binding of P-selectin to monocytes in the area of vascular injury may be important for the initiation of thrombogenesis.

## ACKNOWLEDGMENT

We gratefully acknowledge our collaborators and students who have contributed very significantly to this work. They include A. Celi, R. Gibson, G. Kansas, R. Lorenzet, D. Sanford, G. Larsen, C. Crovello, T. Tedder, J. Yang, S. J. Freedman, J. Galipeau, E. Gottlieb, and T. Palabrica, as well as earlier students who helped to define the biology of P-selectin.

## REFERENCES

1. Springer TA. Adhesion receptors of the immune system. *Nature* 1990;346:425–34.
2. Lasky L. Selectins: interpreters of cell-specific carbohydrate information during inflammation. *Science* 1989;258:964–9.
3. Tedder TF, Isaacs CM, Ernst TJ, Demetri GD, Adler DA, Disteche CM. Isolation and chromosomal localization of cDNAs encoding a novel human lymphocyte cell surface molecule, LAM-1. Homology with the mouse lymphocyte homing receptor and other human adhesion proteins. *J Exp Med* 1989;170:123–33.

4. Lasky LA, Singer MS, Yednock TA, et al. Cloning of a lymphocyte homing receptor reveals a lectin domain. *Cell* 1989;56:1045–55.
5. Bevilacqua MP, Stengalin S, Gimbrone MA Jr, Seed B. Endothelial leukocyte adhesion molecule 1: an inducible receptor for neutrophils related to complement regulatory proteins and lectins. *Science* 1989;243:1160–5.
6. Johnston GI, Cook RG, McEver RP. Cloning of GMP-140, a granule membrane protein of platelets and endothelium: sequence similarity to proteins involved in cell adhesion and inflammation. *Cell* 1989;56:1033–44.
7. Berman CL, Yeo E, Wencel-Drake JD, Furie BC, Ginsberg MH, Furie B. A platelet alpha granule membrane protein that is incorporated into the plasma membrane during activation. Characterization and subcellular localization of PADGEM protein. *J Clin Invest* 1986;78: 130–7.
8. Stenberg PE, McEver RP, Shuman MA, Jacques YV, Bainton DF. A platelet alpha-granule membrane protein (GMP140) is expressed on the plasma membrane after activation. *J Cell Biol* 1985;101:880–6.
9. Bonfanti R, Furie BC, Furie B, Wagner DD. PADGEM is a component of Weibel-Palade bodies in endothelial cells. *Blood* 1989;73:1109–12.
10. McEver RP, Beckstead JH, Moore KL, Marshall-Carlson L, Bainton DF. GMP-140, a platelet alpha-granule membrane protein, is also synthesized by vascular endothelial cells and is localized in Weibel-Palade bodies. *J Clin Invest* 1989;84:92–9.
11. Geng JG, Bevilacqua MP, Moore KL, et al. Rapid neutrophil adhesion to activated endothelium mediated by GMP-140. *Nature* 1990;343:757–60.
12. Larsen E, Celi A, Gilbert G, et al. PADGEM protein: a receptor that mediates the interaction of activated platelets with neutrophils and monocytes. *Cell* 1989;59:305–12.
13. Lawrence MB, Springer TA. Leukocytes roll on a selectin at physiologic flow rates: distinction from and prerequisite for adhesion through integrins. *Cell* 1991;65:859–73.
14. Mayadas TN, Johnson RC, Rayburn H, Hynes RO, Wagner DD. Leukocyte rolling and extravasation are severely compromised in P-selectin-deficient mice. *Cell* 1993;74:541–54.
15. Palabrica T, Lobb R, Furie BC, et al. Leukocyte accumulation which promotes fibrin deposition is mediated in vivo by P-selectin (CD62) on adherent platelets. *Nature* 1992; 359:848–51.
16. Gilbert GE, Drinkwater D, Barter S, Clouse SB. Specificity of phosphatidylserine-containing membrane binding sites for factor VIII. Studies with model membranes supported by glass microspheres (lipospheres). *J Biol Chem* 1992;267:15861–8.
17. Freedman SJ, Furie B, Furie BC, The lectin domain but not the EGF domain is a potent inhibitor of P-selectin-mediated cellular adhesion [Abstract]. *Blood* 1993;82(suppl 1):341A.
18. Kansas GS, Ley K, Zakrzewicz A, et al. A direct role in cell adhesion for the EGF-like domain of P-selectin. *J Cell Biol* 1994;124:609–18.
19. Crovello CS, Furie BC, Furie B. Rapid phosphorylation and selective dephosphorylation of P-selectin accompanies platelet activation. *J Biol Chem* 1993;268:14590–3.
20. Moore KL, Thompson LF. P-selectin (CD62) binds to subpopulations of human memory T lymphocytes and natural killer cells. *Biochem Biophys Res Commun* 1990;186:173–81.
21. Stone JP, Wagner DD. P-selectin mediates adhesion of platelets to neuroblastoma and small cell lung cancer. *J Clin Invest* 1993;92:804–13.
22. Larsen E, Palabrica T, Sajer S, et al. PADGEM-dependent adhesion of platelets to monocytes and neutrophils is mediated by a lineage-specific carbohydrate, LNF III (CD15). *Cell* 1990;63:467–74.
23. Corral L, Singer MS, Macher BA, Rosen SD. Requirement for sialic acid on neutrophils in a GMP-140 (PADGEM) mediated adhesive interaction with activated platelets. *Biochem Biophys Res Commun* 1990;172:1349–56.
24. Larsen G, Sako D, Ahern T, et al. P-selectin and E-selectin: distinct but overlapping leukocyte ligand specificities. *J Biol Chem* 1992;266:11104–10.
25. Moore KL, Varki A, McEver RP. GMP-140 binds to a glycoprotein receptor on human neutrophils: evidence for a lectin-like interaction. *J Cell Biol* 1991;112:491–9.
26. Moore KL, Stults NL, Diaz S, et al. Identification of a specific glycoprotein ligand for P-selectin (CD62) on myeloid cells. *J Cell Biol* 1992;118:445–6.
27. Aruffo A, Kolanus W, Walz G, Fredman P, Seed B. CD62/P-selection recognition of myeloid and tumor cell sulfatides. *Cell* 1991;67:35–44.

28. Picker LJ, Warnock RA, Burns AR, Doerschuk CM, Berg EL, Butcher EC. The neutrophil selectin LECAM-1 presents carbohydrate ligands to the vascular selectins ELAM-1 and GMP-140. *Cell* 1991;66:921–33.
29. Sako D, Chang X-J, Barone KM, et al. Expression cloning of a functional glycoprotein ligand for P-selectin. *Cell* 1993;75:1179–86.
30. Yang J, Furie BC, Furie B. Cloning of a murine homolog of the human P-selectin glycoprotein ligand-1. *Blood* 1994;84(suppl 1):120.
31. Holness CL, Simmons DL. Molecular cloning of CD68, a human macrophage marker related to lysosomal glycoproteins. *Blood* 1993;81:1607–13.
32. Cyster JC, Somoza N, Killeen N, Williams AF. Protein sequence and gene structure for mouse leukosialin (CD43), a T lymphocyte mucin without introns in the coding sequence. *Eur J Immunol* 1990;20:875–81.
33. Galipeau J, Gottlieb E, Yang J, Furie BC, Furie B. Mucin-like CD68 is not an E- or P-selectin ligand [Abstract]. *Circulation* 1994;90:84.
34. Niemetz J, Marcus AJ. The stimulatory effect of platelets and platelet membranes on the procoagulant effect of leukocytes. *J Clin Invest* 1974;54:1437–43.
35. Celi A, Furie BC, Lorenzet R, Pellegrini G, Furie B. P-selectin induces the expression of tissue factor on monocytes. *Proc Natl Acad Sci USA* 1994;91:8767–71.

*Topics in Molecular Medicine, Volume 1,*
edited by Wolfgang Siess, Reinhard Lorenz,
and Peter C. Weber. Raven Press, Ltd.,
New York 1995.

# 10

# P-Selectin–Deficient Mice: A Model to Study Inflammation and Metastasis

Jennifer P. Stone and Denisa D. Wagner

*The Center for Blood Research, Harvard Medical School, Boston, Massachusetts 02115*

During the past 5 years, probably the most amazing development in our knowledge of cell adhesion occurred in the field of vascular selectins. The cloning of three membrane proteins from vascular cells was reported simultaneously in the spring of 1989 (1–4), and it became immediately obvious that these three molecules belong to the same family and that they would share some biologic properties. Because these molecules contain a lectin domain that mediates selective adhesion to specific carbohydrate structures, the family was called the *selectin* family. A capital letter designates each member, based on the cell type in which the selectin was first identified (L, lymphocytes; E, endothelium; P, platelets) (5). All selectins contain an N-terminal calcium-dependent lectin domain, an EGF-like domain, several repeats homologous to those found in complement regulatory proteins, a transmembrane domain, and a short cytoplasmic tail (for review of selectins see 6,7). L-selectin, constitutively present on lymphocytes, mediates their adhesion to high endothelial venules of peripheral lymph nodes (8). It is also present on neutrophils and monocytes, where its suggested role is in mediating adhesion of these cells at sites of inflammation. E-Selectin is found on endothelial cells stimulated by endotoxin or inflammatory cytokines. It is synthesized de novo, and its expression peaks 4 to 6 h after stimulation (9). E-Selectin mediates adhesion of neutrophils and monocytes. P-Selectin (10,11) which also mediates adhesion to neutrophils and monocytes (12,13), is found on activated platelets and endothelial cells. In contrast to E-selectin, P-selectin is constitutively synthesized by the endothelial cells. However, instead of exiting directly to the cell surface, it is delivered to storage granules (14,15) called Weibel–Palade bodies (16). Weibel–Palade bodies also store the adhesive glycoprotein von Willebrand factor (17), and are translocated to the plasma membrane when the cells are stimulated for secretion (18). Similarly in platelets, P-selectin is stored in α-granules and becomes expressed on the platelet surface only on activation (19,20). Therefore, although the circumstances under which the individual selectins may be expressed are very different, they all, among other interactions, mediate the binding of neutrophils and monocytes to activated endothelium. Therefore, it is not surprising that each one has been experimentally implicated in some inflammatory process in vivo (7).

The selectins recognize carbohydrate structures present on glycoproteins or mucin-like molecules. The ligand for L-selectin on the peripheral lymph nodes is a mucin known as CD34 (21). Whether this is also the ligand for L-selectin on inflamed vascular endothelium is not known. Another soluble mucin ligand for L-selectin, Glycam-1 (6,22), can modulate lymphocyte homing by binding to L-selectin, and thus blocks lymphocyte adhesion to CD34. P- and E-selectin appear to share a mucin ligand, PSGL-1, whose *O*-linked carbohydrate appears to be necessary for binding of the selectins (23,24). Another set of ligands was recently identified in the mouse, and these appear to be monospecific for either P- or E-selectin (25). The exact carbohydrate structure(s) that P- or E-selectin recognizes is not known, but it is likely to contain a sialyl Lewis$^{x}$ or sialyl Lewis$^{a}$ moiety (26,27).

## INTERACTION OF CANCER CELLS WITH SELECTINS

Tumor cells express many glycoprotein and glycolipid carbohydrate groups that contain both sialic acid and fucose groups on their cell surfaces. Oncogenic transformation leads both to more carbohydrate branches per core and to higher sialylation of the asparagine-linked complex carbohydrate, as well as to an increased fucose content in cell surface carbohydrates (28–30). This may lead to the formation of carbohydrate ligands for the selectins that are normally present only on vascular cells. Several tumor cells express the carbohydrate motifs sialyl Lewis$^{x}$ or sialyl Lewis$^{a}$ that are recognized by the selectins (31–35). Each of the selectins, however, recognizes with different affinity distinct presentations of these carbohydrate determinants on their target cells (7). Although E-selectin mediates adhesion to several tumors (36–40) and may bind to tumors, to which P-selectin also binds (our unpublished observation), the converse is not true. For example, the melanoma cell line NK1-4 binds to E-selectin but not to P-selectin (41).

We have identified cancer cell lines that bind to P-selectin by screening tumor cells in a rosetting assay with activated platelets (42). In this assay, gel-filtered platelets, which have been either activated with thrombin or maintained undisturbed in a resting state, were incubated with a suspension of tumor cells. In Fig. 1, the binding of activated platelets to HL60 cells (a promyelocytic leukemia cell line) was compared to the binding to NCI-H128 [a small-cell lung cancer (SCLC) cell line]. The binding of the activated platelets to both SCLC and HL60 cells is inhibited by an inhibitory mono-

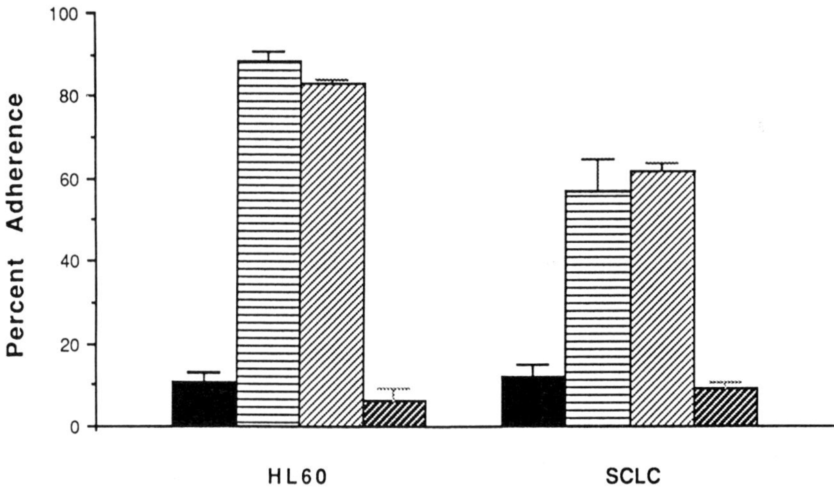


**FIG. 1.** Comparison of HL60 cells and small-cell lung cancer (SCLC) cells (NCI-H128) binding to resting and activated platelets. The percentage of cells binding to two or more platelets was determined by phase-contrast microscopy. Resting platelets, solid bars; activated platelets, horizontal bars; activated platelets incubated with a noninhibitory antibody to P-selectin (AC1.2), hatched bars on white background; activated platelets incubated with an inhibitory antibody to P-selectin (GE12), hatched bars on black background. HL60 cell and SCLC cell binding of platelets was affected to a similar extent by the presence of the monoclonal antibodies to P-selectin. The error bar represents 1 SD. AC1.2 and GE12 were generously provided by Drs. Barbara and Bruce Furie and Biogen Research Corporation. (Adapted from ref. 42.)

clonal antibody (MAb) to P-selectin GE12 and not by the noninhibitory MAb AC1.2 (Fig. 1).

P-Selectin was shown to mediate adhesion of activated platelets to another SCLC cell line (NCI-H345), human neuroblastoma cell lines (NB5-A2 and SK-N-SH), and a mouse teratocarcinoma cell line (CRL 1566) (42). Activated platelets also bind through P-selectin to a colon carcinoma cell line (Colo-205) (35). Three other colon carcinoma cell lines (HCC-P2988, HCT-15, and SW948) were shown to bind to endothelial cells stimulated with thrombin (43). Treatment of these endothelial cells with antibodies to P-selectin or of the colon carcinoma cell with antibodies to Lewis$^x$ and Lewis$^a$ inhibited the binding (43). In our studies, P-selectin expressed on activated platelets did not bind a colon cancer cell line (HT 29) (42) that binds to E-selectin (36), nor to gastric (Hs746T) or breast (MCF7) adenocarcinoma.

To examine if P-selectin can bind to SCLC cells independent of other platelet components, we prepared fluorescent phospholipid vesicles with and without purified P-selectin. Their binding to SCLC cells was evaluated by fluorescent microscopy. Phospholipid vesicles without P-selectin showed no significant binding to the cancer cells. In contrast, phospholipid vesicles containing P-selectin heavily decorated many of the cancer cells (Fig. 2).

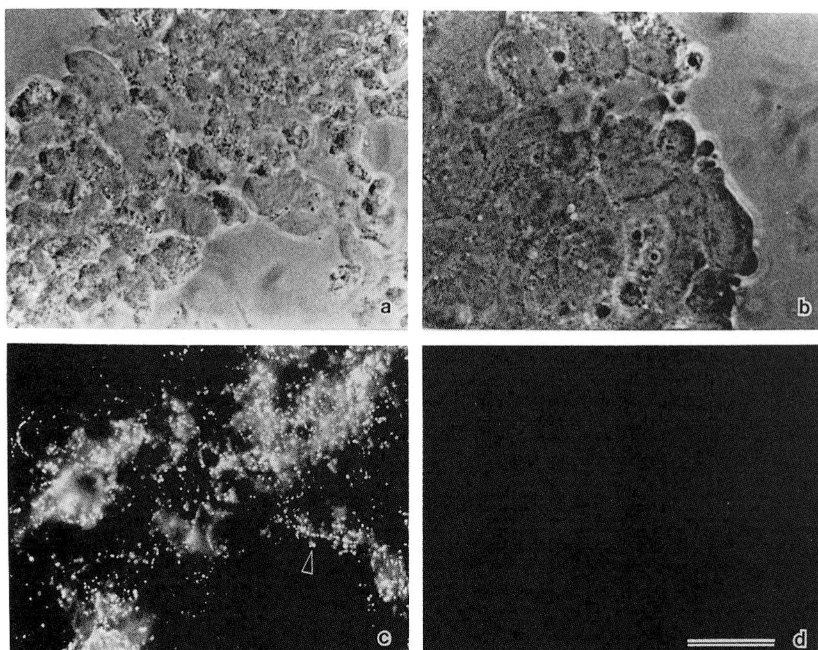


**FIG. 2.** Binding of P-selectin–containing fluorescent phospholipid vesicles to small-cell lung cancer (SCLC) cells (NCI-H128). **a, c:** Adherence of purified P-selectin–containing phospholipid vesicles to SCLC cells after rinsing to remove free vesicles. Cells seen under **(a)** phase-contrast microscopy and identical field observed under **(c)** fluorescence microscopy. **b, d:** Adherence of pure phospholipid vesicles to SCLC cells after removal of unbound vesicles. Cells seen in **(b)** phase-contrast and same field under **(d)** fluorescence microscopy. Only P-selectin–containing vesicles bound to the cancer cells. Bar = 20 μm.

The SCLC cells grew in suspension as floating cell aggregates. Different areas of a small-cell aggregate band with variable intensity to the P-selectin–containing phospholipid vesicles. This suggests that there is heterogeneity within the cell aggregate and that some cells may be expressing more of the ligand or a higher-affinity ligand for P-selectin.

The adherence of platelets to neuroblastoma and SCLC cells is inhibited by EDTA and by pretreatment of these cells with neuraminidase or trypsin (42). These results indicate that the adhesion process is divalent cation-dependent and that the ligand for P-selectin on the neuroblastoma and SCLC cells is a sialylated glycoprotein. In addition, neuroblastoma cells grown in the presence of inhibitors of the biosynthesis or processing of *N*-linked oligosaccharide chains still bind to P-selectin, indicating that the ligand on these cancer cells is an *O*-linked oligosaccharide (42).

To characterize more fully the carbohydrate ligand on the neuroblastoma cells (SK-N-SH), we treated the cancer cells with *Pasteurella haemolytica* glycoprotease (25 μg/4 × $10^5$ cells for 1 h at 37°C), which is specific for *O*-sialoglycoproteins (44). This reaction was performed in the presence of a cocktail of protease inhibitors that included neuraminidase inhibitor (2,3-dehydro-2-deoxy-*N*-acetyl-neuraminic acid) and soybean trypsin inhibitor. We then evaluated the binding of these treated cancer cells to fluorescent phospholipid vesicles containing P-selectin. The results presented in Table 1 show that both the treated HL60 cells and the treated neuroblastoma cells (SK-N-SH) no longer bound to the phospholipid vesicles (PV) containing P-selectin. These data further indicate that the P-selectin ligand on the neuroblastoma cells (SK-N-SH), as well as on the HL60 cells as previously shown (45,46), is a protein-containing *O*-linked carbohydrate side chain. It will be interesting to determine whether the glycoprotein ligand for P-selectin on the neuroblastoma cells is PSGL-1, as it is on the HL60 cells (24).

We have verified that the binding of P-selectin to cancer cells is not only an in vitro phenomenon. P-selectin–containing lipospheres bound in a specific manner to cryostat sections of a human SCLC and two neuroblastoma tumors. Again, with these tumors the binding was abolished by treatment of the sections with trypsin or neuraminidase (42).

In addition to P-selectin binding to tumor cells through a sialic acid-containing glycoprotein, a soluble immunoglobulin chimera of P-selectin appears to bind to myeloid and tumor cell sulfatide, Gal(3-$SO_4$)β1-ceramide (47). The calcium dependence of the binding of the chimeric molecule was not tested, and treating the cancer cells with neuraminidase did not affect the P-selectin immunoglobulin chimera binding (48). This is in contrast to studies showing that platelet binding to leukocytes is fully inhibited by protease or neuraminidase treatment of the myeloid cells (49–51). Aruffo and colleagues (48) demonstrated binding of the P-selectin immunoglobulin chimera to breast and colon carcinoma cell lines and to all carcinoma tissue sections tested (colon, lung, and breast). This soluble form of P-selectin may bind to sulfatides of many cell types. However, the native form of P-selectin incorporated into the membrane of activated platelets either may not recognize sulfatides or may be sterically hindered to reach the sulfatides, as we have found that activated platelets bind only to selected cancer types (42). It will

**TABLE 1.** *Flow cytometry of promyelocytic leukemia (HL60) and neuroblastoma (SK-N-SH) cell lines treated with* P. haemolytica *glycoprotease and then incubated with fluorescent phospholipid vesicles (PV) made with and without P-selectin*

| | Specific linear fluorescent intensity | | |
|---|---|---|---|
| Cell type | PV alone | PV with P-selectin | PV with P-selectin, cells treated with glycoprotease |
| HL60 | 10.1 | 33.0 | 15.7 |
| SK-N-SH | 23.3 | 40.3 | 25.4 |

Representative experiment of three phospholipid-binding assays.

be important to establish whether acquiring a P-selectin ligand gives the cancer cells an advantage for hematogenous metastases.

E-Selectin also binds to selected cancer cells (52) and is probably involved in the homing of specific tumor cells to stimulated vascular endothelium expressing E-selectin in vivo. The colon cancer cell line HT-29 was shown to adhere to TNF-activated human endothelial cells more than to unstimulated endothelial cells (36). This adhesion was inhibited by anti–E-selectin antibodies. In addition, E-selectin on IL-1–activated human umbilical vein endothelial cells mediates binding to adult T-cell leukemia (53) and acute myeloid leukemia blast cells (54). During tumor metastasis, tumor cells circulate in the blood and adhere to the microvasculature before they extravasate into the tissues. Similar to this phenomenon, under laminar flow conditions the binding of colon, ovarian, and breast carcinomas to IL-1–activated human umbilical vein endothelial cells was greater than to unstimulated endothelium (52). Interestingly, the colon carcinoma HT-29M, which was evaluated in more detail in this study, was shown to roll on the stimulated endothelium in a manner similar to leukocytes. The rolling and adhesion were inhibited by an antibody to E-selectin.

E-Selectin also specifically recognizes the sialyl Lewis$^x$ determinant on the colon cancer cell line WiDr (37), as well as the sialyl Lewis$^a$ determinant on pancreatic carcinoma cell lines (39). The presence of these carbohydrates on the cancer cells may provide them with an advantage during metastatic spread (55). Colon cancer cells with a high metastatic potential have been shown to express more lysosomal membrane glycoprotein, lamp-1 and lamp-2 molecules, on their cell surface than their corresponding low-metastatic counterparts (56). These lamp molecules are the major carriers for poly-*N*-acteyl-lactosamines that display sialyl Lewis$^x$ terminals. Sawada and colleagues showed that lamp-1 can efficiently present carbohydrate to E-selectin (57). They further demonstrated that E-selectin expressed on activated endothelial cells binds to high-metastatic colon cancer cells more efficiently and with greater avidity than to low-metastastic colon cancer cells (38).

Because many tumors produce cytokines and other factors, such as vascular permeability factor, that may stimulate granule release from endothelial cells (58), they may cause rapid expression of P-selectin and also a delayed and more prolonged expression of E-selectin. In this manner, these selectins may play a role in the blood-borne metastasis of specific tumors. It is interesting to note that P- and E-selectin bind to human tumors that metastasize frequently, such as the SCLC, neuroblastoma, colon cancer, melanoma, and pancreatic carcinoma. As discussed above, not only do the highly metastatic tumor cells express potential ligands for these selectins, but these cancer cells' ability to bind to the selectins may facilitate the development of hematogenous metastases.

## P-SELECTIN–DEFICIENT ANIMAL MODEL

From the observations presented above it appears evident that all selectins promote binding among similar cell types, e.g., leukocytes and endothelium, and that their function in some processes may be overlapping. To evaluate the role of an individual selectin in inflammation and in cancer metastasis, we prepared a mouse model lacking a functional P-selectin gene (59). This was accomplished through homologous recombination in embryonic stem cells (60) by Tanya Mayadas in collaboration with Richard Hynes (MIT). The P-selectin–deficient animals are fertile, and their general health and viability appear normal. We have verified that disruption of the P-selectin gene did not affect the expression of E- and L-selectin (59), whose genes are in close proximity to that of P-selectin (61).

The peripheral platelet count of the P-selectin–deficient animals was normal, and except for the inability of the mutant platelets to bind to neutrophils after activation (R. Johnson, unpublished observation), we have not identified a platelet defect in these animals. In contrast, the peripheral neutrophil counts were three to four times higher than in wild-type animals (59), probably due to a longer half-life in the circulation (R. Johnson, unpublished observation). The P-selectin–deficient animals appear to have a severe defect in the way their leukocytes interact with the vessel wall. This is reflected by abnormalities in both leukocyte rolling and extravasation.

We studied the leukocyte rolling by intravital microscopy of mesenteric venules (25–35 μm in diameter) in wild-type and homozygous mutant mice (59). Baseline rolling was recorded during the first 10 min after surgery. The number of rolling leukocytes can vary among animals because surgery and exteriorization of the mesentery probably cause some tissue damage leading to variable endothelial cell stimulation. In 10 wild-type mice, the leukocyte rolling index was $10.0 \pm 2.0$ leukocytes/min. However, in eight homozygous mutant mice, no leukocyte rolling was observed in the venules examined (Fig. 3). Although we have not examined heterozygous animals extensively, studies of two heterozygotes showed that rolling occured within the range observed for wild-type mice. Homozygous mutant animals had more neutrophils than wild-type animals and comparable leukocyte counts at both the beginning and the end of the experimental procedure. Therefore, the lack of leukocyte rolling in homozygous mutant animals could not be attributed to systemic neutropenia or leukopenia.

We were able to induce a statistically significant increase in numbers of leukocyte rolling by superfusion of the mesenteric venules with the secretagogue A23187 (59), which causes rapid release of Weibel–Palade bodies in vitro (62). This indicates that it can do so also in vivo, leading to increased expression of P-selectin and increased leukocyte adhesion. In contrast, no significant rolling was observed in the mutant mice even after A23187 treatment (59).

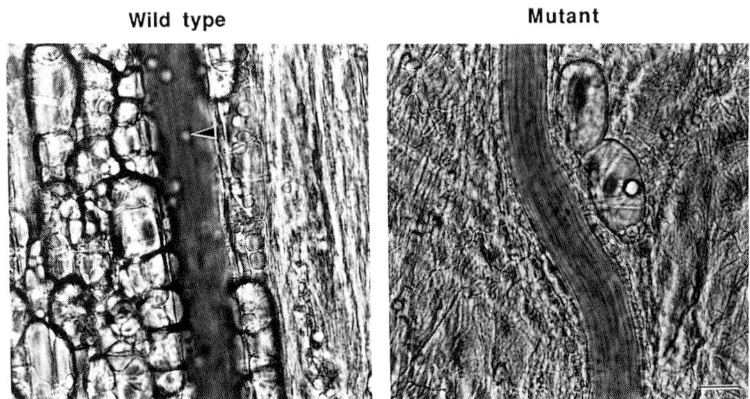


**FIG. 3.** Mesenteric venules of wild-type and P-selectin–deficient (mutant) mice observed by intravital microscopy. Immediately after externalization of the mesentery, rolling leukocytes can be seen as diffractive spheres in the wild-type venules. Under these basal conditions, no rolling leukocytes are observed in the venules of mutant mice. Bar = 20 μm. (From ref. 59 with permission.)

To determine whether neutrophils can be recruited to sites of inflammation in P-selectin–deficient mice, thioglycollate, an inflammatory irritant (63,64), was injected into the peritoneal cavity of wild-type and P-selectin–deficient mice. We observed an approximately 2-h delay in neutrophil extravasation to the inflamed peritoneum in the mutant mice compared with the wild-type animals. After that, the rate of extravasation was similar in both groups of animals (59). Similar delay in neutrophil extravasation was also seen at sites of excisional skin wounds during the first 2 h after infliction of the wound. However, 4 h after wounding the numbers of extravasated neutrophils in the tissue surrounding the injury were comparable (M. Subramaniam, unpublished observations). It appears from these experiments that at later times after an injury or inflammatory stimulus, the endothelium can express other molecules that can mediate leukocyte extravasation in the absence of P-selectin.

To examine whether leukocytes roll in the inflamed venules, R. Johnson (unpublished observation) externalized the mesentery of wild-type and P-selectin–deficient animals 4 h after injection of thioglycollate. Under these inflammatory conditions, many leukocytes rolled on the wall in the wild-type mice and there were also some rolling leukocytes in the P-selectin–deficient venules. However, their numbers in the latter were greatly reduced in comparison with wild-type animals treated in the same way. In addition, the movement of the leukocytes in the mutant vessels was much slower. Recent studies have shown that leukocytes roll more slowly on E-selectin than on P-selectin in vitro (65). This observation argues in favor of the hypothesis that the rolling in inflamed venules, in the absence of P-selectin, is mediated through E-selectin. When animals lacking both P- and E-selectin become available, we will be able to test this hypothesis.

It is clear from our studies with the P-selectin–deficient mice that P-selectin plays an important role in leukocyte recruitment in acute inflammation and injury. These results were expected because of the nature of P-selectin expression. This molecule is stored in granules and can be made available rapidly after granule exocytosis. We have observed in endothelial cells in culture that P-selectin may return from the cell surface back to the storage granules (66), from which it may be reexpressed on the cell surface. Therefore, P-selectin may also be involved in chronic inflammation. Preliminary results obtained in two studies indicate that this is indeed the case. First, examination of the cellular contents of the peritoneum 48 h after thioglycollate injection shows a drastic reduction in the recruitment of macrophages in the P-selectin–deficient animals in comparison to wild-type animals (R. Johnson, unpublished observation). Second, the contact hypersensitivity response to oxazolone appears significantly less prominent in the absence of P-selectin (M. Subramaniam, unpublished observation).

We believe that the P-selectin–deficient animal model will be useful in the identification of pathologic situations in which P-selectin plays a role. In some disease models, especially those closely linked to inflammation, the involvement of this selectin can be predicted. In others, the outcome can be only guessed. For example, what will happen in an experimental cancer metastasis model using these mice? Will the lack of P-selectin inhibit extravasation, similar to that of leukocytes, and thus reduce metastasis? Or, on the contrary, will the reduced surveillance capability of lymphocytes provide the cancer cells with an advantage and thus increase the cancer spread? It is hard to predict. Fortunately, the existence of the P-selectin–deficient animal will enable us to test our predictions experimentally.

## ACKNOWLEDGMENT

We thank Dr. Allan Mellors at the University of Guelph (Ontario, Canada) for generously providing the *P. haemolytica* glycoprotease. This work was supported by National Institutes of Health grant R01HL53756.

## REFERENCES

1. Siegelman MH, van de Rijn M, Weissman IL. Mouse lymph node homing receptor cDNA clone encodes a glycoprotein revealing tandem interaction domains. *Science* 1989;243:1165–72.
2. Lasky LA, Singer MS, Yednock TA, et al. Cloning of a lymphocyte homing receptor reveals a lectin domain. *Cell* 1989;56:1045–55.
3. Johnston GI, Cook RG, McEver RP. Cloning of GMP140, a granule membrane protein of platelets and endothelium: sequence similarity to proteins involved in cell adhesion and inflammation. *Cell* 1989;56:1033–44.
4. Bevilacqua MP, Stengelin S, Gimbrone MA, Seed B. Endothelial leukocyte adhesion mol-

ecule 1: an inducible receptor for neutrophils related to complement regulatory proteins and lectins. *Science* 1989;243:1160–5.
5. Bevilacqua E, Butcher B, Furie B, et al. Selectins: a family of adhesion receptors [Letter]. *Cell* 1991;67:233.
6. Lasky LA, Singer MS, Dowbenko D, et al. An endothelial ligand for L-selectin is a novel mucin-like molecule. *Cell* 1992;69:927–38.
7. Bevilacqua MP, Nelson RM. Selectins. *J Clin Invest* 1993;91:379–87.
8. Gallatin WM, Weissman IL, Butcher EC. A cell-surface molecule involved in organ-specific homing of lymphocytes. *Nature* 1983;304:30–4.
9. Bevilacqua MP, Pober JS, Mendrick DL, Cotran RS, Gimbrone MA. Identification of an inducible endothelial-leukocyte adhesion molecule. *Proc Natl Acad Sci USA* 1987;84:9238–42.
10. Hsu-Lin SC, Berman CL, Furie BC, August D, Furie B. A platelet membrane protein expressed during platelet activation and secretion. *J Biol Chem* 1984;259:9121–6.
11. McEver RP, Martin MN. A monoclonal antibody to a membrane glycoprotein binds only to activated platelets. *J Biol Chem* 1984;259:9799–804.
12. Larsen E, Celi A, Gilbert GE, et al. PADGEM protein: a receptor that mediates the interaction of activated platelets with neutrophils and monocytes. *Cell* 1989;59:305–12.
13. Geng JG, Bevilacque MP, Moore KL, et al. Rapid neutrophil adhesion to activated endothelium mediated by GMP-140. *Nature* 1990;343:757–60.
14. Bonfanti R, Furie BC, Furie B, Wagner DD. PADGEM (GMP-140) is a component of Weibel-Palade bodies of human endothelial cells. *Blood* 1989;73:1109–12.
15. McEver RP, Beckstead JH, Moore KL, Marshall-Carlson L, Bainton DF. GMP-140, a platelet α-granule membrane protein, is also synthesized by vascular endothelial cells and is localized in Weibel-Palade bodies. *J Clin Invest* 1989;84:92–9.
16. Weibel ER, Palade GE. New cytoplasmic components in arterial endothelia. *J Cell Biol* 1964;23:101–12.
17. Wagner DD, Olmsted JB, Marder VJ. Immunolocalization of von Willebrand protein in Weibel-Palade bodies of human endothelial cells. *J Cell Biol* 1982;95:355–60.
18. Wagner DD. The Weibel-Palade body: the storage granule for von Willebrand factor and P-selectin. *Thromb Haemost* 1993;70:105–9.
19. Stenberg PE, McEver RP, Shuman MA, Jacques YV, Bainton DF. A platelet alpha-granule membrane protein (GMP-140) is expressed on the plasma membrane after activation. *J Cell Biol* 1985;101:880–6.
20. Berman CL, Yeo EL, Wencel-Drake JD, Furie BC, Ginsberg MH, Furie B. A platelet alpha granule membrane protein that is associated with the plasma membrane after activation. Characterization and subcellular localization of platelet activation-dependent granule-external membrane protein. *J Clin Invest* 1986;78:130–7.
21. Baumhueter S, Singer MS, Henzel W, et al. Binding of L-selectin to the vasular sialomucin, CD34. *Science* 1993;262:436–8.
22. Imai Y, Singer MS, Fennie C, Lasky LA, Rosen SD. Identification of a carbohydrate-based endothelial ligand for a lymphocyte homing receptor. *J Cell Biol* 1991;113:1213–21.
23. Sako D, Chang X-J, Barone KM, et al. Expression cloning of a functional glycoprotein ligand for P-selectin. *Cell* 1993;75:1179–86.
24. Moore KL, Stults NL, Diaz S, et al. Identification of a specific glycoprotein ligand for P-selectin (CD62) on myeloid cells. *J Cell Biol* 1992;118:445–56.
25. Lenter M, Levinovite A, Isenmann S, Vestweber D. Monospecific and common glycoprotein ligands for E- and P-selectin on myeloid cells. *J Cell Biol* 1994;125:471–81.
26. Foxall C, Watson SR, Dowbenko D, et al. The three members of the selectin receptor family recognize a common carbohydrate epitope, the sialyl Lewis$^x$. *J Cell Biol* 1992;117:895–902.
27. Handa K, Nudelman D, Stroud MR, Shiozawa T, Hakomori S-I. Selectin GMP-140 (CD62: PADGEM) binds to sialosyl-Le$^\alpha$ and sialosyl-Le$^x$, and sulfated glycans modulate this binding. *Biochem Biophys Res Commun* 1991;181:1223–30.
28. Ogata S, Muramatsu T, Kobata A. New structural characteristics of the large glycopeptides from transformed cells. *Nature* 1976;259:580–2.
29. Wagner DD, Ivatt R, Destree AT, Hynes RO. Similarities and differences between the fibronectins of normal and transformed hamster cells. *J Biol Chem* 1981;256:11708–15.

30. Hubbard CS. Differential effects of oncogenic transformation on N-linked oligosaccharide processing at individual glycosylation sites of viral glycoproteins. *J Biol Chem* 1987;262: 16403–11.
31. Magnani JL, Nilsson B, Brockhaus M, et al. A monoclonal antibody-defined antigen associated with gastrointestinal cancer is a ganglioside containing sialylated lacto-N-fucopentaose II. *J Biol Chem* 1982;23:14365–9.
32. Yousefi S, Higgins E, Daoling Z, Pollex-Kruger A, Hindsgaul O, Dennis JW. Increased UDP-GlcNAc:Galβ1-3GalNAc-R(GlcNAc to GalNAc)β-1,6-*N*-acetylglucosaminyltransferase activity in metastatic murine tumor cell lines. *J Biol Chem* 1991;266:1772–82.
33. Fukushima K, Hirota M, Terasaki P, et al. Characterization of sialosylated Lewis$^x$ as a new tumor-associated antigen. *Cancer Res* 1984;44:5279–85.
34. Groves RW, Allen MH, Ross EL, Ahsan G, Barker J, MacDonald DM. Expression of selectin ligands by cutaneous squamous cell carcinoma. *Am J Pathol* 1993;143:1220–5.
35. Polley MJ, Phillips ML, Wayner E, et al. CD62 and endothelial cell-leukocyte adhesion molecule 1 (ELAM-1) recognize the same carbohydrate ligand, sialyl-Lewis x. *Proc Natl Acad Sci USA* 1991;88:6224–8.
36. Rice GE, Bevilacqua MP. An inducible endothelial cell surface glycoprotein mediates melanoma adhesion. *Science* 1989;246:1303–6.
37. Walz G, Kolanus A, Bevilacqua W, Seed B. Recognition by ELAM-1 of the sialyl-LeX determinant on myeloid and tumor cells. *Science* 1990;25;1132–5.
38. Sawada R, Tsuba S, Fukuda M. Differential E-selectin-dependent adhesion efficiency in sublines of a human colon cancer exhibiting distinct metastatic potentials. *J Biol Chem* 1994;269:1425–31.
39. Iwai K, Ishikura H, Kaji M, et al. Importance of E-selectin (ELAM-1) and Sialyl Lewis$^a$ in the adhesion of pancreatic carcinoma cells to activated endothelium. *Int J Cancer* 1993; 54:972–7.
40. Merwin JR, Madri JA, Lynch M. Cancer cell binding to E-selectin transfected human endothelia. *Biochem Biophys Res Commun* 1992;189:315–23.
41. Kunzendorf U, Kruger-Krasagakes S, Notter M, Hock H, Walz G, Diamantstein T. A sialyl-6$^x$-negative melanoma cell line binds to E-selectin but not to P-selectin. *Cancer Res* 1994;54:1109–12.
42. Stone JP, Wagner DD. P-Selectin mediates adhesion of platelets to neuroblastoma and small cell lung cancer. *J Clin Invest* 1993;92:804–13.
43. Martin-Padura I, Metzelaar MJ, Menard S, Mantouani A, Dejana E. Enhanced adhesion of colon carcinoma cells to thrombin stimulated endothelial cells. *Endothelium* 1993;1:41–5.
44. Abdullah KM, Udoh EA, Shewen PE, Mellors A. A neutral glycoprotease of *Pasteurella haemolytica A1* specifically cleaves O-sialoglycoproteins. *Infect Immun* 1992;60:56–62.
45. Kojima N, Handa K, Newman W, Hakomori S. Inhibition of selectin-dependent tumor cell adhesion to endothelial cells and platelets by blocking O-glycosylation of these cells. *Biochem Biophys Res Commun* 1992;182:1288–95.
46. Steininger CN, Eddy CA, Leimgruber RM, Mellors, Welply JK. The glycoprotease of *Pasteurella haemolytica A1* eliminates binding of myeloid cells to P-selectin but not to E-selectin. *Biochem Biophys Res Commun* 1992;188:760–6.
47. Aruffo A, Kolanus W, Walx G, Freedman P, Seed B. CD62/P-selectin recognition of myeloid and tumor cell sulfatides. *Cell* 1991;76:35–44.
48. Aruffo A, Dietsch MT, Wan H, Hellstrom KE, Hellstrom I. Granule membrane protein 140 (GMP140) binds to carcinomas and carcinoma-derived cell lines. *Proc Natl Acad Sci USA* 1992;89:2292–6.
49. Corral L, Singer MS, Macher BA, Rosen SD. Requirement for sialic acid on neutrophils in a GMP-140 (PADGEM) mediated adhesive interaction with activated platelets. *Biochem Biophys Res Commun* 1990;172:1349–56.
50. Moore KL, Varki A, McEver RP. GMP-140 binds to a glycoprotein receptor on human neutrophils: evidence for a lectin-like interaction. *J Cell Biol* 1991;112:491–9.
51. Larsen GR, Sako D, Ahern TJ, et al. P-Selectin and E-selectin: distinct but overlapping leukocyte ligand specificities. *J Biol Chem* 1992;267:11104–10.
52. Giavazzi R, Foppolo M, Dossi R, Remuzzi A. Rolling and adhesion of human tumor cells on vasular endothelium under physiological flow conditions. *J Clin Invest* 1993;92:3038–44.
53. Ishikawa T, Imura A, Tanaka K, Shirane H, Okuna M, Uchiyama T. E-selectin and vas-

cular cell adhesion molecule-1 mediate adult T-cell leukemia cell adhesion to endothelial cells. *Blood* 1993;82:1590–8.

54. Cavenagh JD, Gordon-Smith EC, Gibson FM, Gordon MY. Acute myeloid leukaemia blast cells bind to human endothelium in vitro utilizing E-selectin and vascular cell adhesion molecule-1 (VCAM-1). *Br J Haematol* 1993;85:285–91.
55. Irimura T, Nakamori S, Matsushita Y, et al. Colorectal cancer metastasis determined by carbohydrate-mediated cell adhesion: role of sialyl-Le$^x$ antigens. *Cancer Biol* 1993;4:319–24.
56. Saitoh O, Wang WC, Lotan R, Fukuda M. Differential glycosylation and cell surface expression of lysosomal membrane glycoproteins in sublines of a human colon cancer exhibiting distinct metastatic potentials. *J Biol Chem* 1992;267:5700–11.
57. Sawada R, Lowe JB, Fukuda M. E-selectin-dependent adhesion efficiency of colonic carcinoma cells is increased by genetic manipulation of their cell surface lysosomal membrane glycoprotein-1 expression levels. *J Biol Chem* 1993;268:12675–81.
58. Brock TA, Dvorak HF, Senger DR. Tumor-secreted vascular permeability factor increases cytosolic $CA^{2+}$ and von Willebrand factor release in human endothelial cells. *Am J Pathol* 1991;138:213–21.
59. Mayadas TN, Johnson RC, Rayburn H, Hynes RO, Wagner DD. Leukocyte rolling and extravasation are severely compromised in P-selectin-deficient mice. *Cell* 1993;74:972–7.
60. Capecchi MR. Altering the genome by homologous recombination. *Science* 1989;244:1288–92.
61. Watson ML, Kingsmore SF, Johnston GI, et al. Genomic organization of the selectin family of leukocyte adhesion molecules on human and mouse chromosome 1. *J Exp Med* 1990; 172:263–72.
62. Sporn LA, Marder VJ, Wagner DD. Inducible secretion of large biologically potent von Willebrand factor multimers. *Cell* 1986;46:185–90.
63. Lewinsohn DM, Bargatze RF, Butcher EC. Leukocyte-endothelial cell recognition: evidence of a common molecular mechanism shared by neutrophils, lymphocytes, and other leukocytes. *J Immunol* 1987;138:4313–21.
64. Watson S, Fennie C, Lasky LA. Neutrophil influx into an inflammatory site inhibited by a soluble homing receptor-OgG chimaera. *Nature* 1991;349:164–7.
65. Lawrence MB, Springer TA. Leukocytes roll on a selectin at physiologic flow rates: distinction from the prerequisite for adhesion through integrins. *Cell* 1991;65:859–73.
66. Subramaniam M, Koedam JA, Wagner DD. Divergent fates of P- and E-selectins after their expression on the plasma membrane. *Mol Biol Cell* 1993;4:791–801.

*Topics in Molecular Medicine, Volume 1,*
edited by Wolfgang Siess, Reinhard Lorenz,
and Peter C. Weber. Raven Press, Ltd.,
New York © 1995.

# 11

# Platelet GPIIb/IIIa Inhibition by Monoclonal Antibodies, with Observations on Blood Vessel Wall Passivation After Angioplasty Injury

Barry S. Coller

*Department of Medicine, Mount Sinai Medical Center, New York, New York 10029-6574*

## PLATELET PHYSIOLOGY AND ROLE IN VASCULAR DISEASE

There is abundant evidence that platelets contribute significantly to the vaso-occlusive thrombotic process that produces ischemic damage in cardiovascular, cerebrovascular, and peripheral vascular diseases (1–3). This pathologic role derives from the inability of platelets to differentiate between a damaged normal blood vessel that requires platelet interactions to arrest hemorrhage and a fractured atherosclerotic plaque in which platelet deposition and aggregation may be lethal. Moreover, from an evolutionary standpoint, individuals with enhanced platelet and coagulation reactivity probably enjoyed a selective advantage, because they were less likely to die from hemorrhage before reaching sexual maturity. In contrast, there has probably been little selective pressure against thrombotic disease, because it primarily affects individuals who are beyond their reproductive years. Therefore, our hemostatic system appears maladapted to our modern society, which is composed increasingly of elderly individuals with atherosclerotic vascular disease. As a corollary, however, the effectiveness and redundancy of our hemostatic system permits us to intervene judiciously with relatively high therapeutic doses of anticoagulants, antiplatelet agents, and thrombolytic agents on an acute basis, and with lower doses of anticoagulants and antiplatelet agents on a chronic basis.

## PLATELET PHYSIOLOGY

Platelets adhere to damaged and normal blood vessels and to fractured atherosclerotic plaques because in both cases there is exposure of adhesive glycoproteins (Table 1) for which platelets have specific receptors (Table 2) (1). Additional platelets then interact with the adherent platelets to form extensive aggregates. The first layer of adherent platelets cannot occlude even the smallest blood vessel, but large platelet aggregates can produce frank vaso-occlusion and ischemic infarction. Platelet aggregation is mediated exclusively by the GPIIb/IIIa receptor, which undergoes a transition to a high-affinity ligand-binding state when platelets are suitably activated. The multivalent adhesive glycoproteins fibrinogen and/or von Willebrand factor can then bind to two platelets simultaneously and produce aggregate formation (1).

Control over the activation process must be carefully regulated to avoid ongoing thrombosis. Not unexpectedly, therefore, the agonists that stimulate the GPIIb/IIIa receptor are agents that are likely to be released or generated at sites of injury (Table 3) (1). It is of interest that adhesion itself can activate platelets, as can high shear forces. Most important, some thrombolytic agents have also been shown to activate platelets. It is unknown, however, whether the activation is caused directly by the thrombolytic agent itself or indirectly by generation of plasmin or thrombin (4).

Activation begins with agonist binding or surface perturbation, but this signal must then be transduced by one or more pathways. Two of these pathways have been well defined: release of arachidonic acid from phospholipids, followed by conversion of the arachidonic acid to thromboxane $A_2$ by the enzymes cyclo-oxygenase and thromboxane synthase; and cleavage of phosphatidylinositol 4,5-bisphosphate by phospholipase C, which leads to activation of protein kinase C (by diacylglycerol) and release of internal stores of calcium (by inositol trisphosphate). It is clear, however, that additional transduction pathways exist and that the arachidonic acid pathway is not absolutely required for GPIIb/IIIa activation by any of the agonists (1,4). Therefore, aspirin, which abolishes the arachidonic acid pathway, only partially inhibits platelet aggregation.

The GPIIb/IIIa receptor is also regulated by platelet inhibitors. The best-characterized inhibitors are prostacyclin, which is produced by endothelial

**TABLE 1.** *Components of the blood vessel wall that are hemostatically active*

| Subendothelium | Media |
|---|---|
| von Willebrand factor | Collagen (types I + III) |
| Collagen (types IV, V, VI) | |
| Fibronectin | Adventitia |
| Thrombospondin | Collagen (types I + III) |
| Laminin | Tissue factor |
| Vitronectin | |
| Fibrinogen (fibrin) | |
| Tissue factor (trace amounts) | |

From ref. 1 with permission.

**TABLE 2.** *Platelet receptors*

| Receptor | Ligand | Number of receptors on the platelet surface |
|---|---|---|
| Adhesion | | |
| Integrins | | |
| GPIa/IIa (VLA-2) | Collagen | ~1,000 |
| GPIc/IIa (VLA-6) | Laminin | ~1,000 |
| GPIc*/IIa (VLA-5) | Fibronectin | ~1,000 |
| $\alpha_v$/IIIa | Vitronectin, fibrinogen, von Willebrand factor, thrombospondin | ~100 |
| GPIIb/IIIa | Fibrinogen, fibronectin, von Willebrand factor, vitronectin (?thrombospondin) | 45,000 |
| Others | | |
| GPIb | von Willebrand factor | 25,000 |
| GPIV | Thrombospondin, collagen | 25,000 |
| 67 kDa | Laminin | ? |
| Aggregation | | |
| GPIIb/IIIa | Fibrinogen, fibronectin, von Willebrand factor, vitronectin, (?thrombospondin) | 45,000 |

From ref. 1 with permission.

cells and inhibits the platelet transducing mechanisms by increasing platelet cAMP, and nitric oxide (endothelium-derived relaxation factor), which inhibits the transducing mechanisms by increasing platelet cGMP.

## THE ROLE OF ACTIVATED LUMINAL GPIIb/IIIa RECEPTORS IN HEMOSTASIS AND VASCULAR DISEASE

Data indicate that plaque rupture is not a rare event, whereas myocardial infarction is. Therefore, one of the most important unanswered questions in

**TABLE 3.** *Platelet aggregation*

| Agonists | Transducing mechanisms | | Effectors |
|---|---|---|---|
| Adhesion | Arachidonic acid | ↑ | Activated GPIIb/IIIa |
| Thrombin | Protein kinase C | | |
| Thromboxane $A_2$ | ? | | |
| ADP | | | |
| Epinephrine | | | |
| Serotonin | | cAMP or cGMP | |
| Vasopressin | | ↑ ↑ | |
| Plasmin | | $PGI_2$ NO | |
| t-PA/SK | | | |
| Shear | | | |

t-PA/SK, tissue plasminogen activator/streptokinase; $PGI_2$, prostacyclin; NO, nitric oxide, endothelial-derived relaxation factor (EDRF).
From ref. 28 with permission.

vascular disease is why some plaque injuries lead to vaso-occlusion, ischemia, and infarction, whereas others lead to the deposition of only a monolayer of platelets with no compromise of blood flow.

The recruitment of a second layer of platelets to a surface covered with a monolayer of platelets depends on the availability of activated GPIIb/IIIa receptors on the luminal surface, because these are required for the attachment of the second and subsequent layers of platelets. Many factors can affect this process, including the nature of the subendothelial adhesive glycoproteins, the ability of the adherent platelets to release ADP and to synthesize thromboxane $A_2$, the local generation of thrombin, the availability of the plasma co-factors fibrinogen and von Willebrand factor, and the local synthesis and release of the platelet inhibitors $PGI_2$ and nitric oxide.

The availability of the receptors for interaction with circulating platelets depends on morphologic alterations that enable platelet filopodia to extend outward from the surface into the lumen of the blood vessel. The intrinsic affinity of the receptor for fibrinogen and von Willebrand factor depends on the biochemical transformation of the receptor to a high-affinity state. Finally, rheologic factors affect both the likelihood of a platelet residing long enough in the vicinity of an adherent platelet to interact and the shear forces tending to pull a newly attached platelet from the platelet monolayer.

In model systems, we have created a surface, using high-density fibrinogen, that is extremely effective in binding a monolayer of platelets but is unable to recruit a second layer of platelets because the GPIIb/IIIa receptors appear to have been trapped on the abluminal surface and/or because the platelets adhere to the surface with a "fried egg" appearance and do not extend filopodia into the lumen (5). We are currently studying the biochemical basis of this phenomenon and are trying to exploit it to develop nonthrombogenic surfaces.

## BLOOD VESSEL PASSIVATION

One of the fundamental observations in vascular disease is that soon after injury to a blood vessel, platelets deposit readily on the surface. Less attention has been focused on the observation that within a relatively brief period of time such surfaces stop attracting new platelets. From a phenomenologic standpoint it is useful to define blood vessel passivation as the process or processes that convert platelet-reactive surfaces into platelet-nonreactive surfaces. Table 4 lists potential blood vessel passivation mechanisms, which can be divided into alterations of the blood vessel itself, inhibition of the coagulation mechanism, rheologic effects, and effects on platelets. In animal models of vascular injury, platelet deposition decreases dramatically within 4 h (6) or 8 h (7).

**TABLE 4.** *Potential blood vessel passivation mechanisms*

| |
|---|
| Alteration of blood vessel |
| Re-endothelialization |
| Passivating protein deposition |
| Proteolytic or nonproteolytic alteration of matrix |
| Inhibition of coagulation |
| Inhibition of activators (thrombin-ATIII, thrombin–fibrin, activated protein C) |
| Dilution of activators |
| Negative feedbacks of coagulation (TFPI, substrate depletion) |
| Rheologic effects |
| Shear forces |
| Platelet effects |
| Activation or release of inhibitors ($PGI_2$, NO) |
| Homologous desensitization (refractoriness) |
| Downregulation of GPIb by thrombin |
| Leukocyte inhibition of platelets |
| Loss of luminal GPIIb/IIIa receptors |

## GPIIb/IIIa RECEPTOR BLOCKADE WITH MONOCLONAL ANTIBODIES, PLATELET FUNCTION, AND ANTI-THROMBOTIC EFFECTS

In 1983 and 1985, we reported on the development of monoclonal antibodies (10E5 and 7E3) that blocked the platelet GPIIb/IIIa receptor and inhibited platelet aggregation ex vivo (8,9). A series of animal studies employing fragments of these antibodies lacking the Fc region were then conducted to define their anti-thrombotic effects.

By calculating the number of GPIIb/IIIa receptors blocked by antibody infusions, we were able to correlate the number of free receptors with a variety of platelet functional activities. Decreasing the number of available receptors from approximately 40,000 to approximately 20,000 in both animals and humans had little effect other than mildly decreasing platelet aggregation (10,11). This is in accord with both our in vitro studies (8) and observations that patients who are obligatory heterozygotes for Glanzmann thrombasthenia have approximately 60% of the normal numbers of GPIIb/IIIa receptors (12) but have no hemorrhagic diathesis and essentially normal platelet function.

When antibody infusion decreased the number of available GPIIb/IIIa receptors to less than 20,000 per platelet, platelet aggregation in response to a variety of agonists, including ADP and collagen, showed more significant inhibition (10,11). At approximately 10,000 and 20,000 residual receptors, platelet inhibition was sufficient to prevent occlusive platelet thrombus formation in the animal model developed by Folts, which was designed to simulate the vascular abnormalities found in patients suffering from unstable angina and transient ischemic attacks (11,13,14). The bleeding times in both primates and humans usually were only slightly prolonged compared to the control value when 8,000 and 15,000 GPIIb/IIIa receptors remained unblocked, and rarely exceeded 10 min (10,11). Platelet aggregation responses were usually abolished when the number of residual GPIIb/IIIa receptors decreased below 10,000 (10,11,13), which is consistent with some of the first in vitro studies conducted with antibody 10E5 (8). Bleeding times became progressively prolonged when the residual receptors decreased below 10,000 per platelet.

In animal models of myocardial infarction treated with thrombolytic agents, the addition of 7E3 resulted in more rapid reperfusion and prevention of reocclusion. In addition, 7E3 facilitated the reperfusion of platelet-rich thrombi that resisted lysis by thrombolytic agents alone. Moreover, addition of 7E3 permitted the dose of the thrombolytic agent to be reduced by 50% or more (15–23). In animal models of angioplasty-induced injury, 7E3 was more effective than aspirin (24) or the direct thrombin inhibitor D-Phe-Pro-Arg $CH_2Cl$ (PPACK) (25).

A chimeric version of the Fab fragment of 7E3 (c7E3 Fab) has been developed for human use. When given to humans at a dose of 0.25 mg/kg, it produces approximately 80% blockade of GPIIb/IIIa receptors and almost complete inhibition of platelet aggregation. The EPIC study (26) tested the effects of c7E3 Fab on a 30-day composite end point of death, myocardial infarction, and need for urgent intervention (repeat angioplasty, coronary artery bypass surgery, stent placement, or intra-aortic balloon pump insertion) in patients undergoing angioplasty or atherectomy who were at high risk for one or more of these events. Treatment with a bolus of c7E3 Fab (0.25 mg/kg) followed by a 12-h infusion of c7E3 Fab (10 μg/min) produced a 35% reduction in the composite end point ($p = 0.009$) compared to placebo-treated patients. Patients treated with only a bolus dose of c7E3 Fab also had a reduction in the composite end point, but it was not significantly different from the placebo group.

Data in the EPIC trial indicate that the risk for abrupt occlusion, as judged

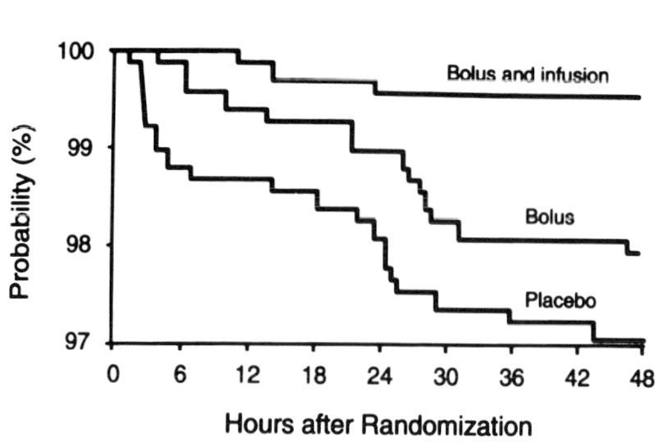


**FIG. 1.** Probability of no urgent repeated percutaneous revascularization procedures in the three treatment groups (Kaplan–Meier plots). Events began to occur shortly after the index procedure in the placebo group, between 6 and 12 h after the procedure in the group given the bolus of c7E3 Fab, and even later in the group given both the bolus and the infusion. The *y* axis is truncated at 97% to demonstrate the difference in this end point, which occurred with low frequency. (From ref. 28 with permission.)

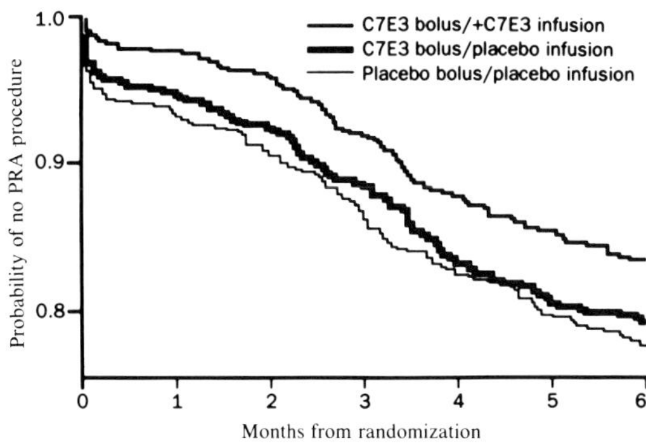


**FIG. 2.** Kaplan–Meier curve of all events (death, myocardial infarction, coronary revascularization) for all patients enrolled. There was a significant reduction of events for the c7E3 bolus/c7E3 infusion group compared with the active bolus only or placebo treatments ($p = 0.001$). A substantial proportion of events occurred after 1 month. (From ref. 30 with permission.)

by the need for emergency repeat angioplasty, is highest during the first 2 to 3 days and diminishes dramatically by the end of 5 to 8 days. The single bolus of c7E3 Fab (0.25 mg/kg), delayed the onset of the need for emergency PTCA by about 4 h and reduced the overall need for emergency repeat angioplasty by approximately 38% by day 5. Treatment with the bolus of c7E3 Fab followed by the constant infusion of c7E3 Fab for 12 h delayed the onset of the need for repeat angioplasty for nearly 12 h and reduced the overall need for emergency repeat angioplasty by approximately 82% at day 5 (Fig. 1).

These data from EPIC suggest that platelet thrombus formation contributes to the vast majority of events that lead to the need for emergency repeat angioplasty. In addition, they indicate that a blood vessel damaged by angioplasty is highly platelet-reactive (thrombogenic) for a period of 2 to 3 days and remains more mildly thrombogenic for up to 5 to 8 days. High-grade GPIIb/IIIa receptor blockade is likely to be needed, therefore, for a period of at least 2 to 3 days. Finally, the 6-month follow-up evaluation of patients in the EPIC study (27) showed that the benefit identified at 30 days was sustained for the 6-month period (Fig. 2).

## ACKNOWLEDGMENT

Supported by grant 19278 from the National Heart, Lung and Blood Institute.

## REFERENCES

1. Coller BS. Platelets in cardiovascular thrombosis and thrombolysis. In: Fozzard HA, Haber E, Jennings RB, Katz AM, Morgan HE, eds. *The heart and cardiovascular system,* 2nd ed. New York: Raven Press, 1992;219–73.
2. Stein G, Fuster V, Halperin JL, Chesebro JH. Antithrombotic therapy in cardiac disease. An emerging approach based on pathogenesis and risk. *Circulation* 1989;80:1501–13.

3. Harker LA. Antithrombotic therapy. In: Williams WJ, Beutler E, Erslev AJ, Lichtman MA, eds. *Hematology*. New York: McGraw–Hill, 1990:1569–81.
4. Coller BS. Platelets and thrombolytic therapy. *N Engl J Med* 1990;322:33–42.
5. Coller BS, Kutor JL, Scudder LE, et al. Studies of activated GPIIb/IIIa receptors on the luminal surface of adherent platelets. Paradoxical loss of luminal receptors when platelets adhere to high density fibrogen. *J Clin Invest* 1993;92:2796–806.
6. Wilentz JR, Sanborn TA, Handenschile CC, Valeri CR, Ryan TJ, Faxon DP. Platelet accumulation in experimental angioplasty: time course and relation to vascular injury. *Circulation* 1987;75:636–42.
7. Groves HM, Kinlough-Rathbone RL, Mustard JF. Development of nonthrombogenicity of injured rabbit aortas despite inhibition of platelet adherence. *Atherosclerosis* 1986;6:189–95.
8. Coller BS, Peerschke EI, Scudder LE, Sullivan CA. A murine monoclonal antibody that completely blocks the binding of fibrinogen to platelets produces a thromobasthenic-like state in normal platelets and binds to glycoproteins IIb and/or IIIa. *J Clin Invest* 1983;72:325–38.
9. Coller BS. A new murine monoclonal antibody reports an activation-dependent change in the conformation and/or microenvironment of the platelet GPIIb/IIIa complex. *J Clin Invest* 1985;76:101–8.
10. Gold HK, Gimple LW, Yasuda T, et al. Pharmacodynamic study of $F(ab')_2$ fragments of murine monoclonal antibody 7E3 directed against human platelet glycoprotein IIb/IIIa in patients with unstable angina pectoris. *J Clin Invest* 1990;86:651–9.
11. Coller BS, Folts JD, Smith SR, Scudder LE, Jordan J. Abolition of in vivo platelet thrombus formation in primates with monoclonal antibodies to the platelet GPIIb/IIIa receptor. *Circulation* 1989;80:1766–74.
12. Coller BS, Seligsohn U, Zivelin A, Zwang E, Lusky A, Modan M. Immunologic and biochemical characterization of homozygous and heterozygous Glanzmann's thrombasthenia in Iraqi-Jewish and Arab populations of Israel: comparison of techniques for carrier detection. *Br J Haematol* 1986;62:723–35.
13. Coller BS, Scudder LE, Beer J, et al. Monoclonal antibodies to platelet glycoprotein IIb/IIIa as antihrombotic agents. *Ann NY Acad Sci* 1991;64:193–213.
14. Coller BS, Folts JD, Scudder LE, Smith SR. Antithrombotic effect of a monoclonal antibody to the platelet glycoprotein IIb/IIIa receptor in an experimental animal model. *Blood* 1986;68:783–6.
15. Yasuda T, Gold HK, Fallon JT, et al. A monoclonal antibody against the platelet GPIIb/IIIa receptor prevents coronary artery reocclusion following reperfusion with recombinant tissue-type plasminogen activator in dogs. *J Clin Invest* 1988;81:1284–91.
16. Gold HK, Coller BS, Yasuda T, et al. Rapid and sustained coronary artery recanalization with combined bolus injection of recombinant tissue-type plasminogen activator and monoclonal anti-platelet GPIIb/IIIa antibody in a dog model. *Circulation* 1988;77:670–7.
17. Yasuda T, Gold HK, Yaoita H, et al. Comparative effects of aspirin, a synthetic thrombin inhibitor and a monoclonal antiplatelet glycoprotein IIb/IIIa antibody on coronary artery reperfusion, reocclusion and bleeding with recombinant tissue-type plasminogen activator in a canine preparation. *J Am Coll Cardiol* 1990;16:714–22.
18. Mickelson JK, Simpson PJ, Cronin M, et al. Antiplatelet antibody [7E3-$F(ab')_2$] prevents rethrombosis after recombinant tissue-type plasminogen activator-induced coronary artery thrombolysis in a canine model. *Circulation* 1990;81:617–27.
19. Fitzgerald DJ, Wright F, FitzGerald GA. Increased thromboxane biosynthesis during coronary thrombolysis. *Circ Res* 1989;65:83–94.
20. Fitzgerald DJ, Hanson M, FitzGerald GA. Systemic lysis protects against the effects of platelet activation during coronary thrombolysis. *J Clin Invest* 1991;88:1589–95.
21. Rote WE, Bates ER, Mu DX, Nedelman MA, Lucchesi BR. Prevention of rethrombosis after coronary thrombolysis in a chronic canine model: I. Adjunctive therapy with monoclonal antibody 7E3 $F(ab')_2$ fragment. *J Cardiovasc Pharmacol* 1994;23;194–202.
22. Yasuda T, Gold HK, Leinbach RC, et al. Lysis of plasminogen activator-resistant platelet-rich coronary artery thrombus with combined bolus injection of recombinant tissue-type plasminogen activator and antiplatelet GPIIb/IIIa antibody. *J Am Coll Cardiol* 1990;7:1728–35.

23. Lu HR, Gold HK, Wu Z, et al. Acceleration and persistence of recombinant tissue plasminogen activator induced arterial eversion graft recanalization with a single bolus injection of $F(ab')_2$ fragments of the antiplatelet GPIIb/IIIa antibody 7E3. *Coronary Artery Dis* 1991;2;1039–46.
24. Bates ER, McGillem MJ, Mickelson JK, Pitt B, Mancini GBJ. A monoclonal antibody against the platelet glycoprotein IIb/IIIa receptor complex prevents platelet aggregation and thrombosis in a canine model of coronary angioplasty. *Circulation* 1991;84:2463–9.
25. Kaplan AV, Leung LL-K, Leung W-H, Grant GW, McDougall IR, Fischell TA. Roles of thrombin and platelet membrane glycoprotein IIb/IIIa in platelet subendothelial deposition after angiplasty in an *ex vivo* whole artery model. *Circulation* 1991;84:1279–88.
26. The EPIC Investigators. Use of monoclonal antibody directed against the platelet glycoprotein IIb/IIIa receptor in high-risk coronary angioplasty. *N Engl J Med* 1994;330:956–61.
27. Topol EJ, Califf RM, Weisman HF, et al. Randomized trial of coronary intervention with antibody against platelet IIb/IIIa integrin for reduction of clinical restenosis: results at six months. *Lancet* 1994;343:881–86.
28. Coller BS. Antiplatelet agents in the prevention and therapy of thrombosis. *Annu Rev Med* 1992;43:171–80.

*Topics in Molecular Medicine, Volume 1,*
edited by Wolfgang Siess, Reinhard Lorenz,
and Peter C. Weber. Raven Press, Ltd.,
New York © 1995.

# 12

# $\beta_2$ Integrins in Neutrophil Adhesion and Their Role in Ischemia–Reperfusion and Trauma

Nicholas B. Vedder, Robert K. Winn, *Sam R. Sharar, and †John M. Harlan

*Departments of Surgery and *Anesthesia, University of Washington, Harborview Medical Center, Seattle, Washington 98104; and †Division of Hematology, Department of Medicine, University of Washington, Seattle, Washington 98195*

A pivotal early event in both host defense and injury repair is the adherence of neutrophils to endothelium. The same mechanisms that mediate this normal process, however, can in certain circumstances contribute to vascular and tissue injury. Neutrophil-mediated endothelial injury has been shown to play a critical role in the pathogenesis of disorders produced by tissue ischemia and reperfusion. Understanding the molecular basis of neutrophil–endothelial interactions, therefore, may yield insights into the basic mechanisms of these disorders and may also suggest new approaches to therapy of a wide range of important disease processes including, stroke, myocardial infarction, organ transplantation, peripheral vascular disease, and shock states (1–3).

## MECHANISMS OF NEUTROPHIL-MEDIATED TISSUE DAMAGE

Neutrophils are equipped with diverse mechanisms for carrying out their normal function of host defense and repair (2,4). In response to a wide variety of inflammatory stimuli, including cytokines, platelet-activating factor, activated complement components, bacterial peptides, or endotoxin, neutrophils can release a multitude of inflammatory mediators. These inflammatory components include reactive oxygen species, proteases, peptides, lipid mediators, and vasoactive substances (3). When this process is appropriately regulated, these endogenous mediators act on microbial organisms or debris brought into the neutrophil by phagocytosis. The result is efficient clearance of pathogens and diseased tissue. However, if regulatory mechanisms fail or if activation is initiated in response to diffuse or systemic inflammatory stimuli, uncontrolled release of toxic substances can ensue, resulting in neutrophil-mediated vascular and tissue injury with damage to otherwise viable host tissues and organs (5).

Neutrophil adherence to endothelium plays a pivotal role in neutrophil-mediated vascular and tissue injury (Fig. 1). Observations of the microcirculation by intravital microscopy have illuminated a sequence of events involved in neutrophil adhesion to endothelium (1,6,7). At the site of inflammation, neutrophils are first seen to leave the laminar flow stream and roll along the endothelium of adjacent postcapillary venules. Next, the neutrophils adhere more firmly to the endothelium and move along the endothelial surface until they encounter an interendothelial junction. There they crawl between the endothelial cells and finally emigrate into the surrounding tissues. With firm adherence to the vessel wall, a protected microenvironment is formed between the adherent neutrophil and the endothelial cell that is sheltered from circulating anti-inflammatory agents. Within this microenvironment, proteases, oxidants, or other inflammatory mediators can cause injury to the endothelium (8,9). This adherence-dependent loss of microvascular integrity can lead to edema formation, hemorrhage, or thrombosis, and such vascular injury then may ultimately be responsible for organ dysfunction. In addition, homotypic adhesion or aggregation of neutrophils within the microcirculation can occlude blood flow, resulting in tissue hypoperfusion and further ischemic damage (10). Finally, once they have emigrated into surrounding tissues, neutrophils can continue to release toxic products that directly injure tissue and cause further organ damage (3,4).

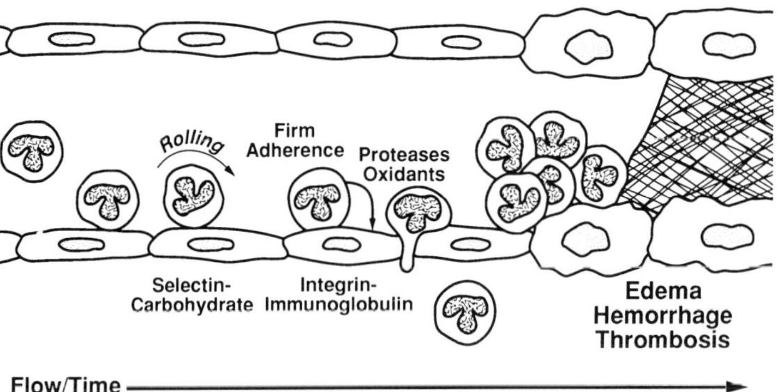


**FIG. 1.** The sequential events at a site of inflammation as neutrophils leave the laminar flow stream of a postcapillary venule. Initial selectin–carbohydrate-mediated rolling along the surface of the endothelium is followed by integrin–immunoglobulin-mediated firm adherence. These two events are required for neutrophil emigration in the setting of neutrophil-mediated endothelial injury in pathologic conditions, such as those associated with ischemia–reperfusion injuries. (Adapted from ref. 67 with permission.)

## MECHANISMS OF NEUTROPHIL ADHESION

In the past decade there has been remarkable progress in the characterization of the specific proteins involved in this adherence interaction. The molecular structures of these proteins have been defined, and much has been learned about their functions through in vitro as well as in vivo study (11–13).

These adherence molecules are currently classified into two major categories: the leukocyte integrins and their ligands on endothelial cells that are members of the immunoglobulin (Ig) superfamily (14), and the selectins and their carbohydrate counterstructures. The integrin–Ig-like ligand interaction mediates the steps of firm adhesion of neutrophils to endothelium at sites of inflammation, diapedesis, and emigration. The selectin receptors are lectin-containing proteins that recognize specific carbohydrate counterstructures expressed on glycoproteins, including mucin-like proteins rich in *O*-linked sugars (15). Selectin-mediated binding, usually of lower affinity, is observed under conditions of flow, whereas integrin-dependent adhesion is sensitive to shear and is optimal under static or low-shear conditions (16). Selectin receptors therefore appear to be responsible for the initial transient adhesion of neutrophils that occurs at sites of inflammation, manifested as rolling. Once slowed by selectin–carbohydrate interactions, local inflammatory stimuli subsequently activate the neutrophils to produce firm adhesion via the integrin–Ig-like ligand interaction.

### Integrin–Ig-Like Ligand Adhesion

The CD11/CD18 complex is a leukocyte membrane glycoprotein complex localized to the leukocyte surface, which plays a major role in mediating leukocyte adhesiveness. It consists of three heterodimeric glycoprotein subunits, each consisting of an identical light or β-chain polypeptide, designated CD18, and a distinct heavy or α-chain polypeptide, designated CD11a, CD11b, or CD11c. These subunits are designated CD11a/CD18 (previously known as LFA-1), CD11b/CD18 (previously known as Mac-1, Mo1 or CR3), and CD11c/CD18 (previously known as p150,95). CD11a/CD18 is found on all leukocytes. CD11b/CD18 and CD11c/CD18 are found primarily on phagocytes and natural killer cells but not on most lymphocytes. The three CD11 α-chains and the CD18 β-chain have been cloned and were shown to be members of the integrin superfamily of adhesion receptors, comprising the $\beta_2$ subclass which is restricted to leukocytes (12,14).

On the basis of studies of cultured human umbilical vein endothelial cells (HUVECs), it has been shown that normal, unstimulated neutrophils exhibit a very low level of basal adherence. After stimulation of neutrophils with tumor necrosis factor (TNFα), C5a, interleukin-8 (IL-8), platelet-activating factor (PAF), or bacterial chemotactic peptide, there is a marked increase in neutrophil adherence to endothelium. Stimulation with some of these substances also causes an increase in homotypic neutrophil adhesion. Addition of monoclonal antibodies (MAbs) directed to functional epitopes of CD11b or CD18 essentially eliminates this increased adhesiveness in vitro (9). Inhibition of tight adherence between the neutrophil and the endothelium prevents formation of the protected microenvironment in which released proteases, oxidants, or other inflammatory mediators can cause injury to the endothelium (8). The results of these in vitro studies suggest that the use of MAbs to inhibit integrin-mediated neutrophil adhesion may have significant advantages over antiproteases or antioxidants, which are generally much less effective once adherence has created a protected microenvironment.

The integrin heterodimer most responsible for firm adhesion in neutrophils appears to be CD11b/CD18 (Mac-1, Mo1, CR3), although CD11a/CD18 (LFA-1) is also involved (17). The mechanism by which CD11b/CD18 augments neutrophil adhesiveness in response to stimulation is not entirely clear. CD11b/CD18 in unstimulated neutrophils exists both on the cell surface and, in far greater quantities, within the secondary or tertiary granules. On stimulation, these granule contents translocate to the cell surface, resulting in a three- to tenfold increase in surface-associated CD11b/CD18. This increase in surface expression, however, is neither necessary nor sufficient to cause increased adhesiveness (18). Instead, the primary mechanism appears to be an activation of surface receptors, producing a conformational change that results in high-avidity ligand binding. The molecular basis for avidity modulation of the $\beta_2$-integrin receptor has not been defined, although allosteric changes induced by intracellular lipids have been suggested (19). Both the activation process and avidity modulation are logical targets for manipulating neutrophil adhesion and neutrophil-mediated injury.

Endothelial counterstructures for CD11a/CD18 include ICAM-1 (intercellular adhesion molecule-1; CD54) and ICAM-2 (CD102), two members of the immunoglobulin superfamily (12,14). ICAM-1 is also recognized by CD11b/CD18, although there appear to be other as yet unidentified endothelial ligands for CD11b/CD18. ICAM-1 is expressed at low levels on resting endothelium in vitro and in vivo and is upregulated over a period of hours in response to inflammatory stimuli, such as lipopolysaccharide (LPS), IL-1$\beta$, or TNF. MAbs to ICAM-1 are effective in inhibiting neutrophil adherence to endothelium in response to inflammatory stimulation in vitro, although in general they are less effective than MAbs to CD11b or CD18.

Animal models of acute inflammation that have used MAbs to CD11b, CD18, or ICAM-1 demonstrate the importance of integrin–Ig-mediated adhesion to the process of neutrophil-mediated inflammation and injury (reviewed in 1). The first in vivo study of this nature used intravital microscopy to examine neutrophil–endothelial adhesion and neutrophil-mediated injury in response to superfusion of a chemoattractant (20). A blocking CD18 MAb was shown to effectively abolish neutrophil adherence to endothelium at the site of inflammatory stimulation. Associated with blockade of adherence, neutrophil emigration and endothelial injury, manifested as plasma leakage, were also effectively inhibited. MAbs to CD11b or ICAM-1 have produced similar findings, further demonstrating the central role of the CD11/CD18–ICAM-1 interactions in neutrophil-mediated inflammation and injury.

Another member of the Ig superfamily, platelet-endothelial cell adhesion molecule-1 [PECAM-1 (CD31)] has recently been immunologically identified and molecularly cloned (21). Found on the surface of platelets, at endothelial intercellular junctions, and on cells of myeloid lineage, it has been shown to play a major role in neutrophil–endothelial adhesion, particularly in the process of neutrophil transendothelial migration (22,23).

### Selectin–Carbohydrate Adhesion

Selectin receptors and their carbohydrate counterstructures are the other major category of adhesion molecules involved in neutrophil–endothelial adhesion (15). Recent in vitro and in vivo evidence suggests that selectin-mediated adhesion plays an important role in the initial rolling of neutrophils along the endothelium at sites of inflammation. The three selectin receptors are L-selectin (CD62L; previously also known as LAM-1, Leu-8, or LECAM-1), E-selectin (CD62E; previously known as ELAM-1), and P-selectin (CD62P; previously known as GMP-140, PADGEM, or CD62). E-Selectin is restricted to endothelial cells, P-selectin is found on platelets and endothelial cells, and L-selectin is expressed only on leukocytes. The function and regulation of selectin function are covered in other chapters, but essentially can be characterized as mediating the weak adherence that is responsible for initial leukocyte rolling along endothelium. Binding occurs between the lectin domain and the specific carbohydrate counterstructure on either the neutrophil (in the case of E- and P-selectin) or the endothelial cell (in the case of L-selectin). For E- and P-selectin receptors, the sialyl Lewis$^x$ (SLe$^x$; CD15s) antigen is a major counterstructure, as well as other sialylated or fucosylated structures (24). The importance of SLe$^x$ in the early events of neutrophil adhesion and subsequent inflammation is well established from in vivo models as well as clinical observations of individuals with clinically dysfunctional SLe$^x$ (25–27).

### Regulation of Neutrophil Adhesion

The mechanisms involved in neutrophil adherence to endothelium and subsequent neutrophil-mediated endothelial injury are complex and varied, depending on the vascular bed, the degree and nature of the inflammatory stimulus, and the time course involved. In the normal healthy state there is minimal or no adherence interaction between circulating neutrophils and endothelium. After an extravascular inflammatory stimulus, there is first rapid induction of selectin-mediated rolling at the site of inflammation, initially involving P- and/or L-selectin and later through the induction and synthesis of E-selectin. Factors released as part of the early inflammatory process, such as thrombin, histamine, and oxidants, can initiate P-selectin expression and thereby induce leukocyte rolling, whereas cytokines generated later, such as IL-1 and TNF$\alpha$, can induce L-selectin ligand or E-selectin. This initial rolling enables the neutrophils to directly contact the endothelium of the inflamed vessels where subsequent activation of integrin receptors by various agonists produces firm adherence (13,16,28).

Lipid mediators, such as PAF, chemotactic peptides, such as $C_{5a}$, or chemokines, such as IL-8, may play important roles in initiating integrin–ICAM-1-mediated adhesion. In vitro, PAF is expressed on the endothelial surface along with P-selectin in response to thrombin, histamine, or oxidizing agents. PAF can in turn cause upregulation and activation of neutrophil integrins, leading to the second step of firm adhesion. This effect can be blocked by inhibiting either P-selectin or the PAF receptor, suggesting that a coordinated or juxtacrine process is involved (29).

Other cytokines (e.g., TNF$\alpha$), bacterial products [e.g., *N*-formyl-methionyl-leucyl-phenylalanine (fMLP), LPS] and chemotaxins [e.g., leukotriene $B_4$ ($LTB_4$), C5a] can all increase the surface expression of $\beta_2$-integrin (CD11/CD18) receptors, which may be an important factor in augmenting adhesion. As mentioned previously, however, simply increasing surface expression of integrin receptors is insufficient to cause adhesion. Receptor activation is required, through a conformational change that increases adhesiveness. Phosphorylation of the cytoplasmic domain of integrin receptor subunits or binding of a lipid mediator to the intracellular region of the receptor have been suggested as potential mechanisms for producing allosteric changes (19). The precise biochemical basis of this "inside-out" signaling, however, remains to be elucidated (30).

It is therefore apparent that many factors are involved in the regulation of neutrophil–endothelial adherence interactions. As our knowledge and understanding of the molecular basis of neutrophil adherence to endothelium increase, the methods for examining and manipulating these processes in vivo will continue to expand, allowing more thorough analysis of neutrophil-mediated injury processes and their role in human disease.

## ANTI-ADHESION THERAPY IN IN VIVO MODELS OF NEUTROPHIL-MEDIATED INJURY

To date, only one phase I clinical trial using a monoclonal antibody to inhibit neutrophil–endothelial adhesion has been completed (31). There have, however, been many animal studies in which MAbs or other inhibitors directed against various adhesion receptors were used to examine the role of neutrophil-mediated injury in models of human disease. These studies have focused on two major groups of disease processes: those of an inflammatory or immune origin and those involving ischemia–reperfusion (1). This chapter focuses on studies of disease models whose basis is ischemia–reperfusion injury.

Ischemia–reperfusion injury forms the basis of many important clinical disorders, including myocardial infarction, stroke, peripheral vascular disease, organ transplantation, and circulatory shock. There is evidence that a significant proportion of the tissue damage triggered by ischemia is often a consequence of events associated with reperfusion of ischemic tissues, i.e., reperfusion injury. Many studies have focused on oxygen-derived free radicals generated at the time of reperfusion as potentially important mediators of this reperfusion injury (32,33). An important source of free radicals is the neutrophil, which in addition can cause injury through generation and release of proteases and phospholipase products. Studies have shown a close association between neutrophil accumulation and tissue injury in this setting and have also demonstrated injury reduction by depletion of circulating neutrophils, suggesting an important role for neutrophils in ischemia–reperfusion injury (34).

There are several mechanisms by which activated neutrophils can cause injury in the setting of ischemia-reperfusion. First, adherent, activated neutrophils can cause direct endothelial injury, resulting in loss of vascular integrity, edema, hemorrhage, and thrombosis (Fig. 1). Another possible mechanism involves occlusion and further ischemia resulting from adherence and accumulation of aggregates of neutrophils within the vessel lumen (10). In this way, a vicious cycle may be triggered in which reperfusion induces neutrophil activation and adherence, leading to leukocyte accumulation and endothelial injury which then results in further ischemia and eventually complete cessation of flow. This may be the basis of what has in the past been called the "no-reflow phenomenon" which, more precisely, might be called a "diminishing reflow phenomenon" (35).

The mechanisms that initiate neutrophil adherence and injury in the setting of ischemia–reperfusion have not been fully elucidated. It is known, however, that oxygen-derived free radicals are generated at the time of reperfusion (32,33). These free radicals can then initiate the process of selectin-mediated adhesion, release of PAF, and subsequent integrin-mediated adhesion (36). The potential role of oxygen free radicals in initiating this process has been demonstrated both in vitro and in vivo (37,38).

MAbs directed against specific adhesion proteins are very effective in inhibiting neutrophil accumulation and neutrophil-mediated injury in a number of animal models of reperfusion injury. The first such study used a CD18 MAb to examine the role of neutrophils in feline gut reperfusion injury (39).

Antibody-mediated adherence blockade significantly reduced the increase in plasma leakage that normally occurs after intestinal ischemia and reperfusion.

A model utilizing the rabbit ear, isolated on its central vascular pedicle, demonstrated that blocking neutrophil adherence with anti-CD18 MAb (35) markedly reduced reperfusion-associated edema formation as well as tissue necrosis (Figs. 2 and 3). Of note in this study was that the degree of protection was similar whether antibody was administered before ischemia or after ischemia but immediately before reperfusion. This finding indicates that neutrophil-mediated injury in this model of ischemia and reperfusion occurs at the time of reperfusion as activated neutrophils flood the vascular bed, causing diffuse endothelial and tissue injury. The injury, therefore, is in fact a true reperfusion injury. Recent studies indicate that significant protection can still be achieved when administration of blocking antibody is delayed for up to 4 h after reperfusion (40,41).

Similar results have also been obtained with anti–P-selectin antibody (42) and with anti–L-selectin antibody (43). In these latter studies, P-selectin surface expression was shown to increase soon after reperfusion, and blockade with P-selectin antibody effectively inhibited neutrophil adherence and accumulation, as well as the associated tissue injury. These results are consistent with the theory that leukocyte rolling precedes firm attachment, that rolling is mediated by either L-selectin and/or P-selectin in ischemia–reper-

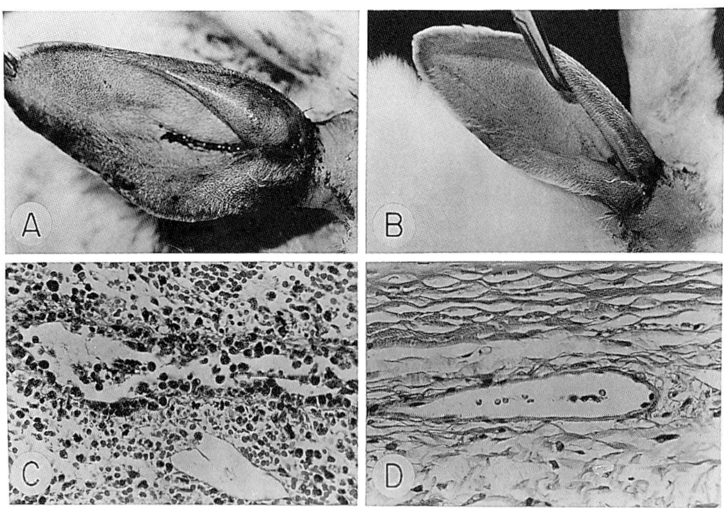


**FIG. 2.** A comparison of tissue injury in rabbits treated with saline (control-ischemic; **A,C**) or with anti-CD18 MAb ("Post"; **B,D**) just before reperfusion after 10 h of complete ischemia. This comparison was made 24 h after initial ischemia. (From ref. 35 with permission.)

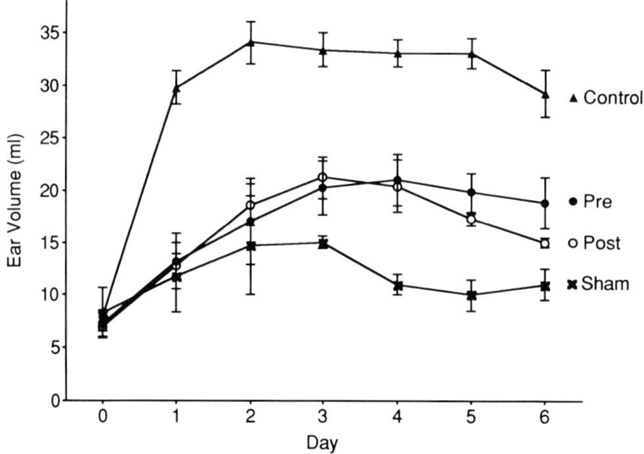


**FIG. 3.** α-CD18 monoclonal antibody attenuates edema formation after 10 h of ischemia followed by reperfusion. Days post reperfusion are indicated on the abscissa and ear volume is indicated on the ordinate. Edema was quantified by volume displacement. Mean values for sham-operated animals are represented by an "x." Values for control-ischemic animals are represented by closed triangles. Values for animals that received antibody before ischemia and again before reperfusion (Pre) are represented by closed circles. Values for animals that received antibody only before reperfusion (Post) are represented by open circles. Ear volumes in both antibody pretreated and antibody post-treated animals were significantly less than in control-ischemic animals ($p < 0.005$ by ANOVA). (From ref. 35 with permission.)

fusion, and that blocking either of these molecules or CD18 is sufficient to prevent reperfusion injury.

Inhibition of neutrophil–endothelial adhesion has also been shown to be effective at attenuating reperfusion injury in myocardium, either by inhibiting CD11/CD18 (44,45) or by inhibiting selectin–carbohydrate-mediated adhesion (46–48).

In models of ischemic brain injury, inhibition of neutrophil–endothelial adhesion has also been effective under certain conditions. In a model of embolic stroke, both CD18 (49) and ICAM-1 (50) inhibition have been shown to reduce experimental brain injury. Both ICAM-1 and P-selectin expression have been shown to increase significantly after focal brain ischemia and reperfusion (51).

Blocking CD18-mediated adherence has also been proven effective in reducing permeability edema in experimental skeletal muscle ischemia–reperfusion models (52). Even the remote effects of neutrophil-mediated reperfusion injury can potentially be inhibited in this way, as has been shown with the reduction of remote lung injury after hindlimb ischemia–reperfusion by blocking either CD18–ICAM-mediated or selectin–carbohydrate-mediated adherence (53–55). This concept of neutrophil-mediated remote organ injury after focal organ ischemia–reperfusion has also been extended to gut ischemia–reperfusion and to remote lung and liver injury (56,57).

Hemorrhagic shock followed by resuscitation can be viewed as a global or generalized ischemia–reperfusion injury. Studies based on this hypothesis have shown that inhibition of CD18-mediated neutrophil adherence and injury effectively reduces organ injury, attenuates the generalized microvascular injury, and improves survival in both small animal and subhuman primate models of hypovolemic shock (58,59). This approach is equally effective whether treatment is given before shock or whether it is begun after shock but before reperfusion, suggesting that the injury that follows shock and resuscitation is also a reperfusion injury (60). The control animals in these studies developed severe hemorrhagic gastritis and diffuse microvascular leakage, requiring large volumes of resuscitation fluid to maintain cardiac output. Administration of a single dose of CD18 MAb at the time of resuscitation almost completely prevented both of these events and led to improved survival. These findings also suggest that neutrophil-mediated injury plays an important role in the development of multiple organ injury and the systemic inflammatory response syndrome (SIRS) which is a common and devastating consequence of severe traumatic shock.

This concept is further supported by a recent epidemiologic study at a level I trauma center, which showed that hypotension is associated with more than 90% of trauma-related deaths and that the subpopulation of trauma patients with hypotension have a 50% mortality. These data compare with a nonhypotensive group matched for multiple criteria, including Injury Severity Score, who had only a 4% mortality, suggesting that hypotension alone is a powerful independent predictor of poor outcome and that blocking the inflammatory consequences of shock and resuscitation has the potential to significantly improve outcome in this setting (Heckbert et al., unpublished data).

## THERAPEUTIC POTENTIAL

In view of the wide array of important disease processes in which anti-adhesion therapy has shown stunning protective efficacy in animal models, it is intriguing to speculate about the therapeutic potential of such an approach in the clinical setting. At present a number of clinical trials are either ongoing or planned that will examine the efficacy and safety of anti-adhesion therapy. These studies will employ either murine or humanized monoclonal antibodies directed to CD11/CD18, ICAM-1, and the selectins or their counterstructures, or will use small molecules to inhibit either integrin–ICAM-1- or selectin-mediated adhesion (61).

In considering such an approach, one must keep in mind that when neutrophil adherence functions are inhibited the risk of infection must be carefully weighed against any potential benefit of this form of therapy. Studies have shown that blocking CD18-mediated functions does not increase mortality or infectious complications in models of bacterial peritonitis (62) and can actually improve outcome in certain septic models, such as bacterial meningitis (63) and Gram-negative sepsis (64). It has also been shown, however, that with extremely high inocula of subcutaneous bacteria there can be an increase in infectious complications and mortality even after a single dose of CD18 MAb, although this effect is not seen with lower, clinically relevant inocula of bacteria (65). Of interest is that this increase in soft tissue infection susceptibility is *not* increased by P-selectin inhibition (66).

Although these studies suggest that it may be safe to inhibit neutrophil adherence functions for a limited period of time in a carefully defined setting, they also raise concern about completely inhibiting such a basic physiologic process. It stands to reason that the more acute disease processes, such as those related to ischemia–reperfusion, may perhaps be more amenable to and more appropriate for controlled, transient inhibition of adherence.

## REFERENCES


1. Harlan JM, Winn RK, Vedder NB, et al. *In vivo* models of leukocyte adherence to endothelium. In: Harlan JM, Lui D, eds. *Adhesion: its role in inflammatory disease*. New York: WH Freeman, 1992:117–50.
2. Malech HL, Gallin JI. Current concepts: immunology. Neutrophils in human diseases. *N Engl J Med* 1987;317:687–94.
3. Weiss SJ. Tissue destruction by neutrophils. *N Engl J Med* 1989;320:365–76.
4. Henson PM, Johnston RB Jr. Tissue injury in inflammation: oxidants, proteinases, and cationic proteins. *J Clin Invest* 1987;79:669–74.
5. Metchnikoff E. Sur lallutte des cellules de l'organisme contre l'invasion des microbes. *Ann Inst Pasteur* 1887;1:321.
6. Grant L. The sticking and emigration of white blood cells in inflammation. In Zweifach BL, McCluskey L, eds. *The inflammatory process,* Vol. 2. Orlando, FL: Academic Press, 1973:205–49.
7. Movat HZ. The role of endothelium in leukocytes emigration: the views of Cohnheim, Metchnikoff and their contemporaries. *Pathol Immunopathol Res* 1989;8:35–41.
8. Harlan JM. Neutrophil-mediated vascular injury. *Acta Med Scand* 1987(suppl 715):123–9.
9. Diener AM, Beatty PG, Ochs HD, et al. The role of neutrophil membrane glycoprotein 150 (GP-150) in neutrophil-mediated endothelial cell injury in vitro. *J Immunol* 1985;135:537–43.
10. Schmid-Schönbein GW. Capillary plugging by granulocytes and the no-reflow phenomenon in the microcirculation. *Fed Proc* 1987;46:2397–2401.
11. Carlos TM, Harlan JM. Membrane proteins involved in phagocyte adherence to endothelium. *Immunol Rev* 1990;114:5–28.
12. Springer TA. Adhesion receptors of the immune system. *Nature* 1990;346:425–34.
13. Butcher EC. Leukocyte-endothelial cell recognition: three (or more) steps to specificity and diversity. *Cell* 1991;67:1033–6.
14. Lobb RR. Integrin-immunoglobulin superfamily interactions in endothelial–leukocyte adhesion. In: Harlan JM, Liu DY, eds. *Adhesion: its role in inflammatory disease.* New York: WH Freeman, 1992:1–18.
15. Bevilacqua MP, Nelson RM. Selectins. *J Clin Invest* 1993;91:379–87.
16. Lawrence MB, Springer TA. Leukocytes roll on a selectin at physiologic flow rates: distinction from and prerequisite for adhesion through integrins. *Cell* 1991;65:859–73.
17. Smith CW, Marlin SD, Rothlein R, et al. Cooperative interactions of LFA-1 and Mac-1 with

intercellular adhesion molecule-1 in facilitating adherence and transendothelial migration of human neutrophils in vitro. *J Clin Invest* 1989;83:2008–17.

18. Vedder NB, Harlan JM. Increased surface expression of CD11b/CD18 (Mac-1) is not required for stimulated neutrophil adherence to cultured endothelium. *J Clin Invest* 1988;81: 676–82.
19. Hermanowski-Vosatka A, Van-Strijp JA, Swiggard WJ, et al. Integrin modulating factor-1: a lipid that alters the function of leukocyte integrins. *Cell* 1992;68:341–52.
20. Arfors K-E, Lundberg C, Lindbom L, et al. A monoclonal antibody to the membrane glycoprotein complex CDw18 (LFA), inhibits PMN accumulation and plasma leakage in vivo. *Blood* 1987;69:338–40.
21. Newman PJ, Berndt MC, Gorski J, et al. PECAM-1 (CD31) cloning and relation to adhesion molecules of the immunoglobulin gene superfamily. *Science* 1990;247:1219–22.
22. Vaporciyan AA, DeLisser HM, Yan HC, et al. Involvement of platelet-endothelial cell adhesion molecule-1 in neutrophil recruitment in vivo. *Science* 1993;262:1580–2.
23. Muller WA, Weigl SA, Deng X, et al. PECAM-1 is required for transendothelial migration of leukocytes. *J Exp Med* 1993;178:449–60.
24. Polley MJ, Phillips ML, Wayner E, et al. CD62 and endothelial cell-leukocyte adhesion molecule 1 (ELAM-1) recognize the same carbohydrate ligand, sialyl-Lewis x. *Proc Natl Acad Sci USA* 1991;88:6224–8.
25. von-Andrian UH, Berger EM, Ramezani L, et al. In vivo behavior of neutrophils from two patients with distinct inherited leukocyte adhesion deficiency syndromes. *J Clin Invest* 1993;91:2893–7.
26. von-Andrian UH, Chambers JD, Berg EL, et al. L-selectin mediates neutrophil rolling in inflamed venules through sialyl LewisX-dependent and -independent recognition pathways. *Blood* 1993;82:182–91.
27. Etzioni A, Frydman M, Pollack S, et al. Brief report: recurrent severe infections caused by a novel leukocyte adhesion deficiency. *N Engl J Med* 1992;327:1789–92.
28. von Andrian UH, Hansell P, Chambers JD, et al. L-selectin function is required for beta 2-integrin-mediated neutrophil adhesion at physiological shear rates in vivo. *Am J Physiol* 1992;263:H1034–44.
29. Lorant DE, Topham MK, Whatley RE, et al. Inflammatory roles of P-selectin. *J Clin Invest* 1993;92:559–70.
30. O'Toole TE, Katagiri Y, Faull RJ, et al. Integrin cytoplasmic domains mediate inside-out signal transduction. *J Cell Biol* 1994;124:1047–59.
31. Haug CE, Colvin RB, Delmonico FL, et al. A phase I trial of immunosuppression with anti-ICAM-1 (CD54) mAb in renal allograft recipients. *Transplantation* 1993;55:766–72.
32. Granger DN, Hollwarth ME, Parks DA. Ischemia-reperfusion injury: role of oxygen-derived free radicals. *Acta Physiol Scand* 1986;548(suppl):47–63.
33. McCord JM. Oxygen-derived radicals: a link reperfusion injury and inflammation. *Fed Proc* 1987;46:2402–6.
34. Romson JL, Hook BG, Kunkel SL, et al. Reduction of the extent of ischemic myocardial injury by neutrophil depletion in the dog. *Circulation* 1983;67:1016–23.
35. Vedder NB, Winn RK, Rice CL, et al. Inhibition of leukocyte adherence by anti-CD18 monoclonal antibody attenuates reperfusion injury in the rabbit ear. *Proc Natl Acad Sci USA* 1990;87:2643–6.
36. Patel KD, Zimmerman GA, Prescott SM, et al. Oxygen radicals induce human endothelial cells to express GMP-140 and bind neutrophils. *J Cell Biol* 1991;112:749–59.
37. Kubes P, Ibbotson G, Russell J, et al. Role of platelet-activating factor in ischemia/reperfusion-induced leukocyte adherence. *Am J Physiol* 1990;259:G300–5.
38. Suzuki M, Inauen W, Kvietys PR, et al. Superoxide mediates reperfusion-induced leukocyte-endothelial cell interactions. *Am J Physiol* 1989;257:H1740–5.
39. Hernandez LA, Grisham MB, Twohig B, et al. Role of neutrophils in ischemia-reperfusion-induced microvascular injury. *Am J Physiol* 1987;253:H699–703.
40. Sharar SR, Mihelcic DD, Han K-T, et al. Ischemia reperfusion injury in the rabbit ear is reduced by both immediate and delayed CD18 leukocyte adherence blockade. *J Immunol* 1994;153:2234–8.
41. Kelly KJ, Williams WW Jr, Colvin RB, et al. Antibody to intercellular adhesion molecule 1 protects the kidney against ischemic injury. *Proc Natl Acad Sci USA* 1994;91:812–6.

42. Winn RK, Liggitt D, Vedder NB, et al. Anti-P-selectin monoclonal antibody attenuates reperfusion injury to the rabbit ear. *J Clin Invest* 1993;92:2042–7.
43. Mihelcic D, Schleiffenbaum B, Tedder TF, et al. Inhibition of leukocyte L-selectin function with a monoclonal antibody attenuates reperfusion injury to the rabbit ear. *Blood* 1994;84:2322–8.
44. Simpson PJ, Todd RF III, Fantone JC, et al. Reduction of experimental canine myocardial reperfusion injury by a monoclonal antibody (anti-Mo1, anti-CD11b) that inhibits leukocyte adhesion. *J Clin Invest* 1988;81:624–9.
45. Ma XL, Tsao PS, Lefer AM. Antibody to CD-18 exerts endothelial and cardiac protective effects in myocardial ischemia and reperfusion. *J Clin Invest* 1991;88:1237–43.
46. Weyrich AS, Ma XY, Lefer DJ, et al. In vivo neutralization of P-selectin protects feline heart and endothelium in myocardial ischemia and reperfusion injury [see comments]. *J Clin Invest* 1993;91:2620–9.
47. Ma XL, Weyrich AS, Lefer DJ, et al. Monoclonal antibody to L-selectin attenuates neutrophil accumulation and protects ischemic reperfused cat myocardium. *Circulation* 1993; 88:649–58.
48. Buerke M, Weyrich AS, Zheng Z, et al. Sialyl Lewis x-containing oligosaccharide attenuates myocardial reperfusion injury in cats. *J Clin Invest* 1994;93:1140–8.
49. Clark WM, Madden KP, Rothlein R, et al. Reduction of central nervous system ischemic injury in rabbits using leukocyte adhesion antibody treatment. *Stroke* 1991;22:877–83.
50. Clark WM, Madden KP, Rothlein R, et al. Reduction of central nervous system ischemic injury by monoclonal antibody to intercellular adhesion molecule. *J Neurosurg* 1991;75: 623–7.
51. Okada Y, Copeland BR, Mori E, et al. P-selectin and intercellular adhesion molecule-1 expression after focal brain ischemia and reperfusion. *Stroke* 1994;25:202–11.
52. Carden DL, Smith JK, Korthuis RJ. Neutrophil-mediated microvascular dysfunction in postischemic canine skeletal muscle. Role of granulocyte adherence. *Circ Res* 1990;66: 1436–44.
53. Seekamp A, Mulligan MS, Till GO, et al. Role of beta 2 integrins and ICAM-1 in lung injury following ischemia-reperfusion of rat hind limbs. *Am J Pathol* 1993;143:464–72.
54. Seekamp A, Till GO, Mulligan MS, et al. Role of selectins in local and remote tissue injury following ischemia and reperfusion. *Am J Pathol* 1994;144:592–8.
55. Welbourn CR, Goldman G, Paterson IS, et al. Pathophysiology of ischaemia reperfusion injury: central role of the neutrophil. *Br J Surg* 1991;78:651–5.
56. Hill J, Lindsay T, Valeri CR, et al. A CD18 antibody prevents lung injury but not hypotension after intestinal ischemia-reperfusion. *J Appl Physiol* 1993;74:659–64.
57. Hill J, Lindsay T, Rusche J, et al. A Mac-1 antibody reduces liver and lung injury but not neutrophil sequestration after intestinal ischemia-reperfusion. *Surgery* 1992;112:166–72.
58. Vedder NB, Winn RK, Rice CL, et al. A monoclonal antibody to the adherence-promoting leukocyte glycoprotein, CD18, reduces organ injury and improves survival from hemorrhagic shock in rabbits. *J Clin Invest* 1988;81:939–44.
59. Mileski WJ, Winn RK, Vedder NB, et al. Inhibition of CD18-dependent neutrophil adherence reduces organ injury after hemorrhagic shock in primates. *Surgery* 1990;108:206–12.
60. Vedder NB, Fouty BW, Winn RK, et al. Role of neutrophils in generalized reperfusion injury associated with resuscitation from shock. *Surgery* 1989;106:509–16.
61. Granert C, Raud J, Xie X, et al. Inhibition of leukocyte rolling with polysaccharide fucoidin prevents pleocytosis in experimental meningitis in the rabbit [see comments]. *J Clin Invest* 1994;93:929–36.
62. Mileski WJ, Winn RK, Harlan JM, et al. Transient inhibition of neutrophil adherence with the anti-CD18 monoclonal antibody 60.3 does not increase mortality rates in abdominal sepsis. *Surgery* 1991;109:497–501.
63. Tuomanen EI, Saukkonen K, Sande S, et al. Reduction of inflammation, tissue damage, and mortality in bacterial meningitis in rabbits treated with monoclonal antibodies against adhesion-promoting receptors of leukocytes. *J Exp Med* 1989;170:959–69.
64. Thomas JR, Harlan JM, Rice CL, et al. Role of leukocyte CD11/CD18 complex in endotoxic and septic shock in rabbits. *J Appl Physiol* 1992;73:1510–6.
65. Sharar SR, Winn RK, Murry CE, et al. A CD18 monoclonal antibody increases the incidence and severity of subcutaneous abscess formation after high-dose Staphylococcus aureus injection in rabbits. *Surgery* 1991;110:213–9.

66. Sharar SR, Sasaki SS, Flaherty LC, et al. P-selectin blockade does not impair leukocyte host defense against bacterial peritonitis and soft tissue infection in rabbits. *J Immunol* 1993;151:4982–8.
67. Vedder NB, Harlan JM. Neutrophil–endothelial cell interactions. In: Shoemaker WC, Ayres SM, Grenvik A, Holbrook PR, eds. *Textbook of critical care*. Philadelphia: WB Saunders, 1995 [*in press*].

*Topics in Molecular Medicine, Volume 1,*
edited by Wolfgang Siess, Reinhard Lorenz,
and Peter C. Weber. Raven Press, Ltd.,
New York © 1995.

# 13

# VCAM-1 Function in Atherogenesis and Embryonic Development

Myron I. Cybulsky

*Vascular Research Division, Departments of Pathology, Brigham and Women's Hospital and Harvard Medical School, Boston, Massachusetts 02115*

There is increasing evidence that the vascular endothelial lining is the site of pathogenically relevant changes in the blood vessel wall that occur in response to conditions such as hypercholesterolemia, viral or bacterial infections, exposure to foreign antigens (immune reactions) or bacterial products, and mechanical perturbation. These changes, termed *endothelial activation* (1), are often the result of local cytokine production and may result in acute or chronic alterations in a variety of vessel wall functions. A characteristic feature associated with endothelial activation in vivo is the emigration of leukocytes from the circulation into the vessel wall or extravascular tissues.

## IDENTIFICATION OF VCAM-1 AND ITS COUNTER-RECEPTOR

Human VCAM-1 (also designated INCAM-110) was identified in activated human umbilical vein endothelial cells by monoclonal antibodies (MAbs) (2,3) and by expression cloning (4). Cytokine activation of endothelium and induction of VCAM-1 expression was key to both strategies. Activated endothelium was used to immunize and screen for MAbs that recognized cytokine-inducible epitopes, unique proteins were identified by immunoprecipitation, and leukocyte adhesive function by MAb adhesion-blocking assays. Expression cloning utilized a subtracted cytokine-activated human umbilical vein endothelial cell library packaged in a eukaryotic expression vector and a novel screening assay in which cDNAs were transfected into COS cells. Cells exhibiting increased leukocyte adhesion were chosen and plasmid DNA was extracted, amplified, and rescreened (4). The immunoglobulin-like structure of VCAM-1 (see below) and similarity to ICAM-1 provided a clue that its counter-receptor may be an integrin. Subsequent experiments determined that VCAM-1 mediates intercellular adhesion via interaction with VLA-4 (CD49d/CD29), an $\alpha_4\beta_1$ integrin, which is expressed on monocytes, lymphocytes, basophils, eosinophils, and certain tumor cells, but not on neutrophils (5).

## VCAM-1 STRUCTURE AND GENOMIC ORGANIZATION

Molecular cloning of human VCAM-1 revealed that this molecule is a transmembrane protein and a member of the immunoglobulin gene superfamily (4). Initially, it was reported to contain six extracellular C2 or H-type immunoglobulin-like domains (4). However, we and others have isolated cDNAs from cytokine-activated endothelium that contained an additional immunoglobulin-like domain, designated domain 4 (the remaining domains were designated 5–7) (6–8). Both transcript forms of VCAM-1 were detected in interleukin-1β (IL-1β)–stimulated human umbilical vein endothelial cells by polymerase chain reaction, although the domain 7 form was markedly predominant and was the only form detected on the activated endothelial cell surface by immunoprecipitation (6). Cloning of the human *VCAM1* gene determined that the 6 and 7 immunoglobulin-like domain forms arise by alternative mRNA splicing (9).

Genomic clones encoding the human and murine *VCAM1* genes revealed that each of the extracellular immunoglobulin-like domains is contained in a separate exon and that all splice junctions occur after the first nucleotide of a codon (type 1) (9,10). The striking conservation of phase 1 introns in the *VCAM1* gene suggests that exons could be spliced in or out of the mRNA without disturbing the reading frame. The structure of the *VCAM1* gene is consistent with the proposal that exon duplication plays a role in the formation of the gene. In VCAM-1, the immunoglobulin-like domains share extensive internal homology. Domains 1, 2, and 3 are highly homologous to domains 4, 5, and 6 (55% to 75% amino acid identity, highest between domains 1 and 4). In addition, the sizes of the introns in this region are similar. This internal homology suggests that the region of the *VCAM1* gene represented by domains 1, 2, and 3 may have been duplicated during evolution to generate domains 4, 5, and 6.

The alternatively spliced forms of VCAM-1 have different adhesive functional properties. The three amino-terminal domains of VCAM-1 are sufficient to support VLA-4–dependent adhesion, determined with a chimeric protein consisting of domains 1 to 3 of VCAM-1 linked to an immunoglobulin heavy chain (11). Because of the homology of domains 1–3 to domains 4–6, the 7 domain form of VCAM-1 has two VLA-4–binding sites, whereas the 6 domain form, lacking domain 4, has only one (12).

The murine *VCAM1* gene contains an additional exon (designated PI exon), unique to rodents, located between domains 3 and 4, which contains a stop codon and a polyadenylation signal but no transmembrane domain.

When utilized by alternative splicing, the resulting VCAM-1 molecule contains three Ig-like domains (domains 1–3), is attached to the cell membrane by a phosphatidylinositol linkage, and binds only one VLA-4 molecule (10, 13,14). In rabbits, VCAM-1 has seven or eight immunoglobulin-like domains, and both forms are expressed at approximately equal levels (M. I. Cybulsky, unpublished observations). The eighth domain is homologous to domains 3, 6, and 7 and is located between the seventh immunoglobulin-like and the transmembrane domain.

## EXPRESSION AND FUNCTION OF VCAM-1 ON CULTURED VASCULAR ENDOTHELIUM

On unactivated, cultured human umbilical vein endothelium, VCAM-1 expression is low or absent. Cell surface expression is rapidly upregulated to maximum at 12 to 24 h by cytokines IL-1, tumor necrosis factor (TNF), and IL-4, or by bacterial lipopolysaccharide (LPS), then gradually declines (2,3). IL-4 induces low levels of VCAM-1 and, in combination with TNF, IL-4 enhances VCAM-1 and partly suppresses ICAM-1 and E-selectin expression (15,16). Recently, we observed that lysophosphatidylcholine, a component of oxidized low-density lipoprotein, upregulates VCAM-1 expression in arterial but not venous endothelium (17). This mechanism for VCAM-1 induction may be relevant to foam-cell–lesion formation in the context of hypercholesterolemia.

In static adhesion assays, VCAM-1–specific MAbs block a significant component of VLA-4–expressing leukocyte adhesion to activated endothelium (monocytes, lymphocytes, basophils, and eosinophils) (2,3,18–23). Similar data are obtained in assays utilizing VCAM-1–transfected cells. VCAM-1 also mediates adhesion of CD11/CD18-deficient lymphocytes (24,25). The contribution of VCAM-1 to monocyte adhesion under nonstatic conditions (in vitro rotation assay) appears less substantial than in static assays and L-selectin interaction with its inducible endothelial counter-receptor plays a major role (26). This may indicate that VLA-4 function is upregulated (27) under static conditions, allowing interaction with VCAM-1, but not when constant motion prevents prolonged monocyte–endothelial contact. A similar phenomenon is observed with the $\beta_2$ integrin CD11a/CD18 (LFA-1), for which activation of lymphocytes by phorbol ester or other treatments increases binding to ICAM-1 (28,29).

## EXPRESSION OF VCAM-1 IN TISSUES

In normal tissues, endothelial cell VCAM-1 expression is usually absent, although occasionally blood vessels show positive immunohistochemical staining (4,30: M. I. Cybulsky, unpublished observations). VCAM-1 is constitutively expressed on some epithelium and on monocyte-derived cells, including dendritic cells in lymphoid tissues and skin and Kupffer cells in liver (30). Cultured bone marrow stromal cells and skeletal muscle cells during myogenesis also express VCAM-1 (31–33). In human inflammatory conditions, including acute appendicitis, acute diverticulitis, sarcoidosis, cat scratch lymphadenitis, a variety of dermatoses (30), rheumatoid and osteoarthritis (35), allograft rejection (35), and atherogenesis (see below), expression of VCAM-1 is induced on vascular endothelium. VCAM-1 expression can also be induced in nonvascular cells, including mesothelial, synovial, and renal tubule epithelial cells (30,34,36).

## POTENTIAL PATHOPHYSIOLOGIC ROLES OF VCAM-1

The temporal and spatial patterns of VCAM-1 expression suggest that this leukocyte adhesion molecule may be important in a variety of pathophysiologic settings. As implied above, VCAM-1 expressed on activated vascular endothelium may participate in the emigration of VLA-4–bearing leukocytes during infections and inflammatory processes. In lymphoid germinal centers, VCAM-1 expressed on dendritic cells can mediate B-lymphocyte adhesion (37) and may participate in the organization of lymphoid follicles. On antigen-presenting cells (dendritic and endothelial cells), VCAM-1 may act as a T-lymphocyte co-stimulatory molecule (38–40). Other potential functions of VCAM-1 include maturation of hematopoietic cells in the bone marrow (31,32) and myogenesis (33).

## VCAM-1 IN ATHEROGENESIS

Careful ultrastructural and immunohistochemical studies in various animal models and human tissues have established that the adherence of blood monocytes and lymphocytes to endothelial cells lining large arteries is one of the earliest detectable events in atherosclerosis (reviewed in 41,42). The subsequent transendothelial migration of monocytes, their accumulation in the intima, and their development into lipid-engorged "foam cells" appear to be important steps in the initiation of atherosclerotic lesions. Monocytes and macrophage foam cells may also contribute to the progression of atherosclerotic lesions by producing cytokines and growth factors (41,42). These, in turn, may amplify mononuclear leukocyte recruitment, induce migration of smooth-muscle cells into the intima, and stimulate cell replication.

The formation of foam-cell–rich lesions during hypercholesterolemia appears to be a highly regulated process during which the vascular endothelium remains intact and may participate in regulating leukocyte recruitment into the intima by expressing leukocyte adhesion molecules. Using a combination of cell biologic and experimental pathologic approaches, we identified a leukocyte adhesion molecule in the rabbit, which appears to be directed to mononuclear leukocytes and is expressed selectively by arterial endothelial cells covering early foam cell lesions of both dietary and Watanabe heritable hyperlipidemic rabbits (43). Immunoaffinity purification and N-terminal sequencing revealed that the molecule was rabbit VCAM-1. VCAM-1 expression in endothelium over foam-cell lesions was not uniform. Expression appeared elevated particularly at edges of lesions and extending several cells beyond the edge. In this region, others demonstrated by scanning electron microscopy that mononuclear leukocyte recruitment through an intact endothelial monolayer was increased and was presumably contributing to lateral expansion of lesions (44). The induction of endothelial VCAM-1 expression was an early event, occurring approximately 1 week after the initiation of a hypercholesterolemic diet in rabbits and preceding detectable intimal monocyte–macrophage accumulation (45). Endothelium not involved by foam-cell lesions did not express VCAM-1 (43,45). In normocholesterolemic rabbits, VCAM-1 was not expressed by aortic endothelium except at sites that are preferential for foam-cell–lesion formation: the aortic arch and downstream of artery ostia. In addition to VCAM-1, ICAM-1 was expressed in endothelium uniformly over lesions. However, unlike VCAM-1, ICAM-1 expression extended widely into noninvolved regions (46). E-Selectin expression was minimal (46).

VCAM-1 was also expressed within neointimal cells with morphologic features consistent with smooth-muscle cells and in the medial smooth-muscle cells adjacent to the internal elastic lamina (46,47). To verify that vascular smooth-muscle cells can express VCAM-1, these cells were cultured from rabbit and human arteries and VCAM-1 expression was induced after activation with appropriate cytokines (47). The pathophysiologic function of VCAM-1 in smooth-muscle cells remains unknown. One possibility is that VCAM-1 may serve as a marker of an activation–differentiation state, without an important biologic function. Alternatively, VCAM-1 may promote the retention of mononuclear leukocytes within atherosclerotic lesions. However, the predominant location of VCAM-1 expression in smooth-muscle cells, near the surface of lesions, at their base, and in the upper media, does not correlate with mononuclear leukocyte distribution. VCAM-1 on smooth-muscle cells may interact with VLA-4 and activate mononuclear leukocytes, thus initiating cytokine cascades and promoting matrix deposition or, alternatively, protease production leading to matrix degradation.

Several recent studies utilized immunohistochemistry to examine the expression patterns of leukocyte adhesion molecules in human atherosclerotic plaques. ICAM-1 expression was found on endothelial cells, smooth-muscle cells, and macrophages (48–53). E-Selectin expression was restricted to endothelium (50–52). In advanced human coronary artery plaques, VCAM-1 was expressed focally by luminal endothelial cells, usually in association with inflammatory infiltrates (52,53). Focal endothelial VCAM-1 expression was also found in uninvolved vessels with diffuse intimal thickening. Within plaques, VCAM-1 was expressed by subsets of smooth-muscle cells and macrophages and by endothelial cells of the neovasculature. The variability of VCAM-1 expression in human atherosclerotic lesions, apart from possible technical difficulties with detection, may reflect states of plaque activity or quiescence with regard to leukocyte recruitment. In contrast to rabbit models, in which relatively high levels of hypercholesterolemia are maintained by an atherogenic diet and intimal lesion growth is likely, humans usually have low levels of hypercholesterolemia, and human plaque expansion as a result of leukocyte recruitment may develop at variable rates.

## VCAM-1 KNOCKOUT MICE

Previously, aspects of VCAM-1 biology have been investigated with function-blocking MAbs. However, this approach has many limitations, particularly in studies of chronic disease processes such as atherosclerosis. An alternative approach is to utilize a genetic strategy and develop mice in which VCAM-1 expression is absent. With this goal in mind, we have produced VCAM-1 knockout mice (54) using the now standard approach of targeted homologous recombination in embryonic stem (ES) cells (55–57). In the targeting construct, the first Ig-like domain of VCAM-1, which is essential for function, was deleted and replaced with a neomycin-resistance cassette. Mice heterozygous for the disrupted VCAM-1 allele developed normally in three backgrounds: pure 129 strain, mixed 129–BALB/c, and mixed 129–C57BL/6. Homozygous mice (VCAM-1 knockout) die during embryonic development at midgestation, indicative of a recessive lethal phenotype.

Timed pregnancies (76 litters comprising more than 650 embryos at gestational ages 7.5 to 12.5 days post-coitus) revealed at 10.5 days an alteration in the expected Mendelian distribution (determined by Southern blotting or polymerase chain reaction of yolk sac DNA). At 9.5 days, a distinctive altered phenotype was present in VCAM-1 knockout mice, consisting of failure in fusion of the allantois to the chorion, with resulting hydropic expansion of the allantois (Fig. 1). After fusion of the allantois and chorion in wild-type mice, these structures form the umbilical cord and fetal placenta. In VCAM-1 knockout mice, the resulting malformations lead to embryonic death and resorption within 1 to 2 days. Immunohistochemical staining of wild-type embryos at 8.0 to 8.5 days revealed expression of VCAM-1 on the tip of the allantois and the integrin $\alpha_4$ (VLA-4), the ligand for VCAM-1, in the chorion.

In approximately 20% of VCAM-1 knockout embryos, allantoic fusion occurs despite the absence of VCAM-1. However, these embryos show developmental delay in many tissues and die before 12.5 days of gestation. This second defect probably results from delayed placental development and

**FIG. 1.** Mouse embryo littermates at 9.5 days of gestation. On the right, a VCAM-1 −/− (knockout) embryo shows failure of allantoic fusion with resulting hydropic expansion. The embryo is turned and no other abnormalities are evident. The left embryo, VCAM-1 +/+ (wild-type), is developing normally. Its allantois was fused to the chorion and was dissected during removal.

insufficient gas or nutrient exchange. We attribute this defect to a role of VCAM-1 in the migration of allantoic mesoderm over the entire chorionic surface. In wild-type embryos, the allantoic mesoderm forms the fetal placenta, including the embryo-derived placental vasculature, and immunohistochemical staining at 9.0 days gestation (after fusion of the allantois) demonstrated VCAM-1 expression on the surface of the allantoic mesoderm in contact with the chorion and $\alpha_4$ integrin expression on the chorion.

Approximately 1% of VCAM-1 knockout embryos survive, becoming healthy adult mice. Analysis of these mice confirmed absence of VCAM-1 expression, suggesting that if embryos can overcome the allantoic–placental defects, VCAM-1 is not necessary for normal development of other tissues. We conclude that VCAM-1 is essential for fusion of the allantois to the chorion and subsequent formation of the umbilical cord and placental embryo-derived vasculature. Therefore, an important new role for VCAM-1 in embryonic development has been identified.

## ACKNOWLEDGMENT

This work was supported by National Institutes of Health grants R01 HL45563 and P01 HL36028.

## REFERENCES

1. Pober JS, Cotran RS. Cytokines and endothelial cell biology. *Physiol Rev* 1990;70:427–51.
2. Rice GE, Bevilacqua MP. An inducible endothelial cell surface glycoprotein mediates melanoma adhesion. *Science* 1989;246:1303–6.
3. Carlos TM, Schwartz BR, Kovach EY, et al. Vascular cell adhesion molecule-1 mediates lymphocyte adherence to cytokine-activated cultured human endothelial cells. *Blood* 1990; 76:965–70.
4. Osborn L, Hession C, Tizard R, et al. Direct expression cloning of vascular cell adhesion molecule 1, a cytokine-induced endothelial protein that binds to lymphocytes. *Cell* 1989; 59:1203–11.
5. Elices MJ, Osborn L, Takada Y, et al. VCAM-1 on activated endothelium interacts with the leukocyte integrin VLA-4 at a site distinct from the VLA-4/fibronectin binding site. *Cell* 1990;60:577–84.
6. Cybulsky MI, Fries JWU, Williams AJ, et al. Alternative splicing of human VCAM-1 in activated vascular endothelium. *Am J Pathol* 1991;138:815–20.
7. Hession C, Tizard R, Vassallo C, et al. Cloning of an alternate form of vascular cell adhesion molecule-1 (VCAM1). *J Biol Chem* 1991;11:6682–5.
8. Polte T, Newman W, Raghunathan G, Gopal TV. Structural and functional studies of full-length vascular cell adhesion molecule-1: internal duplication and homology to several adhesion proteins. *DNA Cell Biol* 1991;10:349–57.
9. Cybulsky MI, Fries, JWU, Williams AJ, Sultan P, Gimbrone MA Jr, Collins T. Gene structure, chromosomal location, and basis for alternative mRNA splicing of the human *VCAM1* gene. *Proc Natl Acad Sci USA* 1991;88:7859–63.
10. Cybulsky MI, Allan-Motamed M, Collins T. Structure of the murine *VCAM1* gene. *Genomics* 1993;18:387–91.
11. Taichman D, Cybulsky MI, Djaffar I, et al. Tumor cell surface $\alpha_4\beta_1$ integrin mediates adhesion to vascular endothelium: demonstration of an interaction with the N-terminal domains of INCAM-110/VCAM-1. *Cell Regul* 1991;2:347–55.
12. Osborn L, Vassallo C, Benjamin CD. Activated endothelium binds lymphocytes through a novel binding site in the alternately spliced domain of vascular cell adhesion molecule-1. *J Exp Med* 1992;176:99–107.
13. Moy P, Lobb R, Tizard R, Olson D, Hession C. Cloning on an inflammation-specific phosphatidyl inositol-linked form of murine vascular cell adhesion molecule-1. *J Biol Chem* 1993;268:8835–41.
14. Terry RW, Kwee L, Levine JF, Labow MA. Cytokine induction of an alternatively spliced murine vascular cell adhesion molecule (VCAM) mRNA encoding a glycosylphosphatidylinositol-anchored VCAM protein. *Proc Natl Acad Sci USA* 1993;90:5919–23.
15. Thornhill MH, Haskard DO. IL-4 regulates endothelial cell activation by IL-1, tumor necrosis factor, or IFN-t. *J Immunol* 1990;145:865–72.
16. Masinovsky B, Urdal D, Gallatin WM. IL-4 acts synergistically with IL-1β to promote lymphocyte adhesion to microvascular endothelium by induction of vascular cell adhesion molecule-1. *J Immunol* 1990;145:2886–95.
17. Kume N, Cybulsky MI, Gimbrone MA Jr. Lysophosphatidylcholine, a component of atherogenic lipoproteins, induces mononuclear leukocyte adhesion molecules in cultured arterial endothelial cells. *J Clin Invest* 1992;90:1138–44.
18. Rice GE, Munro JM, Bevilacqua MP. Inducible cell adhesion molecule 110(INCAM-110) is an endothelial receptor for lymphocytes: a CD11/CD18-independent adhesion mechanism. *J Exp Med* 1990;171:1369–74.
19. Thornhill MH, Kyan-Aung U, Haskard DO. IL-4 increases human endothelial cell adhesiveness for T cells but not for neutrophils. *J Immunol* 1990;144:3060–5.
20. Graber N, Gopal TV, Wilson D, Beall LD, Polte T, Newman W. T cells bind to cytokine-activated endothelial cells via a novel, inducible sialoglycoprotein and endothelial-leukocyte adhesion molecule-1. *J Immunol* 1990;145:819–30.
21. Bochner BS, Luscinskas FW, Gimbrone MA Jr, et al. Adhesion of human basophils, eosinophils, and neutrophils to interleukin 1-activated human vascular endothelial cells: contributions of endothelial cell adhesion molecules. *J Exp Med* 1991;173:1553–6.

22. Schleimer RP, Sterbinsky SA, Kaiser J, et al. IL-4 induces adherence of human eosinophils and basophils but not neutrophils to endothelium. *J Immunol* 1992;148:1086–92.
23. Weller PF, Rand TH, Goelz SE, Chi-Rosso G, Lobb RR. Human eosinophil adherence to vascular endothelium mediated by binding to vascular cell adhesion molecule 1 and endothelial leukocyte adhesion molecule 1. *Proc Natl Acad Sci USA* 1991;88:7430–3.
24. Schwartz BR, Wayner EA, Carlos TM, Ochs HD, Harlan JM. Identification of surface proteins mediating adherence of CD11/CD18-deficient cells to cultured endothelium. *J Clin Invest* 1990;85:2019–22.
25. Vennegoor CJGM, Vandewiel-Vankemenad E, Huijbens RFJ, Sanchez-Madrid F, Melief CJM, Figdor CG. Role of LFA-1 and VLA-4 in the adhesion of cloned normal and LFA-1 (CD11/CD18)-deficient T cells to cultured endothelial cells. *J Immunol* 1992;148:1093–1101.
26. Spertini O, Luscinskas FW, Gimbrone MA Jr, Tedder TF. Monocyte attachment to activated human vascular endothelium in vitro is mediated by leukocyte adhesion molecule-1 (L-selectin) under nonstatic conditions. *J Exp Med* 1992;175:1789–92.
27. Kovach NL, Carlos TM, Yee E, Harlan JM. Monoclonal antibody to β1 integrin (CD29) stimulates VLA-dependent adherence of leukocytes to human umbilical vein endothelial cells and matrix components. *J Cell Biol* 1992;116:499–509.
28. Dustin ML, Springer TA. T-cell receptor cross-linking transiently stimulates adhesiveness through LFA-1. *Nature* 1989;341:619–24.
29. van Kooyk Y, Kemenade VDW, Weder P, Kuijpers TW. Enhancement of LFA-1 mediated cell adhesion by triggering through CD2 or CD3 on T lymphocytes. *Nature* 1989;342:811–3.
30. Rice GE, Munro JM, Corless C, Bevilacqua MP. Vascular and nonvascular expression of INCAM-110. *Am J Pathol* 1991;138:385–93.
31. Miyake K, Medina K, Ishihara K, Kimoto M, Auerbach R, Kincade PW. A VCAM-like adhesion molecule on murine bone marrow stromal cells mediates binding of lymphocyte precursors in culture. *J Cell Biol* 1991;114:557–65.
32. Simmons PJ, Masinovsky B, Longenecker BM, Berenson R, Torok-Storb B, Gallatin WM. Vascular cell adhesion molecule-1 expression by bone marrow stromal cells mediates the binding of hematopoietic progenitor cells. *Blood* 1992;80:388–95.
33. Rosen GD, Sanes JR, Lachance R, Cunningham JM, Roman J, Dean DC. Roles for the integrin VLA-4 and its counter receptor VCAM-1 in myogenesis. *Cell* 1992;69:1107–19.
34. Morales-Ducret J, Wayner E, Elices MJ, Alvaro-Gracia JM, Zvaifler NJ, Firestein GS. α4/β1 integrin (VLA-4) ligands in arthritis. Vascular cell adhesion molecule-1 expression in synovium and on fibroblast-like synoviocytes. *J Immunol* 1992;149:1424–31.
35. Briscoe DM, Schoen FJ, Rice GE, Bevilacqua MP, Ganz P, Pober JS. Induced expression of endothelial-leukocyte adhesion molecules in human cardiac allografts. *Transplantation* 1991;51:537–9.
36. Wuthrich RP. Vascular cell adhesion molecule-1 (VCAM-1) expression in murine lupus nephritis. *Kidney Int* 1992;42:903–13.
37. Freedman AS, Munro JM, Rice GE, et al. Adhesion of human B cells to germinal centers *in vitro* involves VLA-4 and INCAM-110. *Science* 1990;249:1030–3.
38. van Seventer GA, Newman W, Shimizu Y, et al. Analysis of T cell stimulation by superantigen plus major histocompatibility complex class II molecules or by CD3 monoclonal antibody: costimulation by purified adhesion ligands VCAM-1, ICAM-1, but not ELAM-1. *J Exp Med* 1991;174:901–13.
39. Burkly LC, Jakubowski A, Newman BM, Rosa MD, Chi-Rosso G, Lobb RR. Signaling by vascular cell adhesion molecule-1 (VCAM-1) through VLA-4 promotes CD3-dependent T cell proliferation. *Eur J Immunol* 1991;21:2871–5.
40. Damle NK, Aruffo A. Vascular cell adhesion molecule 1 induces T-cell antigen receptor-dependent activation of $CD4^+$ T lymphocytes. *Proc Natl Acad Sci USA* 1991;88:6403–7.
41. Ross R. The pathogenesis of atherosclerosis: a perspective for the 1990s. *Nature* 1993;362: 801–9.
42. Munro JM, Cotran RS. Biology of disease. The pathogenesis of atherosclerosis: atherogenesis and inflammation. *Lab Invest* 1988;58:249–61.
43. Cybulsky MI, Gimbrone MA Jr. Vascular endothelial cells express a monocyte adhesion molecule during atherogenesis. *Science* 1991;251:788–91.

44. Walker LN, Reidy MA, Bowyer DE. Morphology and cell kinetics of fatty streak lesion formation in the hypercholesterolemic rabbit. *Am J Pathol* 1986;125:450–9.
45. Li H, Cybulsky MI, Gimbrone MA Jr, Libby P. An atherogenic diet rapidly induces VCAM-1, a cytokine-regulatable mononuclear leukocyte adhesion molecule, in rabbit aortic endothelium. *Arterioscler Thromb* 1993;13:197–204.
46. Cybulsky MI, Allan-Motamed M, Medoff B, Davis V, Collins T, Gimbrone MA Jr. Endothelial expression of leukocyte adhesion molecules during atherogenesis in the rabbit. *FASEB J* 1992;6:1030A.
47. Li H, Cybulsky MI, Gimbrone MA Jr, Libby P. Inducible expression of vascular cell adhesion molecule-1 (VCAM-1) by vascular smooth muscle cells in vitro and within rabbit atheroma. *Am J Pathol* 1993;143:1551–9.
48. Poston RN, Haskard DO, Coucher JR, Gall NP, Johnson-Tidey RR. Expression of intercellular adhesion molecule-1 in atherosclerotic plaques. *Am J Pathol* 1992;140:665–73.
49. Printseva OY, Peclo MM, Gown AM. Various cell types in human atherosclerotic lesions express ICAM-1. Further immunocytochemical studies employing monoclonal antibody 10F3. *Am J Pathol* 1992;140:889–96.
50. Wood KM, Cadogan MD, Ramshaw AL, Parums DV. The distribution of adhesion molecules in human atherosclerosis. *Histopathology* 1993;23:437–44.
51. van der Wal AC, Das PK, Tigges AJ, Becker AE. Adhesion molecules on the endothelium and mononuclear cells in human atherosclerotic lesions. *Am J Pathol* 1992;141:1427–33.
52. Davis MJ, Gordon JL, Gearing AJH, et al. The expression of the adhesion molecules ICAM-1, VCAM-1, PECAM, and E-selectin in human atherosclerosis. *J Pathol* 1993; 171:223–9.
53. O'Brien KD, Allen MD, McDonald TO, et al. Vascular cell adhesion molecule-1 is expressed in human coronary atherosclerotic plaques. Implications for the mode of progression of advanced coronary atherosclerosis. *J Clin Invest* 1993;92:945–51.
54. Gurtner GC, Davis V, McCoy MJ, Li H, Sharpe A, Cybulsky MI. Targeted disruption of the murine *VCAM1* gene: essential role of VCAM-1 in chorioallantoic fusion and placentation. *Genes Dev* 1995;9:1–14.
55. Thomas KR, Capecchi MR. Site-directed mutagenesis by gene targeting in mouse embryo-derived stem cells. *Cell* 1987;51:503–12.
56. Mansour SL, Thomas KR, Capecchi MR. Disruption of the proto-oncogene int-2 mouse embryo-derived stem cells: a general strategy for targeting mutations to non-selectable genes. *Nature* 1988;336:348–52.
57. McMahon AP, Bradley A. The Wnt-1 (int-1) proto-oncogene is required for development of a large region of the mouse brain. *Cell* 1990;62:1073–85.

*Topics in Molecular Medicine, Volume 1,*
edited by Wolfgang Siess, Reinhard Lorenz,
and Peter C. Weber. Raven Press, Ltd.,
New York © 1995.

# 14

# The Role of Nitric Oxide and Peroxynitrite in Platelet Function: Interactions with Prostacyclin

Marek W. Radomski

*Wellcome Research Laboratories, Beckenham, Kent BR3 3BS, England*

The appreciation of labile mediators as regulators of blood cell–endothelium interactions closely coincides with the discovery of thromboxane $A_2$ and prostacyclin in 1974–1976 (1,2). The biochemistry and pharmacology of these eicosanoids have been the subject of several reviews over the past 20 years (3–5). During the past decade a new platelet-regulatory mediator was born which proved to be even more elusive than prostacyclin and thromboxane $A_2$. This mediator was originally called endothelium-dependent relaxing factor (EDRF) and was subsequently identified as nitric oxide (NO). The gaseous nature of NO enables it to diffuse freely in the surrounding environment and to interact with a number of molecular targets, including heme iron, non-heme iron, oxygen, and oxygen-derived species. The products of these interactions may also have potent biologic activities. One of them is peroxynitrite ($ONOO^-$), which is generated during reaction of NO with superoxide anion ($O_2^-$). This article describes the biologic effects of NO and $ONOO^-$ on platelet function and their interactions with prostacyclin.

## NO ACCOUNTS FOR THE VASODILATOR AND PLATELET-INHIBITORY ACTIVITIES OF EDRF

Although endothelium-dependent relaxation by EDRF of isolated arterial rings was described as early as 1980 (6), NO emerged as a biologic mediator several years later. In 1987 and 1988 it was demonstrated that the generation of NO by NO synthase (NOS) accounts for the biologic activity of EDRF (7–9). NO proved also to be a potent inhibitor of platelet function (10–14).

## NOS IN THE VASCULAR SYSTEM

Gene cloning identifies three isoforms of NOS: endothelial isoform (eNOS), an isoform induced by immunological stimuli (iNOS), and neuronal isoenzyme (nNOS). The two latter isoforms have been isolated and cloned from rodent macrophages and brain (15,16). The eNOS from human umbilical vein and bovine aorta endothelial cells has been also cloned (17,18). The eNOS is a $Ca^{2+}$- NADPH-, flavin-, and tetrahydrobiopterin-dependent enzyme that utilizes the guanido nitrogen atom of L-arginine and incorporates molecular oxygen to generate NO and L-citrulline (19). Platelets also generate NO (20–27). The platelet NOS shows substrate and co-factor dependency similar to that of eNOS (21).

The iNOS that synthesizes NO in macrophages, endothelium, and vascular smooth-muscle cells is an NADPH-, tetrahydrobiopterin-, and flavin-dependent enzyme (19), inducible by bacterial products and/or cytokines, whose expression requires de novo protein synthesis (28,29). The expression of iNOS leads to generation of high amounts of NO over long periods of time and may account for the cytostatic–cytotoxic reactions of nonspecific immunity (30). However, the expression of iNOS may be also responsible at least in part for "self-inflicted" tissue damage; the vascular lesion of septicemia is one example of the cytotoxic reactions mediated by iNOS (29,31). Because platelets have a limited capacity to synthesize new proteins and often acquire them either from the circulation or by transfer from the megakaryocyte, it is rather unlikely that they can express iNOS by themselves. We have recently found that human megakaryoblastic cells (Meg-01) possess NOS similar to that in platelets and express iNOS after stimulation with interleukin-1β and tumor necrosis factor-α (32). Thus, both enzymes are synthesized in megakaryocytes and then may be transferred into platelets. The iNOS is also expressed in human polymorphonuclear leukocytes and peripheral blood monocytes (33,34).

## MOLECULAR TARGETS FOR NO

### The Soluble Guanylate Cyclase

#### *Regulation of the Platelet Transduction Mechanism by cGMP*

NO has a high binding affinity for heme iron and therefore reacts with hemoproteins such as the soluble guanylate cyclase (SGC) (35). Activation of SGC leads to the conversion of magnesium guanosine 5′-triphosphate to guanosine 3′,5′-monophosphate (cGMP), which stimulates cGMP-dependent protein kinase leading to protein phosphorylation, including 46/50-kDa vasodilator-stimulated phosphoprotein (VASP). These effects of cGMP may result in inhibition of phosphorylation of myosin light chains and of protein kinase C and modulation of phospholipase $A_2$- and C-mediated responses (see references in 36). cGMP-regulated responses also decrease intracellular $Ca^{2+}$ levels (37–39).

### *Regulation of Platelet Adhesion Receptors by cGMP*

There is evidence that cGMP regulates the function of platelet adhesion receptors such as IIb/IIIa and P-selectin. It has been found that NO inhibits both platelet contact and spreading on fibrinogen (40), as well as binding of this protein to platelets during aggregation induced by ADP (41,42). These effects are closely correlated with NO-induced increases in cGMP, suggesting that this cyclic nucleotide modulates both adhesion and aggregation mediated by IIb/IIIa. Similarly, both the release of soluble P-selectin and the translocation of membrane-bound protein are inhibited by the NO–cGMP system (42,43).

## Fe-S Groups, Thiols, and Secondary Amines

NO interact also with molecules containing nonheme iron coordinated to sulfur atoms (Fe–S groups), thiols, and secondary amines. The interaction of NO produced by activated macrophages with Fe–S-containing enzymes, such as aconitase and complex I and II of the mitochondrial electron transport chain and of the pathway for the synthesis of DNA, results in inhibition of the activity of these enzymes and may be responsible for NO-mediated cytotoxicity (30). The vascular endothelial cells are known to be susceptible to NO toxicity (44). This cytotoxicity is mainly the result of NO produced in large amounts over long periods by iNOS.

NO may also cause S-nitrosylation of thiol-containing enzymes. Recent evidence suggests that NO can cause the ADP-ribosylation and S-nitrosylation of glyceraldehyde-phosphate dehydrogenase, and enzyme involved in glycolytic formation of ATP. S-Nitrosylation of this enzyme inhibits its activity and might lead to a decrease in glycolysis and gluconeogenesis, which could also contribute to the platelet-inhibitory and cytotoxic effects of NO (45,46). Finally, S-nitrosylation of albumin and other thiol-containing molecules may result in formation of S-nitrosothiols and prolong the biologic half-life of NO in plasma (47–49). Interestingly, S-nitrosothiols are potent inhibitors of platelet aggregation in vitro (47–49) and in vivo in animals (23,24,48) and humans (50). Whether or not S-nitrosylation occurs in vivo remains to be determined.

### The Superoxide Anion

NO generated from endothelial cells or from leukocytes interacts with $O_2^-$, leading to the reduction of its vasorelaxant and platelet antiaggregatory actions (33,51,52). The product of this reaction is the peroxynitrite anion ($ONOO^-$), which is also formed by activated macrophages (53,54). This powerful oxidant, when protonated, decomposes rapidly, resulting in the formation of OH· and $NO_2$, both of which are tissue-damaging species. The pathways of $ONOO^-$-induced cellular injury include inhibition of mitochondrial respiration (55), stimulation of lipid peroxidation (56), nitration of tyrosine-containing proteins (57), and oxidation of reactive thiols (58).

## PHYSIOLOGIC REGULATION OF PLATELET FUNCTION BY NO

### The Platelet NOS

Early observations showed that platelet-aggregating agents caused an increase in the intraplatelet content of cGMP. This, in conjunction with the known platelet-inhibitory role of cAMP, led to the yin–yang hypothesis according to which the function of cGMP was to antagonize the actions of cAMP (59). In 1981, however, it was found that NO inhibited the aggregation of human platelets and that this action was closely associated with platelet cGMP accumulation. This led to a reassessment of the yin–yang hypothesis and to the suggestion that cGMP causes inhibition of platelet aggregation (60).

The formation of NO by platelets (20–27) may explain the mechanism of aggregation-induced increase in cGMP. In resting platelets the synthesis of NO is not detectable. However, the platelet NOS becomes activated during platelet aggregation induced by collagen, ADP, and arachidonic acid (21). Because this enzyme is strictly $Ca^{2+}$-dependent and platelet aggregation is associated with an increase in intraplatelet $Ca^{2+}$, it is possible that this cation controls the activation of the platelet NOS in vivo. However, our recent studies using direct electrochemical measurement of NO released from human platelets have shown that NO is released during collagen- but not thrombin-induced aggregation (61). The reasons for the differential actions of collagen and thrombin on NO release are not clear, as both aggregating agents are known to mobilize $Ca^{2+}$ from intraplatelet compartments. Because differential $Ca^{2+}$ compartments have been described in human platelets (62), the activation of the platelet NOS may be linked to a collagen-sensitive but thrombin-insensitive pool.

## The Endothelial NOS

The amounts of NO available for regulation of platelet function are further increased by its production by the vascular endothelium. Studies with inhibitors of NOS suggest that eNOS generates NO constantly to provide vasodilator tone (63). The physiologic stimuli for generation of NO by the endothelium are not yet fully understood, but flow and shear stresses appear to stimulate the synthesis of NO via activation of the potassium $K_{Ca}$ channel (64).

In 1987 it was also shown that cultured and fresh endothelial cells, when stimulated with bradykinin, release NO in amounts sufficient to inhibit platelet adhesion (13,14,65). Moreover, the coronary and pulmonary vasculature releases NO to inhibit platelet adhesion under constant flow conditions (66–68).

Platelet aggregation in vitro induced by a variety of agonists is inhibited by NO released from fresh or cultured endothelial cells (11,12,69–76). This NO also causes disaggregation of preformed platelet aggregates (12). Moreover, basal release (23,24,77–79) or release of NO stimulated by cholinergic stimuli and substance P (80–82) results in inhibition of platelet aggregation induced by some aggregating agents or endothelial injury in vivo. Thus, a concerted action of endothelial and platelet NOS regulates platelet activation, causing inhibition of adhesion and aggregation and induction of disaggregation.

The contribution of NO released from neutrophils (33) to the regulation of platelet function in vivo remains to be established. However, it is of interest that the NO–cGMP system also inhibits the adhesion and chemotaxis of stimulated neutrophils (83,84) and that inhibition of NO synthesis elicits platelet–leukocyte aggregation that is mediated via expression of P-selectin on the platelet surface (42,85).

In vivo, the synthesis and release of a single inhibitor is unlikely to account for regulation of platelet aggregation. We have shown that NO and prostacyclin synergize with each other as inhibitors of platelet aggregation and inducers of disaggregation (12). In addition, synergistic induction of platelet disaggregation has recently been demonstrated by combining glyceryl trinitrate (an NO donor), prostaglandin $E_1$, and tissue plasminogen activator, which act via cGMP, cAMP and plasmin-dependent mechanisms, respectively (86). Therefore, it is likely that platelet aggregation in vivo is regulated by the synergistic action of several inhibitors of platelet function. The biochemical rationale for the synergistic inhibition of platelet aggregation is unclear. However, for NO and prostacyclin (which increases cAMP levels), it may depend on an NO-induced increase in cGMP with subsequent inhibition of cGMP-inhibited cAMP PDE, leading to an increase in cAMP levels (87). Interestingly, prostacyclin, although a potent inhibitor of aggregation (2), is a weak inhibitor of platelet adhesion and does so only at high doses, at which it also increases cGMP levels (14). This not only explains why there is no synergy between prostacyclin and NO as inhibitors of adhesion but also suggests that this process is controlled by cGMP rather than by cAMP (13,14,88).

## ROLE OF NO IN VASCULAR PATHOLOGY

### Cytotoxic Properties of NO

NO is synthesized in large quantities over a long period of time by iNOS, and present evidence suggests that nonspecific immunity with accompanying inflammation is associated with the expression of this enzyme in leukocytes, macrophages, and the vascular wall (28–30,33). The biologic purpose of these reactions is to contain and eliminate invading organisms. However, NO is not selectively cytotoxic to "non-self" structures, and interacting with cells that produce it can result in "self-inflicted" damage (see references in 89). Indeed, in some models of inflammation pharmacologic inhibition of NO synthesis attenuates the extent of vascular injury (90).

### NO, $ONOO^-$, and Hemostatic–Thrombotic Balance

Septicemia and endotoxic shock are often associated with disturbances in hemostatic–thrombotic balance, such as disseminated intravascular coagulation. There is now evidence that the release of NO during septicemia may attenuate increased thrombogenicity caused by disseminated intravascular coagulation. Indeed, cytokine-stimulated endothelial and vascular smooth-muscle cells produced NO, which caused vasodilatation and inhibition of platelet adhesion and aggregation (28,29,91). This regulating action of NO may be of particular importance in the microvasculature under conditions of enhanced thrombogenic readiness. Schultz and Raij (92) have shown that inhibition of NO synthesis precipitated glomerular thrombosis in endotoxin-treated rats.

Activated macrophages generate a number of oxygen-derived species, including $ONOO^-$ (53,54). We have investigated the effects of $ONOO^-$ on human platelets in vitro to explore the potential of this oxidant to contribute to tissue damage (93). Peroxynitrite caused aggregation of human washed platelets and reversed inhibition of aggregation induced by an NO donor, S-nitroso-*N*-acetyl-DL-penicillamine, and by prostacyclin and indomethacin. However, in the presence of plasma $ONOO^-$ not only did not exert proaggregatory properties but acted as an inhibitor of platelet aggregation. This reversal of the aggregatory effect of $ONOO^-$ by plasma could also be mimicked by endogenous thiols, such as albumin and glutathione, and was accompanied by formation of the respective S-nitrosothiols. Thus, the fate and therefore the actions of $ONOO^-$ in the vascular system are critically dependent on the biologic environment in which this oxidant is present. Indeed, there are pathologic conditions, such as atherosclerosis, for which depletion of thiols may favor the direct interaction of $ONOO^-$ with cell membranes, leading to tissue damage (Fig. 1).

It has been demonstrated that the severe hemodynamic imbalance of septic shock may be controlled by pharmacologic administration of NOS inhibitors (94). This treatment is likely to decrease the generation of $ONOO^-$ from NO generated by iNOS. However, the inhibitors available to date are not selective for iNOS and also inhibit other isoforms of NOS, which may lead to increased thrombogenicity of endothelium (91) and formation of platelet thrombi in the microvasculature (92). Therefore, platelet behavior should be carefully monitored during administration of NOS inhibitors in sepsis.

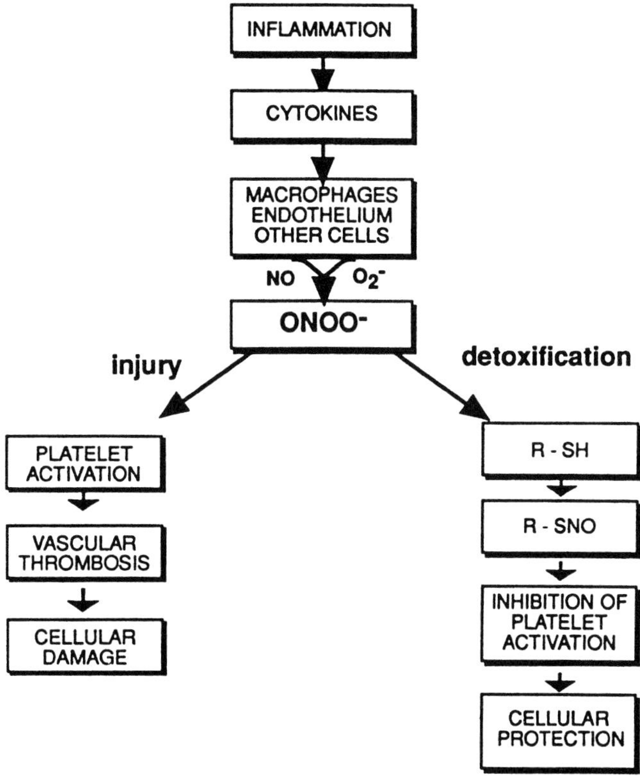


**FIG. 1.** The role of $ONOO^-$ in tissue and organ injury. A hypothesis outlining the pathways of $ONOO^-$-mediated cell damage and protection.

### NO, Atherosclerosis, Hypertension, Vascular Thrombosis, and the Uremic Syndrome

Changes in the generation of NO have been associated with platelet and vascular pathologies. Indeed, the ability of the endothelium to synthesize NO is reduced both in human coronary atherosclerosis (95) and in animal models of this disease (96). This is of importance because NO inhibits mitogen release from stimulated human platelets (97) and decreases the proliferation of vascular smooth-muscle cells (98). Therefore, reduced generation of NO may be one of the contributing factors to atherosclerosis. Interestingly, administration of L-arginine, the substrate for NOS, restores impaired production of endothelium-derived NO, decreases activation of platelets and leukocytes caused by hypercholesterolemia and subsequently attenuates the vascular lesion associated with this disease (99–101).

Furthermore, patients suffering from essential hypertension and insulin-dependent diabetes show impaired formation of NO in the endothelium and platelets (25,102–104). In addition, studies with inhibitors of NOS have shown that decreased generation of NO in vivo can lead to platelet and polymorphonuclear leukocyte activation and thrombosis (23,24,78,79,85). Therefore, it is likely that failure of the endothelium, platelets, or polymorphonuclear leukocytes to produce NO plays a role in the pathogenesis of vascular thromobotic disease.

A hemostatic defect also occurs in uremia and is characterized by prolonged bleeding time and decreased platelet adhesion and aggregation. It has been shown that this platelet defect may be caused by an increased formation of NO in platelets and endothelial cells (27).

All of these observations are consistent with the notion that balanced release and action of NO are necessary for maintenance of vascular homeostasis and that both impaired and exaggerated generation of this mediator results in vascular disease.

## PHARMACOLOGIC CONTROL OF NO GENERATION AND ACTION

### Stimulation of NOS

Studies with NOS have shown that this enzyme requires L-arginine as the substrate and that activation of NOS takes place at concentrations of L-arginine that are several orders of magnitude lower than those present in plasma or in the endothelium and platelets (8,9,20,21). These observations have prompted a discussion of whether the pharmacologic administration of this amino acid may lead to the generation of NO. It appears that there is little effect of L-arginine on the formation of NO by the intact endothelium and resting platelets, showing that eNOS and the platelet NOS are not activated under these conditions. However, exogenous L-arginine can be converted to NO after stimulation of the platelet NOS by aggregation (20,21) or after expression of the iNOS in the vessel wall (29,91,92). Moreover, in human atherosclerosis and in animal models of this disease, L-arginine produces pharmacologic effects such as vasodilatation and inhibition of platelet and leukocyte functions (99–101). Leaving aside the question of whether these effects were due to the formation of NO, further exploration of vascular effects of L-arginine can be of therapeutic importance.

A number of biogenic amines (e.g., acetylcholine and serotonin) and peptides (e.g., bradykinin and calcitonin gene-related peptide) have been shown to activate NOS via a receptor-dependent mechanism (see references in 19). It is unlikely, however, that these could be used as pharmacologic tools because of serious side effects. Recently, nifedipine, a calcium-channel inhibitor, was demonstrated to activate NOS via a mechanism unrelated to the blockade of calcium channels (26). Therefore, it may be possible to develop drugs capable of stimulating endogenous synthesis of NO.

## NO Donors

Nitrovasodilators are drugs whose pharmacologic action depends on the release of NO (105). Although the vasodilator effect of NO donors has been known for many years, appreciation of the platelet-inhibitory activity of these compounds is new.

Organic nitrates (glyceryl trinitrate, isosorbide mononitrate, and dinitrate) are NO donors that release NO after metabolic activation. These compounds are often used for the treatment of coronary artery disease and myocardial infarction, conditions associated with platelet activation. The effects of organic nitrates on platelet function in vitro are weak because platelets lack the appropriate enzymatic system required for the release of NO (106). However, in vivo these compounds have been shown to inhibit platelet function. Indeed, intravenous administration of glyceryl trinitrate for 24 h to patients with acute mycardial infarction significantly inhibited platelet adhesion and aggregation to fibrillar collagen (107). Moreover, oral administration of isosorbide dinitrate decreased platelet reactivity in patients with coronary artery disease (108).

Sydnonimines (molsidomine, its active metabolite SIN-1, and congeners) are the family of NO donors capable of spontaneous generation of NO (105). It is therefore not surprising that they have a profound effect on platelet function both in vitro and in vivo. Indeed, SIN-1 has been shown to inhibit platelet aggregation in patients with acute myocardial infarction (109) and to inhibit deposition of platelets at the site of endothelial denudation (110).

The platelet-inhibitory actions of organic nitrates and sydnonimines cannot be separated from their effects on vascular tone. Therefore, the lack of

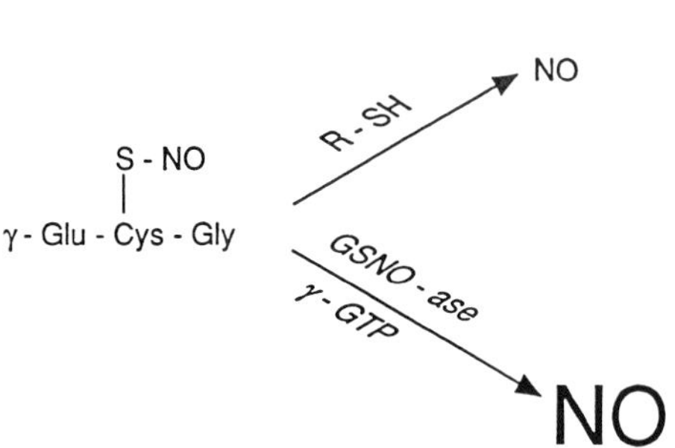


**FIG. 2.** The molecular basis of relative selectivity of S-nitrosoglutathione in inhibiting platelet function. Nitric oxide (NO) is released from S-nitrosoglutathione mainly by the action(s) of an enzyme(s) associated with platelet membranes: S-nitrosoglutathionase (GSNO-ase) and γ-glutamyltranspeptidase (γ-GTP). Transnitrosylation by thiols (R-SH) is a minor pathway for the metabolism of S-nitrosoglutathione in platelets but may be responsible for the release of NO by the vessel wall.

selectivity limits the effectiveness of these drugs as inhibitors of platelet function. The concept of platelet-selective NO donors has arisen from our experiments with S-nitrosoglutathione (GSNO). S-Nitrosoglutathione is a tripeptide S-nitrosothiol formed by S-nitrosylation of glutathione, the most abundant intracellular thiol. We have found that the intravenous administration of GSNO into conscious rat inhibits platelet aggregation at doses that have only small effects on the blood pressure (48). Moreover, similar platelet–vascular differentiation is detected after intra-arterial administration of GSNO into the circulation of the human forearm (50). Finally, we have infused GSNO in the patients undergoing balloon angioplasty and have found that this NO donor effectively protected platelets from activation at the site of angioplastic injury without altering blood pressure (111). This differential effect of GSNO is likely to be due to the presence in platelets of an enzyme(s) that metabolizes GSNO and releases NO (48). Two enzymes have been implicated in the release of NO from GSNO: S-nitrosoglutathionase, an enzyme related to γ-glutamyltransferase (D. J. Meyer et al., personal communication) and γ-glutamyltranspeptidase (S. C. Askew et al., personal communication) (Fig. 2).

## Synergy Between NO Donors and Other Inhibitors of Platelet Function

The synergistic nature of NO has already been explored to enhance the potency and selectivity of this molecule as an inhibitor of platelet function. Both molsidomine and isosorbide dinitrate synergize with prostacyclin and prostaglandin $E_1$ to inhibit platelet activation in peripheral vascular disease (110,112). Potentiation of the platelet-inhibitory activity of NO and its donors can be also achieved by combining them with selective inhibitors of cGMP phosphodiesterase (most of the actions of NO on platelets result from the stimulation of SGC and increase in cGMP levels, an effect that is potentiated by inhibition of phosphodiesterase enzymes) such as M&B 22948 (zaprinast), MY5445, and BY1949 (11–14, 113), and clinically used dipyridamole (114).

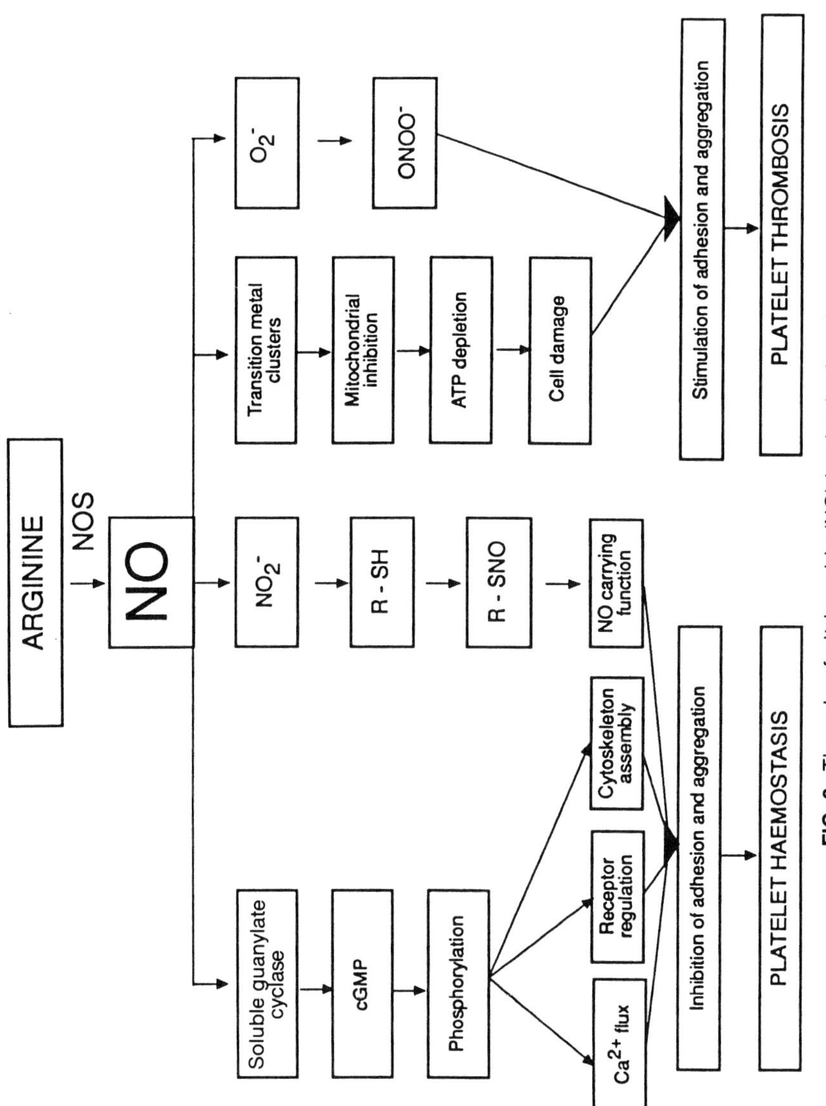


**FIG. 3.** The role of nitric oxide (NO) in platelet function.

Finally, there is a pharmacologic rationale for combination of superoxide dismutase enzymes (SODs), which scavenge superoxide anions ($O_2^-$), with NO donors. Superoxide anions are often co-generated with NO and they contribute to the inactivation of this mediator (51). Moreover, because $O_2^-$ reacts with NO to generate $ONOO^-$, scavenging of $O_2^-$ may protect against the damaging effects of the latter oxidant. Because SOD significantly prolongs the antiadhesive, antiaggregating, and disaggregating activities of NO (11–14), the pharmacologic interactions between NO donors and antioxidants such as SOD deserve to be further explored.

### Inhibition of NO Synthesis

There is little doubt that systemic inhibition of NO synthesis, particularly under conditions of coexisting hypertensive or thrombotic vascular disease, is potentially hazardous. However, both the vasoconstrictor and the platelet-activating effects of inhibitors of NO synthase might be useful in the management of localized bleeding (e.g., gastrointestinal) when other treatments are contraindicated.

## CONCLUSIONS

The hemostatic–thrombotic balance is maintained by a dynamic interplay between mediators of the vascular system, including NO and prostacyclin. NO is an important factor in the network of these regulatory interactions. The molecule shows both physiologic and pathologic characteristics. NO promotes platelet hemostasis by downregulation of adhesion and aggregation. Some prothrombotic actions of NO may result from the secondary formation of tissue-damaging agents such as $ONOO^-$ (Fig. 3). More work is needed to understand further this paradoxical nature of NO. This is likely to be important for the prevention and treatment of vascular pathologies including atherosclerosis, hypertension, vasospasm, thrombosis, and septicemia.

## REFERENCES

1. Hamberg M, Svensson J, Samuelsson B. Thromboxanes—a new group of biologically active compounds derived from prostaglandin endoperoxides. *Proc Natl Acad Sci USA* 1975;72:2994–8.
2. Moncada S, Gryglewski RJ, Bunting S, Vane JR. An enzyme isolated from arteries transforms prostaglandin endoperoxides to an unstable substance that inhibits platelet aggregation. *Nature* 1976;263:663–5.
3. Moncada S, Higgs EA. Prostaglandins in the pathogenesis and prevention of vascular disease. *Blood Rev* 1987;1:141–5.
4. Gryglewski RJ, Botting RM, Vane JR. Mediators produced by the endothelial cell. *Hypertension* 1988;12:530–48.

5. Samuelsson B. An elucidation of the arachidonic acid cascade. Discovery of prostaglandins, thromboxane and leukotriene. *Drugs* 1987;33(suppl 1):2–9.
6. Furchgott RF, Zawadzki JV. The obligatory role of endothelial cells in the relaxation of arterial smooth muscle by acetylcholine. *Nature* 1980;288:373–6.
7. Palmer RMJ, Ferrige AG, Moncada S. Nitric oxide accounts for the biological activity of endothelium-derived relaxing factor. *Nature* 1987;327:524–6.
8. Palmer RMJ, Ashton DS, Moncada S. Vascular endothelial cells synthesize nitric oxide from L-arginine. *Nature* 1988;333:664–6.
9. Palmer RMJ, Rees DD, Ashton DS, Moncada S. L-Arginine is the physiological precursor for the formation of nitric oxide in endothelium-dependent relaxation. *Biochem Biophys Res Commun* 1988;153:1251–6.
10. Azuma H, Ishikawa M, Sekizaki S. Endothelium-dependent inhibition of platelet aggregation. *Br J Pharmacol* 1986;88:411–5.
11. Radomski MW, Palmer RMJ, Moncada S. Comparative pharmacology of endothelium-derived relaxing factor, nitric oxide and prostacyclin in platelets. *Br J Pharmacol* 1987; 92:181–7.
12. Radomski MW, Palmer RMJ, Moncada S. The anti-aggregating properties of vascular endothelium: interactions between prostacyclin and nitric oxide. *Br J Pharmacol* 1987; 92:639–46.
13. Radomski MW, Palmer RMJ, Moncada S. Endogenous nitric oxide inhibits human platelet adhesion to vascular endothelium. *Lancet* 1987;2:1057–8.
14. Radomski MW, Palmer RMJ, Moncada S. The role of nitric oxide and cGMP in platelet adhesion to vascular endothelium. *Biochem Biophys Res Commun* 1987;48:1482–9.
15. Lyons CR, Orloff GJ, Cunningham JM. Molecular cloning and functional expression of an inducible nitric oxide synthase from a murine macrophage cell line. *J Biol Chem* 1992; 267:6470–4.
16. Bredt DS, Hwang PM, Glatt CE, Lowenstein C, Reed RR, Snyder SH. Cloned and expressed nitric oxide synthase structurally resembles cytochrome P-450 reductase. *Nature* 1991;351:714–8.
17. Marsden PA, Shappert KT, Chen HS, et al. Molecular cloning and characterization of human endothelial nitric oxide synthase. *FEBS Lett* 1992;307:287–93.
18. Sessa WV, Harrison JK, Barber CM, et al. Molecular cloning and expression of cDNA encoding endothelial cell nitric oxide synthase. *J Biol Chem* 1992;267:15274–6.
19. Moncada S. The 1991 Ulf von Euler Lecture. The L-arginine:nitric oxide pathway. *Acta Physiol Scand* 1992;145:201–27.
20. Radomski MW, Palmer RMJ, Moncada S. An L-arginine/nitric oxide pathway present in human platelets regulates aggregation. *Proc Natl Acad Sci USA* 1990;87:5193–7.
21. Radomski MW, Palmer RMJ, Moncada S. Characterization of the L-arginine:nitric oxide pathway in human platelets. *Br J Pharmacol* 1990;101:325–8.
22. Pronai L, Ichimori K, Nozaki H, et al. Investigation of the existence and biological role of L-arginine/nitric oxide pathway in human platelets by spin-trapping/EPR studies. *Eur J Biochem* 1991;202:923–30.
23. Golino P, Capelli-Bigazzi M, Ambrosio G, et al. Endothelium-derived relaxing factor modulates platelet aggregation in an in vivo model of recurrent platelet activation. *Circ Res* 1992;71:1447–56.
24. Yao SK, Ober JC, Krishnaswami A, et al. Endogenous nitric oxide protects against platelet aggregation and cyclic flow variations in stenosed and endothelium-injured arteries. *Circulation* 1992;86:1302–9.
25. Cadwgan TM, Benjamin N. Evidence for altered platelet nitric oxide synthesis in essential hypertension. *J Hypertens* 1993;11:417–20.
26. Berkels R, Klaus W, Rosen R. Nifedipine induced inhibition of platelet aggregation—a nitric oxide mediated process. *Endothelium* 1993;1:S69.
27. Noris M, Benigni A, Boccardo P, et al. Enhanced nitric oxide synthesis in uremia: implications for platelet dysfunction and dialysis hypotension. *Kidney Int* 1993;44:445–50.
28. Radomski MW, Palmer RMJ, Moncada S. Glucocorticoids inhibit the expression of an inducible but not the constitutive, nitric oxide synthase in vascular endothelial cells. *Proc Natl Acad Sci USA* 1990;87:10043–7.
29. Rees DD, Cellek S, Palmer RMJ, Moncada S. Dexamethasone prevents the induction by

endotoxin of a nitric oxide synthase and the associated effects on vascular tone: an insight into endotoxin shock. *Biochem Biophys Res Commun* 1990;173:541–7.

30. Hibbs JB Jr, Taintor RR, Vavrin Z, et al. Synthesis of nitric oxide from a terminal guanidino nitrogen atom of L-arginine: a molecular mechanism regulating cellular proliferation that targets intracellular iron. In: Moncada S, Higgs EA, eds. *Nitric oxide from L-arginine: a bioregulatory system* Amsterdam: Elsevier, 1990:189–224.
31. Kilbourn RG, Gross SS, Adams J, et al. $N^G$-methyl-L-arginine inhibits tumor necrosis factor-induced hypotension: implications for the involvement of nitric oxide. *Proc Natl Acad Sci USA* 1990;87:3629–32.
32. Lelchuk R, Radomski MW, Martin JF, Moncada S. Constitutive and inducible nitric oxide synthases in human megakaryoblastic cells. *J Pharmacol Exp Ther* 1992;262:1220–4.
33. McCall TB, Boughton-Smith NK, Palmer RMJ, Whittle BJR, Moncada S. Synthesis of nitric oxide from L-arginine by neutrophils. *Biochem J* 1989;261:293–6.
34. Hunt NCA, Goldin RD. Nitric oxide production by monocytes in alcoholic liver disease. *J Hepatol* 1992;4:146–50.
35. Craven PA, De Rubertis FR. Restoration of the responsiveness of purified guanylate cyclase to nitrosoguandine, nitric oxide, and related activators by heme and hemoproteins. Evidence for involvement of the paramagnetic nitrosyl-heme complex in enzyme activation. *J Biol Chem* 1978;253:8433–43.
36. Walter U. Physiological role of cGMP and cGMP-dependent protein kinase in the cardiovascular system. *Rev Physiol Biochem Pharmacol* 1989;13:41–88.
37. Geiger J, Nolte C, Butt E, Sage SO, Walter U. Role of cGMP and cGMP-dependent protein kinase in nitrovasodilator inhibition of agonist-evoked calcium elevation in human platelets. *Proc Natl Acad Sci USA* 1992;89:1031–5.
38. Morgan RO, Newby AC. Nitroprusside differentially inhibits ADP-stimulated calcium influx and mobilization in human platelets. *Biochem J* 1989;258:447–54.
39. Nakashima S, Tohmatsu T, Hattori H, Okano Y, Nozawa Y. Inhibitory action of cyclic GMP on secretion, phosphoinositide hydrolysis and calcium mobilization in thrombin-stimulated human platelets. *Biochem Biophys Res Commun* 1986;35:1099–104.
40. Shahbazi T, Jones N, Radomski M, Gingell D. Nitric oxide donors inhibit platelet spreading on surfaces coated with fibrinogen but not with fibronectin. *Abstracts, UCLA Conference on Nitric Oxide,* July, 1994.
41. Mendelsohn M, O'Neill S, George D, Loscalzo J. Inhibition of fibrinogen binding to human platelets by S-nitroso-N-acetylcysteine. *J Biol Chem* 1990;265:19028–34.
42. Salas E, Moro MA, Askew S, et al. Comparative pharmacology of analogues of S-nitroso-N-acetyl-DL-penicillamine on human platelets. *Br J Pharmacol* 1994;112:1071–6.
43. Rosen P, Schwippert B, Kaufman L, Tschope D. Expression of adhesion molecules on the surface of activated platelets is diminished by $PGI_2$ analogues and an NO (EDRF)-donor: a comparison between platelets of healthy and diabetic subjects. *Platelets* 1994;5:45–52.
44. Palmer RMJ, Bridge L, Foxwell NA, Moncada S. The role of nitric oxide in endothelial cell damage and its inhibition by glucocorticoids. *Br J Pharmacol* 1992;105:11–2.
45. Ivanova K, Schaefer M. Drummer C, Gerzer R. Effects of nitric oxide-containing compounds on increases in cytosolic ionized calcium and on aggregation of human platelets. *Eur J Pharmacol* 1993;244:37–47.
46. Molina y Vedia L, McDonald B, Reep B, et al. Nitric oxide-induced S-nitrosylation of glyceraldehyde-3-phosphate dehydrogenase inhibits enzymatic activity and increases endogenous ADP-ribosylation. *J Biol Chem* 1992;257:24929–32.
47. Mellion BT, Ignarro LJ, Myers CB, et al. Inhibition of human platelet aggregation by S-nitrosothiols. Heme-dependent activation of soluble guanylate cyclase and stimulation of cyclic GMP accumulation. *Mol Pharmacol* 1983;23:653–64.
48. Radomski MW, Rees DD, Dutra A, Moncada S. S-nitrosoglutathione inhibits platelet activation in vitro and in vivo. *Br J Pharmacol* 1992;107:745–9.
49. Stamler JS, Jaraki O, Osborne, J, et al. Nitric oxide circulates in mammalian plasma primarily as an S-nitroso adduct of serum albumin. *Proc Natl Acad Sci USA* 1992;89: 7674–7.
50. De Belder A, MacAllister RA, Radomski MW, Moncada S, Vallance P. Effects of S-nitrosoglutathione in the human forearm circulation. Evidence for selective inhibition of platelet activation. *Cardiovasc Res* 1994;28:691–54.

51. Gryglewski RJ, Palmer RMJ, Moncada S. Superoxide anion is involved in the breakdown of endothelium-derived vascular relaxing factor. *Nature* 1986;320:454–6.
52. Rubanyi GM, Vanhoutte PM. Superoxide anions and hyperoxia inactivate endothelium-derived relaxing factor. *Am J Physiol* 1986;250:H822–7.
53. Beckman JS, Beckman TW, Chen J, Marshall PA, Freeman BA. Apparent hydroxyl radical production by peoxynitrite: implications for endothelial injury from nitric oxide and superoxide. *Proc Natl Acad Sci USA* 1990;87:1620–4.
54. Ischiropoulos H, Zhu L, Beckman JS. Peroxynitrite release from macrophage-derived nitric oxide. *Arch Biochem Biophys* 1992;298:446–51.
55. Radi R, Rodriguez M, Castro L, Telleri R. Inhibition of mitochondrial electron transport by peroxynitrite. *Arch Biochem Biophys* 1994;308:89–95.
56. Radi R, Beckman JS, Bush KM, Freeman BA. Peroxynitrite-induced membrane lipid peroxidation: the cytotoxic potential of superoxide and nitric oxide. *Arch Biochem Biophys* 1991;222:481–7.
57. Van der Vliet A, O'Neil CA, Halliwell B, Cross CE, Kaur H. Aromatic hydroxylation and nitration of phenylalanine and tyrosine by peroxynitrite. Evidence for hydroxyl radical production from peroxynitrite. *FEBS Lett* 1994;339:89–92.
58. Radi R, Beckman JS, Bush KM, Freeman BA. Peroxynitrite oxidation of sulfhydryls. *J Biol Chem* 1991;266:4244–50.
59. Goldberg ND, Haddox MK, Nicol SE, et al. Biological regulation through opposing influences of cyclic GMP and cyclic AMP: the yin yang hypothesis. In: Drummond GI, Greengard P, Robinson GA, eds. *Advances in cyclic nucleotide research,* Vol 5. New York: Raven Press, 1975:307–30.
60. Mellion BT, Igmarro LJ, Ohlstein EH, Pontecorvo EG, Hyman AL, Kadowitz PJ. Evidence for the inhibitory role of guanosine 3′,5′-monophosphate in ADP-induced human platelet aggregation in the presence of nitric oxide and related vasodilators. *Blood* 1981; 57:946–55.
61. Malinski T, Radomski MW, Taha Z, Moncada S. Direct electrochemical measurement of nitric oxide released from human platelets. *Biochem Biophys Res Commun* 1993;194: 960–5.
62. Brune B, Volker U. Different calcium pools in human platelets and their role in thromboxane $A_2$ formation. *J Biol Chem* 1991;266:19232–7.
63. Rees DD, Palmer RMJ, Moncada S. Role of endothelium-derived nitric oxide in the regulation of blood pressure. *Proc Natl Acad Sci USA* 1989;86:3375–8.
64. Cooke JP, Rossitch E Jr, Andon NA, Loscalzo J, Dzau VJ. Flow activates an endothelial potassium channel to release an endogenous nitrovasodilator. *J Clin Invest* 1991;88:1663–71.
65. Sneddon JM, Vane JR. Endothelium-derived relaxing factor reduces platelet adhesion to bovine endothelial cells. *Proc Natl Acad Sci USA* 1988;85:2800–4.
66. Venturini CM, Del Vecchio PJ, Kaplan JE. Thrombin-induced platelet adhesion to endothelium is modified by endothelial derived relaxing factor (EDRF). *Biochem Biophys Res Commun* 1989;159:349–54.
67. Pohl U, Busse R. EDRF increases cyclic GMP in platelets during passage through the coronary vascular bed. *Circ Res* 1989;65:1798–1803.
68. De Graaf JC, Banga JD, Moncada S, Palmer RMJ, de Groot PG, Sixma JJ. Nitric oxide functions as an inhibitor of platelet adhesion under flow conditions. *Circulation* 1992;85: 2284–90.
69. Furlong B, Henderson AH, Lewis MJ, Smith JA. Endothelium-derived relaxing factor inhibits in vitro platelet aggregation. *Br J Pharmacol* 1987;90:687–92.
70. Busse R, Luckhoff A, Bassenge E. Endothelium-derived relaxing factor inhibits platelet activation. *Naunyn Schmiedebergs Arch Pharmacol* 1987;336:566–71.
71. Macdonald PS, Read MA, Dusting GJ. Synergistic inhibition of platelet aggregation by endothelium-derived relaxing factor and prostacyclin. *Thromb Res* 1988;49:437–49.
72. Hawkins DJ, Meyrik BO, Murray JJ. Activation of guanylate cyclase and inhibition of platelet aggregation by endothelium-derived relaxing factor released from cultured cells. *Biochim Biophys Acta* 1988;969:289–96.
73. Bult H, Fret HRL, van den Bossche RM, Herman AG. Platelet inhibition by endothelium-derived relaxing factor from the rabbit perfused aorta. *Br J Pharmacol* 1988;95:1308–14.

74. Alheid U, Reichwehr I, Forstermann U. Human endothelial cells inhibit platelet aggregation by separately stimulating platelet cyclic AMP and cyclic GMP. *Eur J Pharmacol* 1989;164:103–10.
75. Broekman MJ, Eiroa AM, Marcus AJ. Inhibition of human platelet reactivity by endothelium-derived relaxing factor from human umbilical vein endothelial cells in suspension: blockade of aggregation and secretion by an aspirin-insensitive mechanism. *Blood* 1991; 78:1033–40.
76. Houston DS, Robinson P, Gerrard JM. Inhibition of intravascular platelet aggregation by endothelium-derived relaxing factor: reversal by red blood cells. *Blood* 1990;76:953–8.
77. Rosenblum WI, Nelson GH, Povlishock JT. Laser-induced endothelial damage inhibits endothelium-dependent relaxation in the cerebral microcirculation of the mouse. *Circ Res* 1987;60:169–76.
78. May GR, Crook P, Moore PK, Page CP. The role of nitric oxide as an endogenous regulator of platelet and neutrophil activation within the pulmonary circulation. *Br J Pharmacol* 1991;102:759–63.
79. Herbaczynska-Cedro K, Lembowicz K, Pytel B. $N^G$-monomethyl-L-arginine increases platelet deposition on damaged endothelium in vivo. A scanning electron microscopy study. *Thromb Res* 1991;64:1–9.
80. Bhardwaj R, Page CP, May GR, Moore PK. Endothelium-derived relaxing factor inhibits platelet aggregation in human whole blood in vitro and in the rat in vivo. *Eur J Pharmacol* 1988;157:83–91.
81. Hogan JC, Lewis MJ, Henderson AH. In vivo EDRF activity influences platelet function. *Br J Pharmacol* 1988;94:1020–2.
82. Humphries RG, Tomlinson W, O'Connor SE, Leff P. Inhibition of collagen- and ADP-induced platelet aggregation by substance P in vivo: involvement of endothelium-derived relaxing factor. *J Cardiovasc Pharmacol* 1990;16:292–7.
83. Kubes P, Suzuki M, Granger DN. Nitric oxide: an endogenous modulator of leukocyte adhesion. *Proc Natl Acad Sci USA* 1991;88:4651–5.
84. Moilanen E, Vuorinen P, Metsa-Ketela T, Vapaatalo H. Inhibition by nitric oxide donors of human polymorphonuclear leucocyte functions. *Br J Pharmacol* 1993;109:852–8.
85. Kurose I, Kubes P, Wolf R, et al. Inhibition of nitric oxide production. Mechanisms of vascular albumin leakage. *Circ Res* 1993;73:164–71.
86. Stamler JS, Vaughan DE, Loscalzo J. Synergistic disaggregation of platelets by tissue-type plasminogen activator, prostaglanding $E_1$ and glyceryl trinitrate. *Circ Res* 1989;65: 796–804.
87. Maurine DH, Haslam RJ. Molecular basis of the synergistic inhibition of platelet function by nitrovasodilators and activators of adenylate cyclase: inhibition of cyclic AMP breakdown by cyclic GMP. *Mol Pharmacol* 1990;37:671–81.
88. Venturini CM, Weston LK, Kaplan JE. Platelet cGMP but not cAMP, inhibits thrombin-induced platement adhesion to pulmonary vascular endothelium. *Am J Physiol* 1992;32: H606–12.
89. Moncada S, Higgs EA. The L-arginine–nitric oxide pathway. *N Engl J Med* 1993;329: 2002–12.
90. Ialenti A, Moncada S, Di Rosa M. Modulation of adjuvant arthritis by endogenous nitric oxide. *Br J Pharmacol* 1993;110:701–6.
91. Radomski MW, Vallance P, Whitley G, Foxwell N, Moncada S. Platelet adhesion to human vascular endothelium is modulated by constitutive and cytokine induced nitric oxide. *Cardiovasc Res* 1993;27:1380–2.
92. Schultz PJ, Raij L. Endogenously synthesized nitric oxide prevents endotoxin-induced glomerular thrombosis. *J Clin Invest* 1992;90:1718–25.
93. Moro MA, Darley-Usmar VM, Goodwin DA, et al. Paradoxical fate and biological action of peroxynitrite on human platelets. *Proc Natl Acad Sci USA* 1994;91:6702–6.
94. Petros A, Bennett D, Vallance P. Effect of nitric oxide synthase inhibitors on hypotension in patients with septic shock. *Lancet* 1991;338:1557–8.
95. Chester AH, O'Neil GS, Moncada S, Tadjkarimi S, Yacoub M. Low basal and stimulated release of nitric oxide in atherosclerotic epicardial coronary arteries. *Lancet* 1990;336: 897–900.
96. Verbeuren TJ, Jordaens FH, Zonnekeyn LL, van Hove CE, Coene MC, Herman AG.

Effect of hypercholesterolemia on vascular reactivity in the rabbit. I. Endothelium-dependent and endothelium-independent contractions and relaxations in isolated arteries of control and hypercholesterolemic rabbits. *Circ Res* 1986;58:552–64.
97. Barrett ML, Willis AL, Vane JR. Inhibition of platelet-derived mitogen release by nitric oxide (EDRF). *Agents Actions* 1989;27:488–91.
98. Scott-Burden T, Vanhoutte PM. Regulation of vascular smooth muscle proliferation: role of endothelium-derived relaxing factor (nitric oxide). *J Vasc Biol* 1991;3:445–6.
99. Drexler H, Zeiher AM, Meinzer K, Just H. Correction of endothelial dysfunction in coronary microcirculation of hypercholesterolaemic patients by L-arginine. *Lancet* 1991; 338:1546–50.
100. Craeger MA, Gallagher SJ, Girerd XJ, Coleman SM, Dzau VJ, Cooke JP. L-Arginine improves endothelium-dependent vasodilatation in hypercholesterolemic humans. *J Clin Invest* 1992;90:1248–53.
101. Cooke JP, Tsao P. Cellular mechanisms of atherogenesis and the effects of nitric oxide *Curr Opin Cardiol* 1992;7:799–804.
102. Calver A, Collier J, Moncada S, Vallance P. Effect of local infusion on $N^G$-monomethyl-L-arginine in patients with hypertension. The nitric oxide dilator mechanism appears abornal. *J Hypertens* 1992;10:1025–31.
103. Calver A, Collier J, Vallance P. Inhibition and stimulation of nitric oxide synthesis in the human forearm arterial bed of patients with insulin-dependent diabetes. *J Clin Invest* 1992;90:2548–54.
104. Amado JA, Salas E, Botana MA, Poveda JJ, Berrazueta JR. Low levels of intraplatelet cGMP in IDDM. *Diabetes Care* 1993;16:809–11
105. Feelisch M, Noack EA. Correlation between nitric oxide formation during degradation of organic nitrates and activation of guanylate cyclase. *Eur J Pharmacol* 1987;139:19–30.
106. Gerzer R, Karrenbrock B, Siess W, Heim JM. Direct comparison of the effects of nitroprusside SIN 1, and various nitrates on platelet aggregation and soluble guanylate cyclase activity. *Thromb Res* 1988;52:11–21.
107. Gebalska J. Platelet adhesion and aggregation in relation to clinical course of acute myocardial infarction. MD thesis, Warsaw, 1990 [in Polish].
108. Sinzinger H, Virgolini I, O'Grady J, Rauscha F, Fitscha P. Modification of platelet function by isosorbide dinitrate in patients with coronary artery disease. *Thromb Res* 1992; 65:323–35.
109. Wautier JL, Weill D, Kadeva H, Maclouf J, Soria C. Modulation of platelet function by SIN-1A. J *Cardiovasc Pharmacol* 1989;14(suppl 11):S111–4.
110. Sinzinger H, Rauscha F, O'Grady J, Fitscha P. Prostaglandin $I_2$ and the nitric oxide donor molsidomine have synergistic effect on thromboresistance in man. *Br J Clin Pharmacol* 1992;33:289–92.
111. Langford EJ, Brown AS, Wainwright RY, et al. Inhibition of platelet activity by S-nitroglutathione during coronary angioplasty. *Lancet* 1994;344:1458–60.
112. Sinzinger H, Fitscha P, O'Grady J, Rauscha F, Rogatt W, Vane JR. Synergistic effect of prostaglandin $E_1$ and isosorbide dinitrate in peripheral vascular disease. *Lancet* 1990;335: 627–8.
113. Aono J, Sugawa M, Koide T, Takato M. The role of cGMP in the anti-aggregating properties of BY1949, a novel dibenzoxazepine derivative. *Eur J Pharmacol* 1991;195:225–31.
114. Bult H, Fret HRL, Jordaens FH, Herman, AG. Dipyridamole potentiates the antiaggregating and vasodilator activity of nitric oxide. *Eur J Pharmacol* 1991;199:1–8.

*Topics in Molecular Medicine, Volume 1,*
edited by Wolfgang Siess, Reinhard Lorenz,
and Peter C. Weber. Raven Press, Ltd.,
New York © 1995.

# 15

# Endothelial Cell 13-HODE and Blood Cell Adhesion to the Vessel Wall

Michael R. Buchanan and Stephanie J. Brister

*McMaster University Health Sciences Centre, and the Hamilton Civic Hospitals, Hamilton, Ontario, Canada L8L 2X2*

Under normal conditions the intact monolayer of endothelial cells lining the vascular wall is not reactive to the circulating blood cells (1,2). However, when the wall becomes exposed to cytokines, such as interleukin-1 (IL-1) or tumor necrosis factor (TNF), or to procoagulants such as thrombin, the endothelium becomes reactive to circulating blood cells at the sites of stimulation or perturbation (3–7). A key event that is central to these pathologic responses is adhesion of circulating blood cells (e.g., platelets, leukocytes, macrophages, monocytes, and tumor cells) to the vascular endothelium and/or to any exposed underlying extracellular matrices. Fundamental to this adhesive phenomenon is the expression of a number of adhesion molecule receptors (7,8). Each adhesion molecule receptor possesses affinity for specific ligands, thereby facilitating specific blood cell–vessel wall interactions during the course of inflammation, hemostasis–thrombosis, or metastasis. Recognition of the unique properties of each adhesion molecule receptor has led to a better identification of the various structural properties of these receptors and a better understanding of their specific roles in specific pathogenic conditions. However, it is still unclear what mechanism(s) regulate(s) the expression of these adhesion molecule receptors and what regulates their ability to recognize specific ligands. A number of studies performed by our laboratory and others over the last decade suggest that intracellular monohydroxides derived from the lipoxygenase pathway play important roles in this process under both basal and stimulated conditions.

## FATTY ACID METABOLITES AND THE ENDOTHELIUM

From the perspective of fatty acid metabolites, most investigators over the past few decades have focused on the importance of prostaglandin synthesis in circulating blood cells and endothelial cells in regulating inflammation, thrombosis, and metastasis. It is apparent that a variety of stimuli generated in response to cell injury, such as thrombin, endotoxin, and cytokines, can induce the synthesis and subsequent release of cyclo-oxygenase–derived arachidonic metabolites (9–13). Specifically, cell membrane arachidonic acid is released from the phospholipid stores by a plasma membrane-associated phospholipase after stimulation. In the endothelial cell, the free arachidonic acid is rapidly isomerized, then hydrolyzed to a labile endoperoxide which, in turn, is metabolized into prostacyclin or $PGI_2$ (9,10). $PGI_2$ is a potent inhibitor of platelet function (10) and achieves its effect by increasing intracellular cAMP (14,15). This, in turn, renders platelets, tumor cells, and other circulating blood cells (as well as endothelial cells) less responsive to further stimulation. It has also been demonstrated that endothelial cells per se not only metabolize their own endogenous endoperoxides into $PGI_2$ but also metabolize exogenously-derived endoperoxides into $PGI_2$ (16). In addition, endothelial cells contribute to the conversion of the stable but inactive end-product 6-keto-$PGF_{1\alpha}$ into a potent and stable inhibitor, 6-keto-$PGE_1$, in concert with platelets or leukocytes (17–21). The observations that both 6-keto-$PGE_1$ and $PGI_2$ are potent inhibitors of platelet function (mediating their effect by enhancing cAMP) have suggested to some that these prostanoids play key roles both in regulating blood–vessel wall interactions after perturbation or injury and in regulating the biocompatibility or thromboresistance of the endothelium under homeostatic conditions (10). The latter possibility, however, is questionable, because cell stimulation is a requirement for the synthesis of both $PGI_2$ and 6-keto-$PGE_1$.

## 13-HODE METABOLISM AND THE ENDOTHELIUM

Other investigators have reported that under basal conditions (when endothelial cells are thromboresistant), endothelial cells continuously turn over triglycerides (22) and release linoleic acid, which, in turn, is synthesized to another fatty acid metabolite via the 15-lipoxygenase pathway, i.e., 13-hydroxyoctadecadienoic acid (13-HODE) (23–25). Triglyceride turnover is dependent on the level of cAMP (25). Linoleic acid is continuously released from this cytosolic triglyceride pool and is metabolized by the endothelial cell enzyme 15-lipoxygenase into the monohydroxide, 13-HODE. 13-HODE remains intracellular and is not released into the ambient surroundings, unlike $PGI_2$, which is released from endothelial cells after stimulation or injury. When endothelial cells are stimulated and their cAMP level falls, 13-HODE synthesis ceases (e.g., after stimulation by thrombin, endotoxin, or a number of cytokines) (23–26). Finally, the levels of intracellular 13-HODE in vascular cells can be elevated by increasing the intracellular level of cAMP as measured both in vitro and in vivo (23–27). This effect is associated with an increase in linoleic acid release from the triglyceride pool, thus providing a greater amount of substrate for 13-HODE synthesis via the lipoxygenase pathway.

Other investigators have also reported that endothelial cells can metabolize exogenously added linoleic acid into 13-HODE and into 9-HODE via the cyclo-oxygenase pathway (28–30). These observations were seen in studies in which the cells were exposed to high concentrations of the exogenous substrate for a short duration in vitro. The significance of these observations is not yet clear, as 18 carbon fatty acids with two double bonds, such as linoleic acid, are poor substrates for the cyclo-oxygenase enzyme. Consequently, high concentrations of such substrate are required to generate cyclo-oxygenase-derived 9- and 13-HODE. It is unclear whether such high concentrations can actually be generated in vivo. Moreover, it requires 12 to 18 h for the linoleic acid to become incorporated into the triglyceride stores under steady-state conditions (25), indicating that cyclo-oxygenase–derived 9-HODE and 13-HODE in the above studies are generated under nonphysiologic conditions.

## MECHANISM OF ACTION OF 13-HODE

It has been well established that the expression of adhesion molecules on the surface of cells in the blood vessel wall is required for cell–cell interactions after stimulation or injury. Many of these adhesion molecules are composed of two glycoprotein chains that belong to the super family of adhesion molecules called integrins. Integrins have a specific adhesive domain in their β-chain, which binds ligands containing the RGD sequence. The RGD-binding domain must be expressed for receptor–ligand binding and subsequent cell–cell adhesion (8). What is not as well established is the mechanism that regulates the expression of these integrins in their adhesive form from within the cell.

A number of studies suggest that the intracellular 13-HODE and monohydroxides derived from arachidonic acid (5-, 12-, and/or 15-hydroxyeicosatetaenoic acid) regulate adhesion in endothelial cells and in other cells (23–26,31–37). For example, when the intracellular level of 13-HODE is increased in either tumor cells or endothelial cells, there is a corresponding decrease in the ability of tumor cells to adhere to the endothelial cells (26, 33,37). In contrast, when the tumor cells or the endothelial cells are stimulated (e.g., by naturally occurring cytokines), 13-HODE synthesis decreases and tumor cell–endothelial cell adhesion increases. These observations prompted us to test the hypothesis that 13-HODE downregulates vessel wall reactivity to circulating blood cells. Therefore, we examined the relationship between a specific integrin, the vitronectin receptor, and endothelial cells under basal (nonadhesive) and perturbed (adhesive) conditions (38) (the vitronectin receptor is a two-chain glycoprotein belonging to the integrin family, which is expressed on endothelial cells). We found, using immunohisto-fluorescence techniques, that in unstimulated (i.e., nonadhesive) endothelial cells, the vitronectin receptor co-localizes with 13-HODE in specific vesicles immediately below the plasma membrane. Neither 13-HODE nor vitronectin receptor could be detected on the apical surface of these nonadhesive endothelial cells. Both 13-HODE and the vitronectin receptor were detected in a punctate pattern within the endothelial cells. The patterns of localization of both 13-HODE and the vitronectin receptor were identical, indicating that both moieties were co-localized. We also confirmed that the vitronectin receptor was in its nonadhesive form in the unstimulated endothelial cells, because the RGD-binding site could not be detected by a specific monoclonal antibody that recognized only the RGD-binding site. In contrast, when the endothelial cells were stimulated with IL-1, 13-HODE disappeared and the vitronectin receptor re-localized to the apical surface of the endothelial cells and was now expressed in its adhesive form (the latter was confirmed by detection of the receptor with a specific monoclonal antibody to the RGD site). These data supported the hypothesis that the inverse relationship between 13-HODE and endothelial cell adhesiveness is due to a physiochemical interaction of 13-HODE with the vitronectin receptor within endothelial cells. This possibility is also consistent with observations that other receptors in their nonadhesive form are located in specific vesicles in cells, such as the laminin and fibronectin receptors in specific adhesomes in leukocytes under resting conditions (39), and that the lipid environment in which the vitronectin receptor is located influences its ability to recognize its ligands (40).

## RECENT ONGOING STUDIES

More recently, we transfected human umbilical cord endothelial cells with an adenoviral plasmid containing either sense or antisense 15-lipoxygenase cDNA + T cDNA (41). Endothelial cell message 15-lipoxygenase was measured by Western blotting, endothelial cell 13-HODE was measured by HPLC, and endothelial cell reactivity was measured by determining the number of radiolabeled platelets adhering to the endothelial cell monolayer, using an adhesion assay. When endothelial cells were transfected with the sense adenoviral 15-lipoxygenase vector, endothelial cell 15-lipoxygenase mRNA increased. In contrast, when the endothelial cells were transfected with the antisense 15-lipoxygenase vector, 15-lipoxygenase mRNA was decreased (41). The changes in the 15-lipoxygenase activity were associated with corresponding changes in endothelial cell 13-HODE synthesis and with inverse changes in endothelial cell reactivity ($p < 0.001$) (Table 1). These data provide direct proof for a causal relationship. Therefore, endothelial cell 13-HODE appears to regulate the conformational presentation of the vitronectin receptor, thereby influencing its ability to recognize its ligands.

**TABLE 1.** *15-Lipoxygenase, 13-HODE, and platelet–endothelial cell interactions of endothelial cells containing sense or antisense 15-lipoxygenase cDNA + T cDNA*

| | 15-Lipoxygenase mRNA | 13-HODE | PLT/EC ADH |
|---|---|---|---|
| Sense | +40 (4) | +17 ± 1 (4) | −9 ± 2 (8) |
| | | * | ** |
| Antisense | ND (4) | −9 ± 2 (4) | +18 ± 4 (8) |

PLT, platelets; EC, endothelial cells.
Data are expressed as a percent change from control, mean ± SEM (*n*).
** $p < 0.001$; *$p < 0.01$; ND, not detected.

This was confirmed by study of the relative effects of 13-HODE and 15-HETE (15-hydroxyeicosatetraenoic acid) on the binding affinity of the vitronectin receptor for its ligands (15-HETE is the monohydroxide derived from arachidonic acid via 15-lipoxygenase). Each monohydroxide was inserted into liposomes containing purified vitronectin receptor, after which vitronectin receptor–liposome binding to vitronectin-, fibronectin-, or laminin-coated substrates was measured. The vitronectin receptor–liposomes bound more avidly to vitronectin than to fibronectin than to laminin (Fig. 1).When 15-HETE was inserted into the liposomes containing the vitronectin receptor, vitronectin receptor–liposome binding to the three ligands increased (Fig. 2). In contrast, when 13-HODE was inserted into the liposomes containing the vitronectin receptor, vitronectin receptor–liposome binding decreased (41). The former observations indicate a causal relationship between the 15-lipoxygenase pathway and the adhesiveness of the endothelial cell vitronectin receptor. The latter results indicate that the mechanism by which the 15-lipoxygenase monohydroxides regulate endothelial cell vitronectin receptor adhesion is a direct physiochemical interaction, thereby altering the ability of the vitronectin receptor to recognize its ligands.

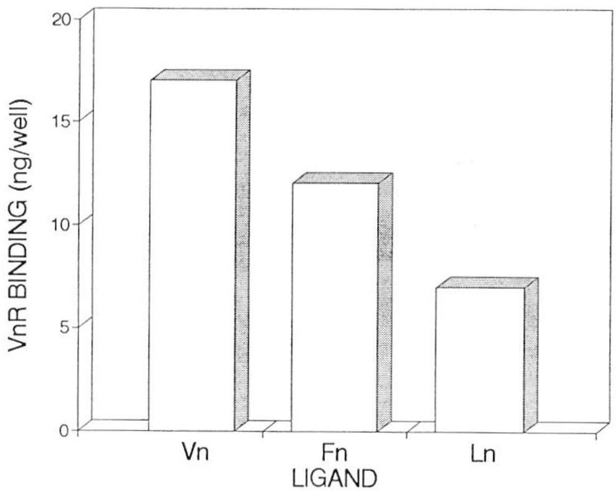


**FIG. 1.** Differential binding of vitronectin–liposomes to vitronectin-, fibronectin-, and laminin-coated substrates.

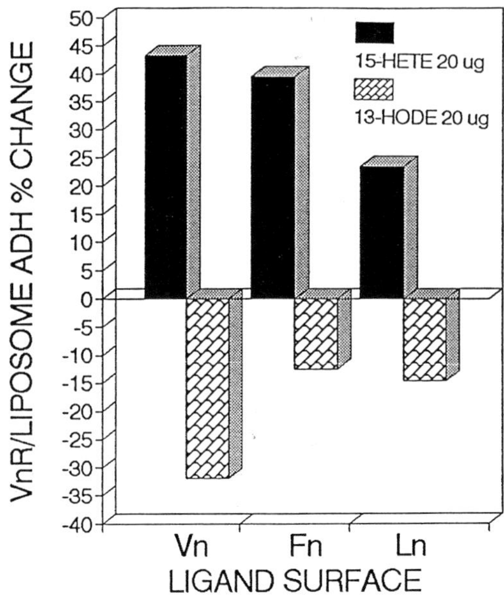


**FIG. 2.** Effect of 13-HODE and 15-HETE on vitronectin receptor–liposome binding to vitronectin-, fibronectin-, and laminin-coated substrates. Data are expressed as a percent change from control.

## POSSIBLE RELEVANCE IN VIVO

These in vitro and cell culture studies suggest that the blood–vessel wall interactions in vivo may be mediated by modulation of the vascular 15-lipoxygenase pathway. For example, up- or downregulation of endothelial cell reactivity may influence either hemostasis or thrombosis. To test this possibility, a number of experiments were performed using strategies designed to alter the amount of substrate available for metabolism into 13-HODE (27). Rabbits were pretreated orally with salicylate or dipyridamole for 1 week. The salicylate was given to block the peroxidation of 13-HPODE to 13-HODE. The dipyridamole was given to inhibit the phosphodiesterase activity, thereby increasing vessel wall cAMP and facilitating the release of linoleic acid. After 1 week of oral treatment with either drug, both carotid arteries of each animal were isolated and the endothelium was selectively damaged. Three hours later, all animals were injected with radiolabeled platelets and the number of platelets adhering to the injured vessel wall surface over the following hour was determined as a measure of vessel wall thrombogenicity.

Platelet adhesion to the injured carotid arteries increased twofold in the salicylate-treated animals. This increased thrombogenicity was associated with a significant decrease (greater than 60%) in 13-HODE synthesis. Platelet adhesion to the injured vessel wall in the dipyridamole-treated animals decreased by 50% and was associated with a significant increase in 13-

HODE synthesis. The increase in 13-HODE synthesis also correlated with the circulating dipyridamole plasma level and vessel wall cAMP. Neither compound had any significant effect on platelet function as assessed ex vivo (27). Similar relations between vascular 13-HODE levels and vessel wall thrombogenicity were seen in animals fed semipurified diets that were linoleic acid-rich or linoleic acid-depleted (42). These observations provided direct evidence that the amount of 13-HODE produced by vessel wall correlates with vessel wall thromboresistance in vivo. This gives credence to the possibility that manipulating vessel wall thromboresistance per se may be an effective approach to inhibit platelet–vessel wall interactions after vessel wall injury in the clinical setting.

## HUMAN STUDIES

Human vessel walls also synthesize 13-HODE (43). Their reactivity to platelets also correlates inversely with their level of 13-HODE. For example, segments of internal mammary arteries and saphenous veins obtained from patients undergoing elective coronary artery bypass grafting (CABG) produced 13-HODE (43). The arteries produced six times more 13-HODE than the veins. Interestingly, 13-HODE synthesis in both veins and arteries decreased with age. This age-dependent decrease in 13-HODE was paralleled by an increase in vessel wall thrombogenicity (43). Preliminary results from an ongoing study in patients undergoing elective CABG suggest that specific dietary supplements can alter 13-HODE synthesis (44). Thus, artery and vein segments obtained from individuals who received a dietary supplement rich in linoleic acid for 1 month before surgery produced more 13-HODE than arteries and veins obtained from patients who received a placebo control (Table 2). Moreover, vessel wall 13-HODE decreased with age ($r = 0.7620$; $p < 0.005$), and this age-dependent decrease was associated with an age-dependent increasse in vessel wall reactivity ($r = 0.9480$; $p < 0.01$), consistent with our previous studies (43). Thus, upregulating 13-HODE in the clinical setting can decrease vessel wall thrombogenicity. The impact on subsequent clinical end points, however, has yet to be elucidated.

**TABLE 2.** *Effect of dietary linoleic acid supplements on human internal mammary artery 13-HODE and thrombogenicity*

| | Treatment | |
|---|---|---|
| | +Linoleic acid | Placebo |
| VW 13-HODE | 1344 ± 416 | 734 ± 206 |
| VW THROMB | 2.65 ± 0.9 | 3.09 ± 0.5 |

VW, vessel wall; THROMB, thrombogenicity.
Data are expressed as ng/g/cm$^2$, mean ± SEM.

## SUMMARY

A number of experimental data indicate that intracellular 13-HODE varies inversely with cell surface adhesiveness, not only in endothelial cells but also in smooth-muscle cells and a variety of tumor cells. Moreover, a number of these in vitro studies demonstrate that increasing endothelial cell 13-HODE decreases endothelial cell reactivity for platelets and tumor cells, and vice versa.

More recent studies have confirmed that modulation of purified vitronectin receptor by 13-HODE decreases vitronectin receptor adhesion by reducing the receptor's ability to recognize its specific ligands. These data suggest that this lipoxygenase metabolite plays a direct causal role in regulation of integrin expression and in recognition of its ligands via a direct physiochemical relationship.

The clinical relevance of these observations is supported by the observation that both human and animal arteries and veins also produce 13-HODE in vivo and that the same inverse relationships exist between intravascular 13-HODE and vessel wall thrombogenicity, which affect subsequent blood cell–vessel wall interactions. Moreover, 13-HODE synthesis decreases with age, coincidental with the progressive onset of cardiovascular disease. It is therefore suggested that therapeutic modulation of this lipoxygenase pathway may have a beneficial impact on thrombogenesis.

## ACKNOWLEDGMENT

The studies performed by us that are cited herein were supported by the Heart and Stroke Foundation of Ontario and by the Medical Research Council of Canada.

## REFERENCES

1. Fuster V, Chesebro JH. Current concepts of thrombogenesis: Role of platelets. *Mayo Clin Proc* 1981;56:102.
2. Ross R. Atherosclerosis: a problem of the biology of arterial wall cells and their interactions with blood components. *Arteriosclerosis* 1981;1:293.
3. Hekman CM, Loskutoff DJ. Endothelial cells produce a potent inhibitor of plasminogen activators that can be activated by denaturants. *J Biol Chem* 1985;260:11581.
4. Bevilacqua MP, Gimbrone MA Jr. Inducible endothelial functions in inflammation and coagulation. *Semin Thromb Hemost* 1987;13:425.
5. Preissner KT, Delvos U, Muller-Berghaus G. Binding of thrombin to thrombomodulin accelerates inhibition of the enzyme by antithrombin III. Evidence for a heparin-independent mechanism. *Biochemistry* 1987;26:2521.
6. Lafrenie RM, Podor TJ, Buchanan MR, Orr FW. Up-regulated biosynthesis and expression of endothelial cell vitronectin in receptor enhances cancer cell adhesion. *Cancer Res* 1992; 52:2202.
7. Van Mourik JA, Von Dem Borne AEG, Giltay JC. Pathophysiological significance of integrin expression by vascular endothelial cells. *Biochem Pharmacol* 1990;39:233.

8. Hynes RO. Integrins: a family of cell surface receptors. *Cell* 1987;48:549.
9. Hamberg M, Svensson J, Samuelsson B. Prostaglandin endoperoxides: a new concept concerning the mode of action and release of prostaglandins. *Proc Natl Acad Sci USA* 1974;71:3824.
10. Moncada S, Vane JR. Arachidonic acid metabolites and the interactions between platelets and blood vessel walls. *N Engl J Med* 1979;300:1142.
11. Defreyn G, Deckmyn H, Vermylen J. A thromboxane synthetase inhibitor agonist reorients endoperoxides metabolism in whole blood towards prostacyclin and prostaglandin $E_2$. *Thromb Res* 1982;26:389.
12. Mehta J, Mehta P, Lawson DL, Ostrowski N, Brigmon L. Influence of selective thromboxane synthetase blocker CGS-13080 on thromboxane and prostacyclin biosynthesis in whole blood: evidence for synthesis of prostacyclin by leukocytes from platelet-derived endoperoxides. *J Lab Clin Med* 1985;106:246.
13. Burrows FJ, Haskard DO, Hart IR, Marshall JF, Selkirk S, Poole S. Influence of tumor-derived interlukin 1 on melanoma-endothelial cell interactions *in vitro*. *Cancer Res* 1991; 51:4768.
14. Hajjar DP, Weksler BB, Falcone DJ, Hefton JM, Taik-Goldman K, Minick CR. Prostacyclin modulates cholesteryl hydrolytic activity by its effects on cyclic adenosine monophosphate in rabbit aortic smooth muscle cell. *J Clin Invest* 1982;70:479.
15. Pohlman T, Yates J, Needleman P, Klahr S. Effects of prostacyclin on short circuit current and water flow in the toad urinary bladder. *Am J Physiol* 1983;244:270.
16. Marcus AJ, Broekman MJ, Weksler BB, et al. Arachidonic and metabolism in endothelial cells and platelets. *Ann NY Acad Sci* 1982;401:195.
17. Quilley CP, Wong PYK, McGiff JC. Hypotensive and renovascular actions of 6-keto-prostaglandin $E_1$: a metabolite of prostacyclin. *Eur J Pharmacol* 1979;57:273.
18. Wong PY-K, Lee WH, Chao PH-W, Reiss RF, McGiff JC. Metabolism of prostacyclin by 9-hydroxyprostaglandin dehydrogenase in human platelets. *J Biol Chem* 1980;225:9021.
19. Wong PY-K, Malik KU, Desiderio DM, McGiff JC, Sun FF. Hepatic metabolism of prostacyclin ($PGI_2$) in the rabbit: formation of a potent novel inhibitor of platelet aggregation. *Biochem Biophys Res Commun* 1980;93:486.
20. Berry CN, Hoult JRS. 6-Keto-prostaglandin $E_1$: its formation by platelets from prostacyclin and resistance to pulmonary degradation. *Pharmacology* 1983;26:324.
21. Buchanan MR. Inhibition of thrombosis by leukocytes: role of endogenous fatty acid metabolites. *Sanofi Foundation for Thrombosis Research Bulletin #4,* 3rd Sanofi Prix 14, 1989.
22. Denning GM, Figard PH, Koduce TL, Spector A. Role of triglycerides in endothelial cell arachidonic acid metabolism. *J Lipid Res* 1983;24:993.
23. Buchanan MR, Butt RW, Magas Z, Van Ryn J, Hirsh J, Nazir DJ. Endothelial cells produce a liopoxygenase-derived chemorepellant which influences platelet/endothelial cell interactions: effect of aspirin and salicylate. *Thromb Haemost* 1985;53:306.
24. Buchanan MR, Haas T, Lagarde M, Guichardant M. 13-Hydroxy-octadecadienoic acid is the vessel wall chemorepellant factor, LOX. *J Biol Chem* 1985;260:16056.
25. Haas TA, Bertomeu MC, Bastida E, Buchanan MR. Cyclic AMP regulation of endothelial cell triglyceride turnover, 13-hydroxyoctadecadienoic acid (13-HODE) synthesis and endothelial cell thrombogenicity. *Biochim Biophys Acta* 1990;1031:174.
26. Haas TA, Bastida E, Nakamura K, Hullin F, Almirall L, Buchanan MR. Binding of 13-HODE, 5-, 12- and 15-HETE to endothelial cells and subsequent platelet, neutrophil and tumor cell adhesion. *Biochim Biophys Acta* 1988;961:153.
27. Weber E, Haas TA, Mueller TH, et al. Relationship between vessel wall 13-HODE synthesis and vessel wall thrombogenicity following injury. Influence of salicylate and dipyridamole treatment. *Thromb Res* 1990;57:383.
28. Funk CD, Powell WS. Release of prostaglandins and monohydroxy and trihydroxy metabolites of linoleic and arachidonic acids by adult and fetal aortae and ductus arteriosus. *J Biol Chem* 1985;260:7481.
29. Kaduce TL, Figard PH, Leifur R, Spector AA. Formation of 9-hydroxyoctadecadienoic acid linoleic acid in endothelial cells. *J Biol Chem* 1989;264:6830.
30. Kuhn H, Belkner J, Wiesner R, Alder L. Occurrence of 9- and 13-keto-octadecadienoic

acid in biological membranes oxygenated by the reticulocyte lipoxygenase. *Arch Biochem Biophys* 1990;270:218.

31. Cook J, Delebassee S, Aldigier JC, Gaulde N, Kazatchkine M. 15-HETE "modulates" expression of C3b receptor (CR1) antigen on peripheral blood B-lymphocytes. *Prost Leuko Med* 1986;23:201.
32. Ryn-McKenna JV, Buchanan MR. Relative effects of flurbiprofen on platelet 12-hydroxy-eicosatetraenoic acid and thromboxane $A_2$ production: influence on collagen-induced platelet aggregation and adhesion. *Prost Leuko Essent Fatty Acids* 1989;36:171.
33. Grossi IM, Fitzgerald LA, Umbarger LA, et al. Bidirectional control of membrane expression and/or activation of tumor cell IRGpIIb/IIIa receptor and tumor cell adhesion by lipoxygenase products of arachidonic acid and linoleic acid. *Cancer Res* 1989;49:1029.
34. Lin B, Timor J, Howlett J, Diglio DA, Honn KV. Lipoxygenase metabolites of arachidonic and linoleic acids modulate the adhesion of tumor cells to endothelium via regulation of protein kinase C. *Cell Regul* 1991;2:104.
35. Valles J, Santos MT, Marcus AJ, et al. Downregulation of human platelet reactivity by neutrophils. Participation of lipoxygenase derivative and adhesive proteins. *J Clin Invest* 1993;92:1357.
36. Hernandez R, Alemany M, Bozzo J, Buchanan MR, Ordinas A, Bastida E. Platelet adhesivity to subendothelium is influenced by polymorphonuclear leukocytes: studies with aspirin and salicylate. *Haemostasis* 1993;23:1.
37. Bastida E. Bertomeu MC, Haas TA, et al. Regulation of tumor cell adhesion by intracellular 13-HODE:15HETE ratio. *J Lipid Mediator* 1990;2:281.
38. Buchanan MR, Bertomeu MC, Haas TA, Orr FW, Eltringham-Smith LJ. Localization of 13-HODE and the vitronectin receptor in human endothelial cells and subsequent *in vitro* endothelial cell platelet interactions. *Blood* 1993;81:3303.
39. Singer II, Scott S, Kawka W, Kazazis DM. Adhesomes: specific granules containing receptors for laminin, C3bi/fibrinogen, fibronectin, and vitronectin in human polymorphonuclear leukocytes and monocytes. *J Cell Biol* 1989;109:3169.
40. Conforti G, Zanetti A, Pasquali-Ronchetti I, Quaglino D Jr, Neyroz P, Dejana E. Modulation of vitronectin receptor binding by membrane lipid composition. *J Biol Chem* 1990; 265:4011.
41. Buchanan MR, Wale C, Smith LJ, Gallo S, Lafrenie RM, Haas TA. Regulation of vitronectin receptor ligand binding by the 15-lipoxygenase monohydroxides, 15-HETE and 13-HODE. *Proceedings of the VIIIth Symposium on the Biology of Vascular Cells,* Heidelberg, 1994.
42. Bertomeu MC, Crozier GL, Haas TA, Fleith M, Buchanan MR. Selective effects of dietary fats on vascular 13-HODE synthesis and platelet/vessel wall interactions. *Thromb Res* 1990;59:819.
43. Brister SJ, Haas TA, Bertomeu MC, Austin J, Buchanan MR. 13-HODE synthesis in internal mammary arteries and saphenous veins: implications in cardiovascular surgery. *Adv Prost Thromb Leuko Res* 1990;21:667.
44. Brister SJ, Buchanan MR, Horrobin D. Altering vessel wall 13-HODE and vessel wall reactivity with dietary fatty acid supplements in patients undergoing cardiac surgery. *Proceedings VIIIth Symposium on the Biology of Vascular Cells,* Heidelberg, 1994.

*Topics in Molecular Medicine, Volume 1,*
edited by Wolfgang Siess, Reinhard Lorenz,
and Peter C. Weber. Raven Press, Ltd.,
New York © 1995.

# 16

# The Cell Matrix Adhesion Regulator Gene

Marco R. Novelli, Helga Durbin, and Walter F. Bodmer

*Imperial Cancer Research Fund, London WC2A 3PX, England*

The cell matrix adhesion regulator (CMAR) gene was first discovered during a search for genes that increase cell adhesion to collagen. It had been noted in previous work that there is great variability among colon carcinoma cell lines in their binding to collagen type I (1). One of the cell lines investigated, SW1222, bound extremely well to collagen type I and, unlike most other human colon carcinoma cell lines, could differentiate into glandular structures when cultured in collagen gels (2). Further experimentation showed that the adhesion of SW1222 to collagen type I was dependent on the $\alpha_2\beta_1$ integrin (2,3) and that gland formation also depended on the same $\alpha_2\beta_1$ integrin and on E-cadherin (4).

Examination of other human colon carcinoma cell lines showed that although they also expressed similar levels of the $\alpha_2\beta_1$ integrin they bound more weakly, or not at all, to collagen type I, and were unable to form glandular structures in collagen gels. This suggested that another factor was needed to activate collagen I binding and differentiation induction. The mouse colorectal cell line CMT93 is unable to form glandular structures when grown in collagen gels. However, when somatic cell hybrids were made between SW1222 and an HRGPT-deficient variant of CMT93, a number of these hybrids were able to form glandular structures, this ability apparently segregating with human chromosome 15 (2). This suggested that a dominant genetic factor was responsible for gland formation, possibly mapping to chromosome 15.

In an attempt to determine what gene or genes might be causing these changes, an adaptation of the expression cloning technique described by Seed and Aruffo was used (5). Initially, a cDNA library was constructed from the cell line SW1222 in the expression vector pCDM8 (6). The library was then transfected by electroporation into the mouse transformed fibroblast cell line WOP, which does not normally adhere to collagen type I. WOP cells carry the SV40 T antigen and so are able to support extra chromosomal replication and expression of the pCDM8 vector. After transfection, cells were incubated on a collagen-coated plate and the plate gently washed to remove nonadherent cells. Any adherent cells remaining were then lysed and the plasmids they contained harvested. This procedure was repeated several times to enrich for any plasmids containing cDNAs that induce cell adhesion. After four rounds of "panning" on collagen type I, several resulting cDNA clones were isolated, transfected into WOP cells, and adhesion assays performed to determine which clones increased cell adhesion to collagen type I. The clone inducing maximal adhesion, clone 27, increased the binding of WOP cells to collagen type I by three to fivefold. This clone was sequenced and was initially named CAR (cell adhesion regulator) (7). However a gene already existed with this name, so the gene was renamed cell matrix adhesion regulator (CMAR).

Inspection of the CMAR sequence shows that it is a small intronless gene encoding an 81-amino-acid protein (Fig. 1A). This protein has several interesting features. First, it contains two myristoylation consensus signatures, suggesting an intracytoplasmic submembranous location within the cell. Second, there is a terminal tyrosine lying in a region highly homologous to regions at which tyrosine phosphorylation is commonly found.

The importance of the terminal tyrosine was investigated by constructing a mutated CMAR clone in which tyrosine was replaced with a stop codon. Transfection of this mutated CMAR clone into WOP cells failed to increase binding to collagen, suggesting that the terminal tyrosine is essential for the cell adhesion function of CMAR (7).

The small size of the CMAR protein and the fact that its protein structure contains no *trans*-membrane regions makes it very unlikely to act as an adhesion molecule in itself. However, its probable submembranous location and its terminal tyrosine suggest that it could act as a signaling molecule to activate binding function, interacting with the cytoplasmic tails of adhesion molecules such as the integrins and possibly with the cytoskeleton.

For most epithelial malignancies, the degree of tumor differentiation generally correlates well with the clinical stage of the tumor and the overall prognosis; the more poorly differentiated the tumor, the higher the clinical stage and poorer the prognosis. Because the CMAR gene appears to regulate cell adhesion to collagen, and thus possibly the state of differentiation of the tumor cells, it was suggested that CMAR might act as a tumor suppressor gene.

Screening of somatic cell hybrids, followed by FISH studies using a CMAR-containing cosmid, showed a 16q.24 subtelomeric localization for the gene, which is a region showing high levels (45% to 55%) of allelic loss in prostate (8,9), liver (10), and breast cancers (11).

Studies to detect mutations in the CMAR gene were performed using SSCP and direct sequencing on 25 colon carcinoma cell lines (12) and by direct sequencing alone on hepatocellular and prostate cancers (13). No mutations of the CMAR gene have been found, but a polymorphism was

```
(A) GCATGGAACACTTCGAGTTCCCAGGGTTATAGACAGTCGTTCCCAGTGTGG    51
    CTGAGGCCACCCAGAGGCAGCAGAGCATTCAGACTCCAAACAGACCCCTGT   102
    TCATGCCGACGCTTGCACGACCGCCCCAGTTCCTGTGGCTCCCTCGGAATG   153
                                                      M     1
    CTAAGGGGATCGGACATGAAAGGACCCTGTGAGCCGATTGTCCTATCTCCA   204
    L  R  G  S  D  M  K  G  P  C  E  P  I  V  L  S  P      18

    GCGGCCCTGTCATCCAGCTCACTCATCAATGGGGCCAGTCAGGCCCAGGCA   255
    A  A  L  S  S  S  S  L  I  N  G  A  S  Q  A  Q  A      35
                                         1

    CTGGGCTCCGGAGGACTCACCACTGCCCCCTGCTGCCATGTGGACTGGTGC   306
    L  G  S  G  G  L  T  T  A  P  C  C  H  V  D  W  C      51
                      2

    AAGTTGAGGACTTCTTGCTGGTCTAGTCACGCATGCAGTGTTGGGGATGCC   357
    K  L  R  T  S  C  W  S  S  H  A  C  S  V  G  D  A      68

    TTGGTTTTTACTGCTCTGAGAATTGTTGAGATACTTTACTAATAAACTGTG   408
    L  V  F  T  A  L  R  I  V  E  I  L  Y  .  .            81
                               3

    TAGTTGGAAAAAAAAAAAAAA

(B)                      AAGGACCCTGTGAGCCGATTGTCCTATCTCCAGCGG    36
    CCCTGTCATCCAGCTCACTCATCAATGGGGCCACACAGTCAGGCCCAGGCA   187
                                M  G  P  H  S  Q  A  Q  A       9

    CTGGGCTCCGGAGGACTCACCACTGCCCCCTGCTGCCATGTGGACTGGTGC   238
    L  G  S  G  G  L  T  T  A  P  C  C  H  V  D  W  C      26
                      1

    AAGTTGAGGACTTCTTGCTGGTCTAGTCACGCATGCAGTGTTGGGGATGCC   289
    K  L  R  T  S  C  W  S  S  H  A  C  S  V  G  D  A      43

    TTGGTTTTTACTGCTCTGAGAATTGTTGAGATACTTTACTAATAAACTGTG   340
    L  V  F  T  A  L  R  I  V  E  I  L  Y  .  .            56
                               2

    TAGTTGGAAAAAAAAAAAAAA
```

**FIG. 1. A:** Nucleotide and amino acid sequence of the cell matrix adhesion regulator (CMAR) gene product (1 and 2, myristoylation sites; 3, tyrosine phosphorylation site). **B:** Nucleotide and amino acid sequence of the polymorphic variant of the CMAR gene product (CMAR +4) (1, myristoylation site; 2, tyrosine phosphorylation site).

detected for a 4-base-pair (bp) CACA insertion within the coding region of the gene. Initially this seemed rather surprising, as the 4 bp insertion changes the reading frame of the cDNA distal (3 prime) to the insertion. This variant has since been recovered from a human melanoma cDNA library DX3, known to be heterozygous for the +4 bp polymorphism. Functional studies have also been performed and show that both the original CMAR and its polymorphic +4-bp variant can increase the adhesion of WOP cells to collagen on transient transfection (Fig. 2).

Closer examination of the sequence of the polymorphic variant shows that if the reading frame distal (3 prime) to the 4-bp insertion is maintained and applied to the entire sequence there is a further ATG/methionine before (5 prime to) the 4-bp insertion, which could act as the initiation site of a shorter protein of 56 amino acids. This retains both a myristoylation consensus signature and the terminal tyrosine and tyrosine phosphorylation sequence (Fig. 1B).

To study this polymorphism further, the segregation of the +4 bp variant in the pedigree of a family from CEPH (the Centre for the Study of Human Polymorphism, Paris) was investigated. In this family, both parents are known to be heterozygous for the +4 bp polymorphism. Results show that the polymorphic variant of CMAR segregates as expected, producing a classical Hardy–Weinberg fit (Fig. 3; Table 1). The population frequency of this polymorphism has also been investigated, the +4 bp polymorphic variant being present in approximately 20% of English Caucasians. To determine if the polymorphic variant could be linked to a predisposition to cancer, the frequencies of the two alleles were compared between the normal English Caucasoid subjects, patients with colorectal cancer, and human colon carcinoma cell lines (see Table 1). No significant differences were seen among the three groups, suggesting that the polymorphic variant of CMAR is not associated with a predisposition to colon cancer.

To facilitate further functional studies, CMAR has been cloned into the plasmid vector pMAMneo, which allows dexamethasone induction of gene expression (14). With neomycin selection, this CMAR.pMAMneo construct has been stably transfected into both the mouse colon carcinoma cell line CMT93 and a human colon carcinoma cell line, LS174T. Results from these studies show that CMAR transfection is able to increase markedly the binding of CMT93 to collagen type I (Fig. 4), and preliminary results suggest that an LS174T CMAR transfectant is able to form primitive glandular structures in collagen gels, unlike the control plasmid transfectant or parental cell line (Fig. 5). CMAR has proven to be an extremely interesting if sometimes rather elusive gene. There still remains much research to be undertaken and many problems to be solved. Great difficulty has been encountered in demonstrating the messenger RNA for CMAR, suggesting that this message is particularly unstable. Polyclonal antisera and monoclonal antibodies are being raised against CMAR fusion proteins, to establish CMAR's cellular and tissue distributions. Immunoprecipitations will also be performed to discover with which other proteins it interacts.

The processes of cell adhesion are becoming increasing implicated in both the initial development and progression of tumours. Previous work has shown that reduced expression in the $\beta_1$ integrins is a common finding in

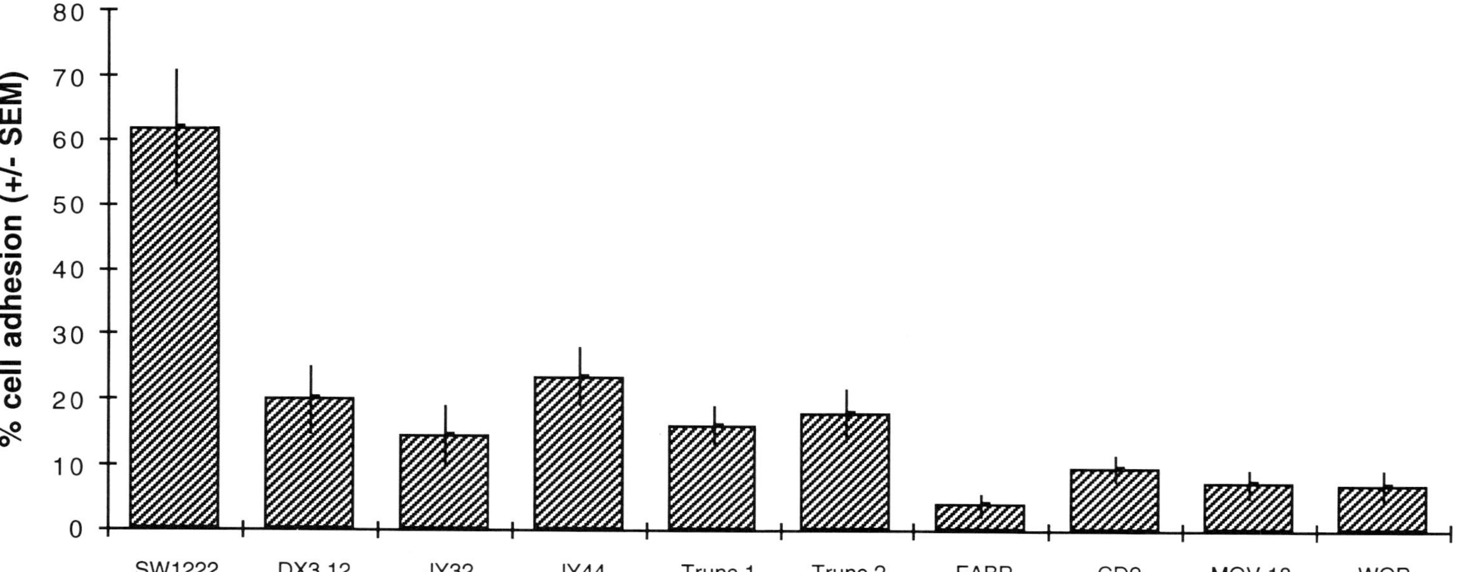


**FIG. 2.** Binding of CMAR transfected WOP cells to Vitrogen 100. WOP cells were transfected by electroporation with pCDM8 plasmid constructs containing full-length CMAR clones (JY32, JY44, and DX3.12), truncated clones containing only the CMAR coding sequence (Trunc 1, Trunc 2), or control clones [fatty acid-binding protein (FABP) CD2, and MOV 18]. After transfection the cells were grown under standard tissue culture conditions for 48 h, then harvested. Adhesion assays were performed on collagen type I (Vitrogen 100, bovine dermal collagen; Celtrix Laboratories, Palo Alto, CA) coated Terasaki plates after labeling the cells with a fluorescent dye (3). Untransfected WOP cells and SW1222 cells were used as negative and positive controls, respectively.

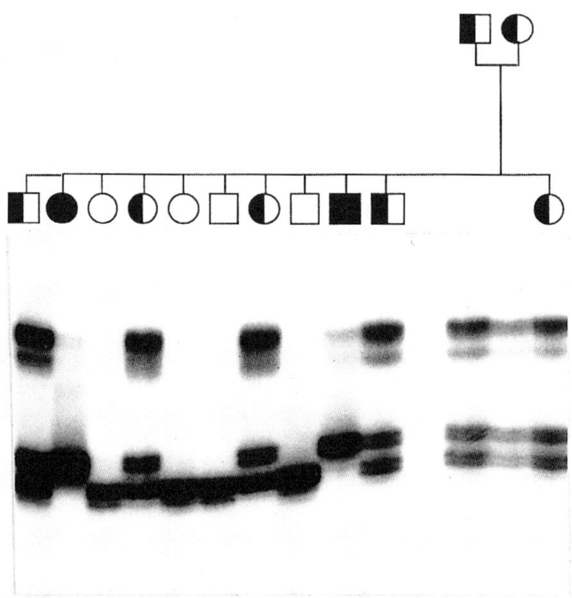

**FIG. 3.** Segregation of CMAR polymorphism in a CEPH family pedigree. CEPH Amish family pedigree and the results from a denaturing sequencing gel showing the segregation of the two polymorphic variants of CMAR. In the pedigree, the CMAR allele is shown in white, the CMAR +4 variant in black. Beneath the pedigree, corresponding bands can be seen on the sequencing gel. In the bottom set of bands, the lower band represents the CMAR allele and the upper band the slower-migrating CMAR +4 variant.

colorectal tumors, this being particularly associated with poorly differentiated tumors (15). Although mutations of the APC gene have been shown to be an early step in the development of colorectal cancer, recent evidence shows that the APC gene product associates with β-catenin (16,17), suggesting that it may also be involved in the processes of cell adhesion. DCC, another of the genes implicated in colorectal carcinogenesis (18), has a protein sequence closely related to the neural cell adhesion molecule (NCAM) and other related surface glycoproteins, suggesting that it probably acts as an adhesion molecule. Thus, particularly in the case of colorectal cancer, the

**TABLE 1.** *Frequency of CMAR and CMAR +4 alleles (%) in unaffected patients, cancer patients, and colon carcinoma cell lines*

| | Total | Heterozygote | CMAR Homozygote | CMAR +4 Homozygote |
|---|---|---|---|---|
| Unaffected | 77 (100) | 17 (22) | 57 (74) | 3 (4) |
| Colorectal cell lines | 28 (100) | 9 (32) | 18 (64) | 1 (4) |
| Cancer patients | 26 (100) | 7 (27) | 19 (73) | 0 (0) |
| Total cancer | 54 (100) | 16 (30) | 37 (68) | 1 (2) |
| Total samples | 131 (100) | 33 (25) | 94 (72) | 4 (3) |

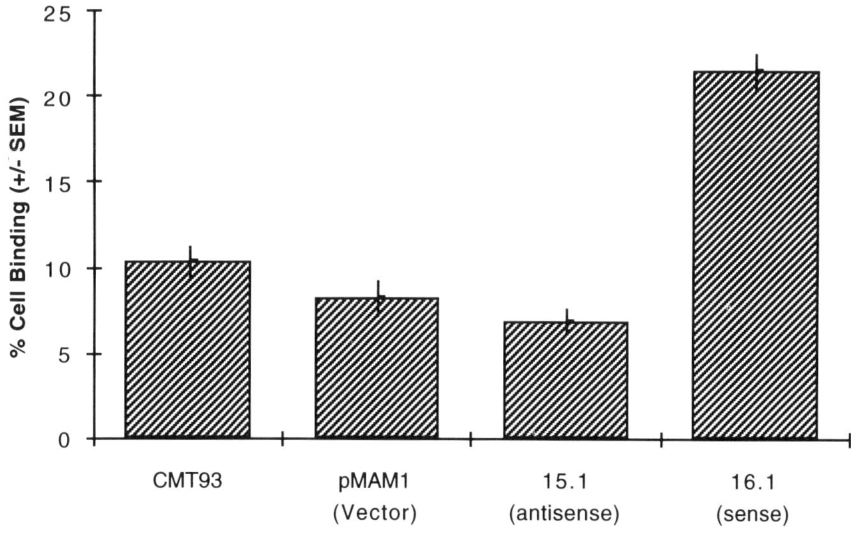


**FIG. 4.** Binding of CMT93 CMAR transfectants to Vitrogen 100. CMT93 cells were transfected by lipofection with either pMAMneo (vector), CMAR.pMAMneo (clone 16.1), or reverse.CMAR.pMAMneo (clone 15.1) constructs. Resulting transformants were selected with geneticin and cloned. Adhesion assays to Vitrogen 100 (bovine dermal collagen; Celtrix Laboratories, Palo Alto, CA) were performed as previously described (3).

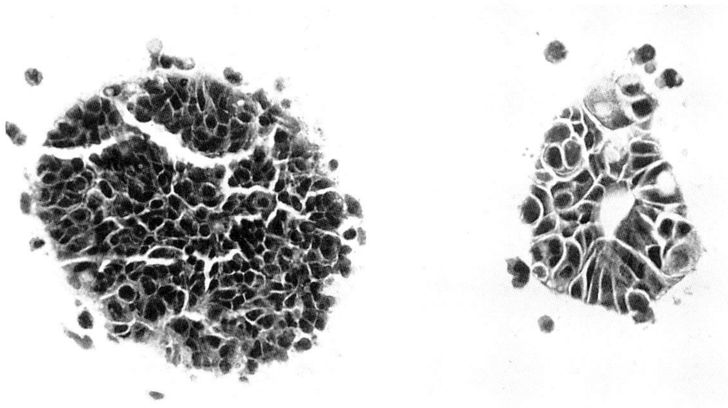

**FIG. 5.** The morphological appearance of LS174T transfectants. **Left:** pMAMneo (vector) transfectant. **Right:** CMAR.pMAMneo transfectant. LS174T cells were transfected with either the vector pMAMneo or the construct CMAR.pMAMneo using geneticin selection. Resulting transformants were cloned, then seeded in Vitrogen 100 gels (bovine dermal collagen; Celtrix Laboratories, Palo Alto, CA) and cultured under standard tissue culture conditions. After 14 days the collagen gels were fixed in formal saline and paraffin-embedded. Serial 4-μm sections were cut and stained with hematoxylin and eosin. Magnification ×500.

control of cell adhesion is increasingly seen as being central to the initial development of a tumor.

With further study, the function of the CMAR gene and its protein product will be better understood and its place in these processes of cell adhesion and possibly in tumor formation resolved.

## REFERENCES

1. Pignatelli M, Bodmer WF. Integrin-receptor-mediated differentiation and growth inhibition are enhanced by transforming growth factor-beta in colorectal tumour cells grown in collagen gel. *Int J Cancer* 1989;44:518–23.
2. Pignatelli M, Bodmer WF. Genetics and biochemistry of collagen binding-triggered glandular differentiation in a human colon carcinoma cell line. *Proc Natl Acad Sci USA* 1988; 85:5561–5.
3. Novelli MR, Durbin H, Bodmer WF. A fluorescence based cell adhesion assay using Terasaki plates. *In Vitro Cell Dev Biol* 1995;31:83–5.
4. Pignatelli M, Nasim MM, Stamp GWH, Hirano, S, Takeichi M. Morphoregulatory activities of E-cadherin and beta-1 integrins in colorectal tumour cells. *Br J Cancer* 1992;66: 629–34.
5. Seed B, Aruffo A. Molecular cloning of the CD2 antigen, the T-cell erythrocyte receptor, by a rapid immunoselection procedure. *Proc Natl Acad Sci USA* 1987;84:3365–9.
6. Seed B. An LFA-3 cDNA encodes a phospholipid-linked membrane protein homologous to its receptor CD2. *Nature* 1987;329:840–2.
7. Pullman, WE, Bodmer WF. Cloning and characterization of a gene that regulates cell adhesion. *Nature* 1992;356:529–32.
8. Carter BS, Ewing CM, Ward WS, et al. Allelic loss of chromosomes 16q and 10q in human prostate cancer. *Proc Natl Acad Sci USA* 1990;87:8751–5.
9. Kunimi K, Bergerheim, US, Larsson IL, Ekman P, Collins VP. Allelotyping of human prostatic adenocarcinoma. *Genomics* 1991;11:530–6.
10. Fujimori M, Tokino T, Hino O, et al. Allelotype study of primary hepatocellular carcinoma. *Cancer Res* 1991;51:89–93.
11. Sato T, Tanigami A, Yamakawa K, et al. Allelotype of breast cancer: cumulative allele losses promote tumor progression in primary breast cancer. *Cancer Res* 1990;50:7184–9.
12. Durbin H, Novelli M, Bodmer W. Detection of a 4-bp insertion (CACA) functional polymorphism at nucleotide 241 of the cellular adhesion regulatory molecule CMAR (formerly CAR). *Genomics* 1994;19:181–2.
13. Koyama K, Emi M, Nakamura Y. The cell adhesion regulator (CAR) gene, TaqI and insertion/deletion polymorphisms, and regional assignment to the peritelomeric region of 16q by linkage analysis. *Genomics* 1993;16:264–5.
14. Lee F, Mulligan R, Berg P, Ringold G. Glucocorticoids regulate expression of dihydrofolate reductase cDNA in mouse mammary tumour virus chimaeric plasmids. *Nature* 1981;294: 228–32.
15. Pignatelli M, Smith ME, Bodmer WF. Low expression of collagen receptors in moderate and poorly differentiated colorectal adenocarcinomas. *Br J Cancer* 1990;61:636–8.
16. Adams JC, Watt FM. Expression of beta 1, beta 3, beta 4, and beta 5 integrins by human epidermal keratinocytes and nondifferentiating keratinocytes. *J Cell Biol* 1991;115:829–41.
17. Chammas R, Brentani R. Integrins and metastases: an overview. *Tumour Biol* 1991;12:309–20.
18. Vogelstein B, Fearon ER, Hamilton SR, et al. Genetic alterations during colorectal-tumor development. *N Engl J Med* 1988;319:525–32.

*Topics in Molecular Medicine, Volume 1,*
edited by Wolfgang Siess, Reinhard Lorenz,
and Peter C. Weber. Raven Press, Ltd.,
New York © 1995.

# 17

# Distinct Functions of Integrins $\alpha_4\beta_7$, $\alpha_4\beta_1$, and $\alpha_{IEL}\beta_7$ in Lymphocyte Recirculation and Matrix Adhesion

Ulrike G. Strauch, Uwe Gosslar, Alexandra Lifka, and
Bernhard Holzmann

*Institute of Medical Microbiology and Hygiene, Technical University,
D-81675 Munich, Germany*

The recruitment of leukocytes to lymphoid organs and sites of inflammation is critical to the function of the immune system. The physiologic recirculation of lymphocytes is regulated by selective interactions of lymphocytes with specialized postcapillary venules that differentiate to high endothelial venules (HEVs) in lymphoid organs such as Peyer's patches (PP) or lymph nodes (1). Alternatively, lymphocytes may extravasate through endothelium in peripheral tissues, such as skin or intestine, and enter lymph nodes through afferent lymphatic vessels. Native lymphocytes migrate almost exclusively through HEVs, whereas memory T lymphocytes extravasate preferentially through peripheral tissue endothelia (2). Lymphocyte adhesion to endothelial cells was shown to display a remarkable organ selectivity that may result from the adhesive specificity of individual lymphocyte or endothelial adhesion receptors and also from the combinatorial specificity of sequential adhesion and activation events (3–7).

L-Selectin was defined as a murine lymphocyte receptor mediating in vitro and in vivo adhesion to peripheral lymph node HEVs (8). In addition, L-selectin is involved in lymphocyte migration to PP in vivo (9). L-Selectin interacts with two distinct mucin-like counter-receptors on lymph node HEVs: GlyCAM-1 is secreted by HEVs, whereas CD34 is cell surface-bound (10,11). Binding of L-selectin is dependent on the sialylation and sulfation of GlyCAM-1 and CD34 *O*-linked carbohydrate side chains (12). Both L-selectin ligands express carbohydrate determinants recognized by the monoclonal antibody (MAb) MECA-79, and the isolated MECA-79 antigen directly binds L-selectin (13).

The mucosal vascular addressin MAdCAM-1 is a novel immunoglobulin family member expressed on PP HEVs, on blood vessels in the lamina propria of the small intestine, in the marginal sinus of the spleen, and on lymph node HEVs of mice early in development (1,14–16). Antibodies to MAdCAM-1 selectively inhibit in vitro lymphocyte adhesion to PP HEVs as well as in vivo lymphocyte migration to mucosal sites (15). Recently, it was shown that lymphocyte adhesion to MAdCAM-1 is mediated by the $\alpha_4\beta_7$ integrin LPAM-1 (17). Interestingly, in some tissues the MAdCAM-1 adhesion receptor is modified by carbohydrate structures that are recognized by L-selectin (18).

Previous work using MAbs directed against the $\alpha_4$ subunit revealed a crucial role for $\alpha_4$ integrins in lymphocyte adhesion to PP HEVs (19,20) and for in vivo homing of lymphocytes to mucosal lymphoid organs such as PP (21,22). Cloning and expression of the $\beta_7$ subunit directly demonstrated the function of the murine integrin LPAM-1 ($\alpha_4\beta_7$) as a lymphocyte adhesion receptor for PP HEVs (23). The integrin VLA-4 ($\alpha_4\beta_1$) is structurally related to LPAM-1 and interacts with VCAM-1 as well as with the CS-1 peptide derived from fibronectin (24–28). Adhesive interactions of VLA-4 appear to be important for leukocyte recruitment to inflammatory sites in vivo (21,29), for lymphocyte differentiation and activation (30–33), and for the development of autoimmune disorders (34–36) and tumor metastasis (37,38).

The integrin $\alpha_{IEL}\beta_7$ is also structurally related to LPAM-1 (39–41). Whereas LPAM-1 is expressed by most lymphocytes, expression of $\alpha_{IEL}\beta_7$ is more restricted and is detected mainly on intraepithelial and lamina propria lymphocytes, as well as on dendritic cells (39,42,43). Functional analysis revealed that the integrin $\alpha_{IEL}\beta_7$ mediates adhesion of lymphocytes to epithelial cells (44,45) and that binding of antibodies to $\alpha_{IEL}$ enhances CD3-induced activation of human intraepithelial lymphocytes (46). Owing to the restricted distribution of $\alpha_{IEL}\beta_7$ on lymphocytes, it was also speculated that this integrin is involved in lymphocyte migration to epithelial sites (40).

In the present study, the function of integrins $\alpha_4\beta_1$, $\alpha_4\beta_7$, and $\alpha_{IEL}\beta_7$ in mediating lymphocyte endothelial interactions was studied and the binding specificity for the endothelial adhesion molecules MAdCAM-1, VCAM-1, and fibronectin was critically examined. We report that lymphocyte adhesion to PP HEVs is dependent on LPAM-1 but not on VLA-4. The functional difference appears to result from a higher binding activity of LPAM-1 compared to VLA-4 for the mucosal addressin MAdCAM-1. By contrast, LPAM-1 and VLA-4 were equally efficient in mediating adhesion to VCAM-1 and fibronectin. The integrin $\alpha_{IEL}\beta_7$ did not show detectable binding activity for MAdCAM-1, VCAM-1, or fibronectin, and was not involved in lymphocyte binding to several endothelioma cell lines.

## LPAM-1 BUT NOT VLA-4 MEDIATES LYMPHOCYTE ADHESION TO MUCOSAL HEV

The murine B-cell lymphoma 38C13 expresses cell surface–associated integrin $\alpha_4$ subunits but no detectable mRNA transcripts for $\beta_1$ or $\beta_7$ chains (23). 38C13 cells were therefore infected with Lβ1SN (38-$\beta_1$ cells) or Lβ7SN (38-$\beta_7$ cells) recombinant retroviruses or with a virus carrying the wild-type vector pLXSN (38-LXSN cells). Immunoprecipitation with an MAb directed against the $\alpha_4$ subunit and Western blot analysis with rabbit antisera specific for $\beta_1$ or $\beta_7$ showed that both 38-$\beta_1$ and 38-$\beta_7$ cells expressed de novo synthesized β subunits as heterodimers with $\alpha_4$ (47). In addition, flow cytometric analysis revealed that 38-$\beta_1$ and 38-$\beta_7$ transfectants expressed similar levels of integrin $\alpha_4$ subunits. Analysis of 38-$\beta_1$ and 38-$\beta_7$ cells therefore allowed a functional comparison of integrins LPAM-1 and VLA-4 in an identical cellular environment.

The binding capacity of 38C13 transfectants was examined in an in vitro HEV binding assay. As is shown in Fig. 1, adhesion of 38-$\beta_7$ cells to PP HEVs was four- to fivefold greater than binding of normal mesenteric node lymphocytes, whereas 38-$\beta_1$ and 38-LXSN cells did not exhibit significant binding. The mucosal addressin MAdCAM-1 has been shown to be a counter-receptor for integrin $\alpha_4\beta_7$ (17). To define the role of MAdCAM-1 in 38-$\beta_7$ cell adhesion to PP HEVs in situ, MAb MECA-367 was tested for binding inhibition. It was also shown that MAb MECA-367 completely blocked binding of 38-$\beta_7$ transfectants to PP HEVs, whereas the VCAM-1–specific MAb 16HB5 had no detectable effect (47).

## BINDING TO MAdCAM-1 REVEALS DISTINCT FUNCTIONS FOR LPAM-1 AND VLA-4

To compare the interactions of LPAM-1 and VLA-4 directly with those of MAdCAM-1, adhesion of lymphoma lines to immunoaffinity-purified

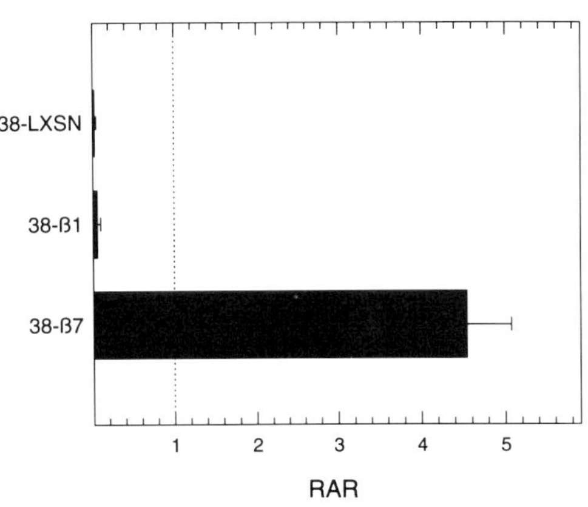


**FIG. 1.** Adhesion of 38C13 transfectants to PP HEVs. 38C13 transfectants were mixed with fluorescence-labeled normal mesenteric node lymphocytes and incubated with frozen sections of mouse PPs for 30 min at room temperature. Bound cells were fixed to the tissue sections and adhesion was analyzed microscopically. The RAR represents the number of lymphoma cells bound to HEV per normal mesenteric node lymphocyte. The binding of unlabeled mesenteric node lymphocytes defines an RAR of unity (dotted line).

MAdCAM-1 was investigated. As shown in Fig. 2, 38-$\beta_7$ cells strongly bound to purified MAdCAM-1 but not to control protein. Similar to the results obtained with the in vitro HEV binding assay, no significant adhesion was detected with unstimulated 38-$\beta_1$ or 38-LXSN cells (Fig. 2). To confirm binding specificity, MAb inhibition experiments were performed. MAb PS/2 (anti-integrin $\alpha_4$ subunit) and MAb MECA-367 (anti–MAdCAM-1) completely inhibited binding, whereas MAb M18/2.a.8 (anti-integrin $\beta_2$ subunit) did not affect adhesion (47).

Integrin functions can be upregulated by treatment of cells with $Mn^{2+}$. In contrast to unstimulated cells, 38-$\beta_1$ transfectants treated with $Mn^{2+}$ showed weak adhesion to purified MAdCAM-1 (Fig. 2). Binding was about four-fold less efficient than adhesion of unstimulated 38-$\beta_7$ cells (Fig. 2).

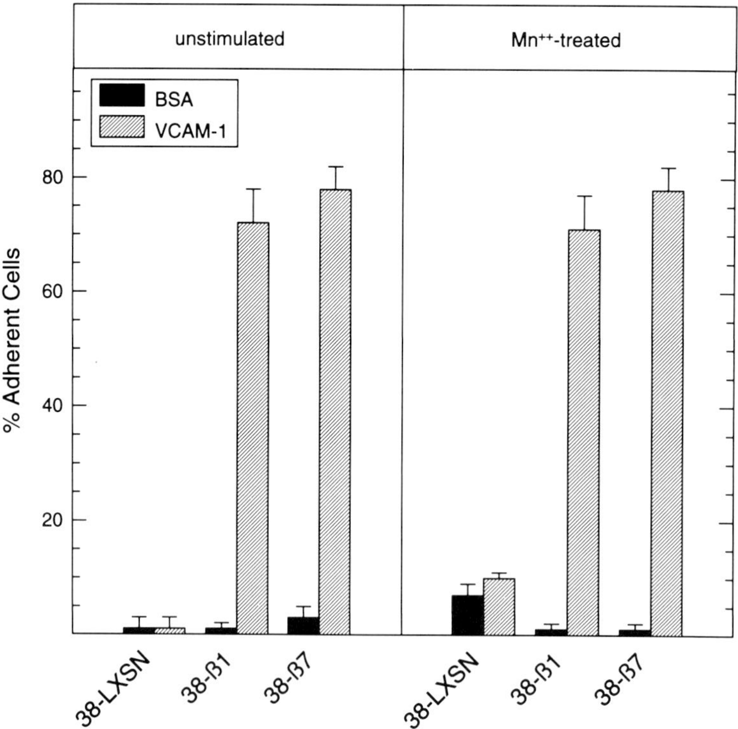


**FIG. 2.** Adhesion of 38C13 transfectants to the purified MAdCAM-1 protein. Immunoaffinity-purified MAdCAM-1 was adsorbed to glass slides for 16 h at 4°C. Slides were blocked with RPMI containing 1% BSA and incubated with 38C13 transfectants for 30 min at room temperature. The cells were used unstimulated or were treated for 15 min at 37°C with 1 m*M* $MnCl_2$. Unbound cells were removed by washing the slides in PBS and bound cells were fixed to the slides in PBS 2.5% glutaraldehyde. The number of cells bound per unit area was determined microscopically. The unit area was defined as the whole field covered by a grid subdivided ocular at 160× magnification. Data represent mean ± SD three or four independent experiments.

Specificity of binding was demonstrated by inhibition with MAb PS/2 and MECA-367 (47). Interestingly, binding of 38-$\beta_7$ cells was neither dependent on nor enhanced by $Mn^{2+}$ treatment. Immunofluorescence staining and FACS analysis revealed that $Mn^{2+}$ stimulation of cells did not alter expression of LPAM-1 or VLA-4 (data not shown). It should be noted that in a previous report VLA-4/MAdCAM-1 interactions were not detected (17). However, 38-$\beta_1$ cells were not used in this study; also, consistent with the results reported by Berlin et al. (17), we did not detect binding of $\alpha_4\beta_1{}^+$ RAW112 cells to purified MAdCAM-1 (data not shown).

In summary, these results indicate that lymphocyte homing to mucosa-associated lymphoid organs is regulated by a direct interaction of $\alpha_4\beta_7$ integrin with MAdCAM-1. Although weak binding of VLA-4 to MAdCAM-1 was demonstrated with $Mn^{2+}$-treated 38-$\beta_1$ cells, several lines of evidence suggest that LPAM-1 is the functionally dominant receptor for regulating lymphocyte homing to mucosal lymphoid organs: binding of $Mn^{2+}$-activated normal lymphocytes ($\alpha_4{}^+$ $\beta_1{}^+$ $\beta_7{}^+$) to purified MAdCAM-1 is inhibited by MAb to $\beta_7$ with the same efficiency as with MAb to $\alpha_4$ (17); MAbs specific for the $\alpha_4\beta_7$ heterodimer block the in vivo homing of normal lymphocytes to mucosal sites as effectively as MAbs to $\alpha_4$ (22); and $Mn^{2+}$ treatment did not induce adhesion of 38-$\beta_1$ cells to PP HEVs in vitro (data not shown). Nevertheless, it cannot be excluded that the activated VLA-4 may act in conjunction with LPAM-1 to enhance the migration of lymphocytes to mucosal sites or to promote the adhesion of lymphocytes to MAdCAM-1 expressed in other tissues during the early stages of development (i.e., in peripheral lymph nodes) or the marginal sinus of the spleen (1,14).

## BOTH LPAM-1 AND VLA-4 MEDIATE ADHESION TO VCAM-1 AND FIBRONECTIN

To further compare the functions of integrins LPAM-1 and VLA-4 in cell adhesion, binding of 38C13 transfectants to the inducible endothelial adhesion receptor VCAM-1 was examined. The results in Fig. 3 demonstrate that both 38-$\beta_7$ and 38-$\beta_1$ cells strongly bound to recombinant soluble VCAM-1, whereas 38-LXSN cells did not exhibit significant adhesion. Interestingly, LPAM-1–dependent adhesive interactions with VCAM-1 were as efficient as binding mediated by VLA-4. Activation with $Mn^{2+}$ did not increase the adhesive capacity of 38C13 transfectants for VCAM-1. Incubation of 38-$\beta_1$ and 38-$\beta_7$ cells with MAb PS/2, but not with control antibodies against CD45 or integrin $\beta_2$, completely inhibited binding (47).

In additional experiments, VCAM-1 and MAdCAM-1 were independently transfected into CHO cells and it was shown that both receptors were expressed at similar densities (47). Thus, the efficiency of LPAM-1 in interacting with different endothelial counter-receptors could be compared. Ad-

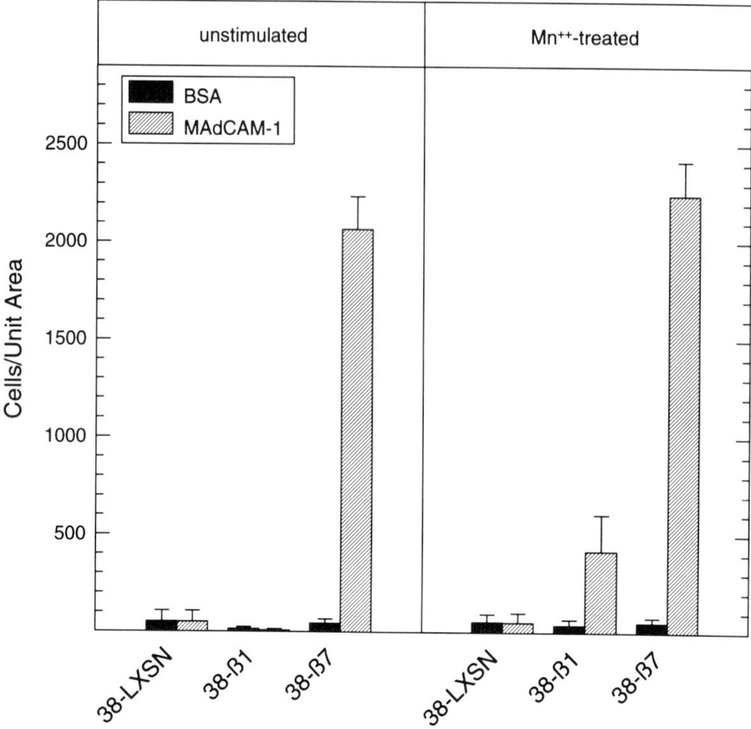


**FIG. 3.** Adhesion of 38C13 transfectants to VCAM-1. Microwell plates were coated with recombinant soluble VCAM-1 (0.45 μg/well) or BSA (1 μg/well) and incubated with fluorescence-labeled 38C13 transfectants for 30 min at 37°C. Cells were used untreated or were preincubated for 15 min at 37°C with 1 m*M* $MnCl_2$. Unbound cells were removed by centrifugation of plates in an inverted orientation and adherence was determined by fluorimetry. Results are expressed as percent of bound cells and represent the mean ± SD from four to six independent experiments.

hesion assays with CHO transfectants revealed that the integrin LPAM-1 mediates adhesion of 38-$\beta_7$ to VCAM-1 and MAdCAM-1 with similar efficiency (Fig. 4).

We next examined interactions of 38C13 transfectants with the extracellular matrix protein fibronectin. The results in Fig. 5 demonstrate that both unstimulated 38-$\beta_1$ and 38-$\beta_7$ transfectants, but not 38-LXSN cells, strongly bound to fibronectin. $Mn^{2+}$ treatment of 38-$\beta_1$ and 38-$\beta_7$ cells further increased their adhesive capacity. Interestingly, addition of $Mn^{2+}$ appeared to stimulate LPAM-1 more efficiently than the VLA-4 to bind fibronectin (Fig. 5). Fibronectin adhesion of 38-$\beta_1$ and 38-$\beta_7$ cells was completely inhibited by MAb PS/2, confirming that binding was indeed mediated by $\alpha_4$ integrins and not by other fibronectin receptors expressed on these lymphoma cells (47).

VLA-4 interacts with the endothelial adhesion molecule VCAM-1 and

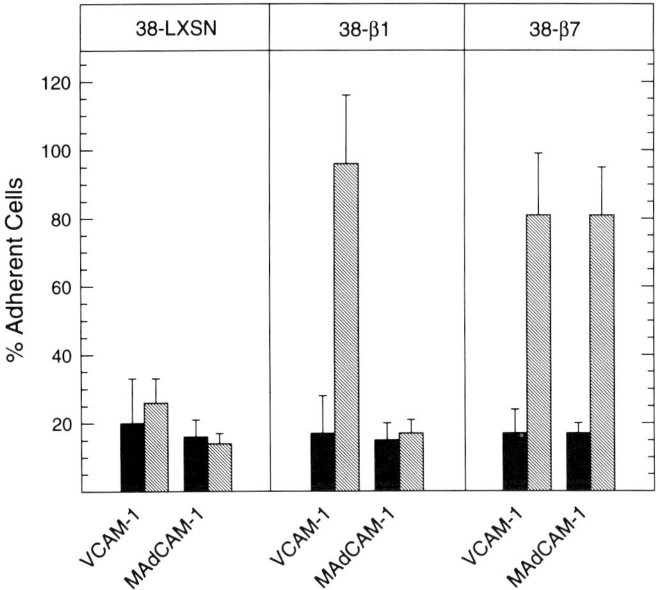


**FIG. 4.** Adhesion of 38C13 transfectants to VCAM-1- or MAdCAM-1-transfected CHO cells. Confluent monolayers of VCAM-1– or MAdCAM-1–transfected CHO cells (shaded bars) or control transfectants (solid bars) were incubated for 30 min at 37°C with fluorescence-labeled 38C13 transfectants. Unbound cells were removed by centrifugation of plates in an inverted orientation and adherence was measured by fluorimetry. Results are expressed as percent of bound cells and represent the mean ± SD from four to six independent experiments. Fluorescence flow cytometry revealed that CHO–VCAM-1 and CHO–MAdCAM-1 transfectants expressed similar levels of the vascular adhesion molecules (data not shown).

with the CS-1 peptide derived from fibronectin, playing an important role in various cell adhesion-dependent physiologic and pathologic processes. The binding function of the $\alpha_4\beta_7$ integrin for these ligands remained controversial, however, as the activated LPAM-1 integrin expressed on TK1 cells was shown to bind the same ligands as VLA-4 (48), whereas on the human B-cell line JY it was much less effective than VLA-4 (49). In the present study, adhesive interactions of LPAM-1 with VCAM-1 transfectants and recombinant soluble VCAM-1, as well as with fibronectin, were demonstrated, and LPAM-1 was found to be equally effective as VLA-4 when expressed in 38C13 cells. Moreover, the LPAM-1 integrin on 38-$\beta_7$ cells did not display any preference for interacting either with MAdCAM-1 or VCAM-1, as was tested with vascular adhesion proteins expressed in same CHO cell environment. Confirming these observations, it was recently reported that the integrin $\alpha_4\beta_7$ functions as a receptor for VCAM-1 and fibronectin on human B lymphocytes (50). Together, these data indicate that the integrins LPAM-1 and VLA-4 have the same intrinsic capacity to mediate lymphocyte inter-

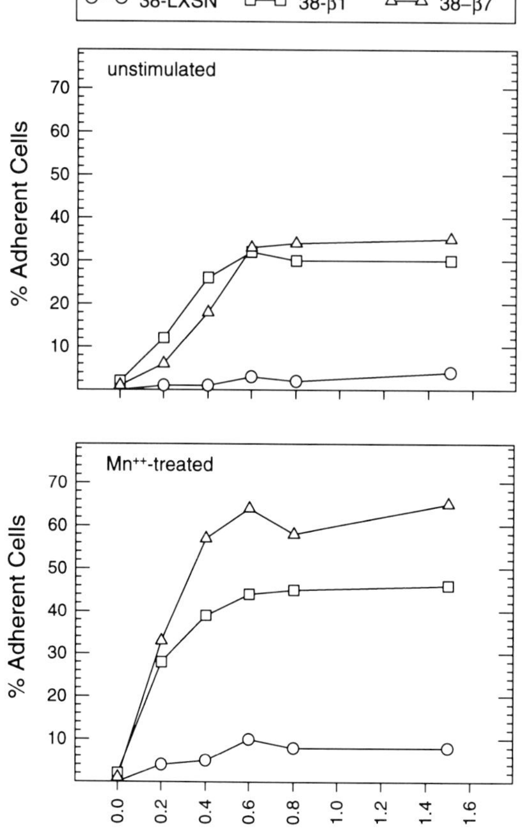


**FIG. 5.** Adhesion of 38C13 transfectants to fibronectin. Microwell plates were coated with increasing concentrations of fibronectin or BSA (1 μg/well) and incubated with fluorescence-labeled 38C13 transfectants for 30 min at 37°C. Cells were used untreated or were preincubated for 15 min at 37°C with 1 m*M* $MnCl_2$. Unbound cells were removed by centrifugation of plates in an inverted orientation and adherence was determined by fluorimetry. Results are expressed as percent of bound cells and represent the mean ± SD from four to six independent experiments.

actions with VCAM-1 and fibronectin and that LPAM-1 may contribute significantly to cellular functions previously attributed to VLA-4.

It is well established that the binding of lymphocyte integrins to their ligands can be modulated by divalent cations, activating antibodies to integrin subunits, treatment of cells with PMA, or crosslinking of lymphocyte surface receptors (51). In addition, the basal activation state appears to be dependent on the cell type expressing the integrin receptor. The data presented here therefore suggest that in 38C13 transfectants the integrin $\alpha_4\beta_7$ is expressed at a high-activation state. By contrast, it has been reported that normal mouse lymphocytes as well as sublines of the T-cell lymphoma TK1 require stimulation with PMA or $Mn^{2+}$ for efficient VCAM-1 binding, but not for adhesion to MAdCAM-1 (17). These experiments therefore indicate that the binding specificity of integrin $\alpha_4\beta_7$ for MAdCAM-1 versus VCAM-1 is activation-dependent and that at a lower activation state LPAM-1 preferentially interacts with MAdCAM-1 (17).

## DISTINCT BINDING SPECIFICITY OF INTEGRIN $\alpha_{IEL}\beta_7$

Because of the restricted distribution of $\alpha_{IEL}\beta_7$, it was speculated that this integrin may be involved in lymphocyte migration to epithelial sites. To analyze the binding specificity of integrin $\alpha_{IEL}\beta_7$ expression variants of the T-cell hybridoma MTC-1 were selected by repeated cell sorting using MAb M290 and the sublines MTC-$1^{lo}$ and MTC-$1^{hi}$ were generated. The T-cell lymphocyte LB constitutively expresses high levels of $\alpha_{IEL}\beta_7$ and was included in the functional analysis of $\alpha_{IEL}\beta_7$. Both MTC-$1^{hi}$ and LB cells did not express cell surface $\alpha_4$ subunits (data not shown). Binding activity of $\alpha_{IEL}\beta_7$ expressed on MTC-$1^{hi}$ and LB cells was demonstrated by adhesion to the lung carcinoma cells CMT64/61 and inhibition of binding by MAb M290 (data not shown). When tested for fibronectin binding, weak adhesion of LB lymphoma cells was detected, whereas MTC-$1^{hi}$ cells did not bind to fibronectin (Table 1). However, the binding to fibronectin was not inhibited by the function-blocking antibody M290 (Table 1). As shown in Table 1, MTC-$1^{hi}$ and LB lymphoma cells also did not adhere to purified MAdCAM-1 or VCAM-1, and binding activity was not induced by treatment of cells with $Mn^{2+}$. These results therefore suggest that the $\alpha_{IEL}\beta_7$ integrin does not interact with MAdCAM-1, VCAM-1, or fibronectin.

In additional experiments, binding of MTC-1 expression variants to endothelioma cell lines was tested. As is shown in Fig. 6, the expression of integrin $\alpha_{IEL}\beta_7$ did not correlate with the binding capacity of MTC-1 sublines for the endothelioma lines TME, eEND.2, and mlEND. Treatment of endo-

**TABLE 1.** *Ligand binding specificity of integrin $\alpha_{IEL}\beta_7$*

| Ligand | Treatment | MTC-$1^{hi}$ | LB | 38-$\beta_7$ |
|---|---|---|---|---|
| Fibronectin[a] | Medium | 3 ± 1 | 9 ± 3 | 29 ± 10 |
| | $Mn^{2+}$ | ND | 6 ± 2 | ND |
| | M290 | ND | 8 ± 4 | ND |
| VCAM-1[a] | Medium | 1 ± 1 | 2 ± 2 | 64 ± 4 |
| | $Mn^{2+}$ | ND | 1 ± 1 | ND |
| MAdCAM-1[b] | Medium | 9 ± 15[c] | 9 ± 9[c] | 926 ± 239 |
| | $Mn^{2+}$ | 15 ± 20 | 19 ± 16 | ND |

ND, not determined

[a] Microwell plates were coated with fibronectin (1 μg/well), recombinant soluble VCAM-1 (0.45 μg/well), or BSA (1 μg/well) and incubated with fluorescence-labeled MTC-$1^{hi}$, LB, or 38-β7 cells for 30 min at 37°C. Adherence was determined as described for Fig. 3. The results represent the mean from four to six independent experiments and are given as % adherent cells. Background binding to BSA was <4%.

[b] Immunoaffinity-purified MAdCAM-1 was adsorbed to glass slides for 16 h at 4°C. Slides were blocked with RPMI containing 1% BSA and incubated with 38C13 transfectants for 30 min at room temperature. Adherence was determined as described for Fig. 2. The results represent the mean ± SD from three or four independent experiments and are given as number of cells bound per unit area.

[c] Binding of MTC-$1^{hi}$ and LB cells to MAdCAM-1 was not significantly different from the BSA control.

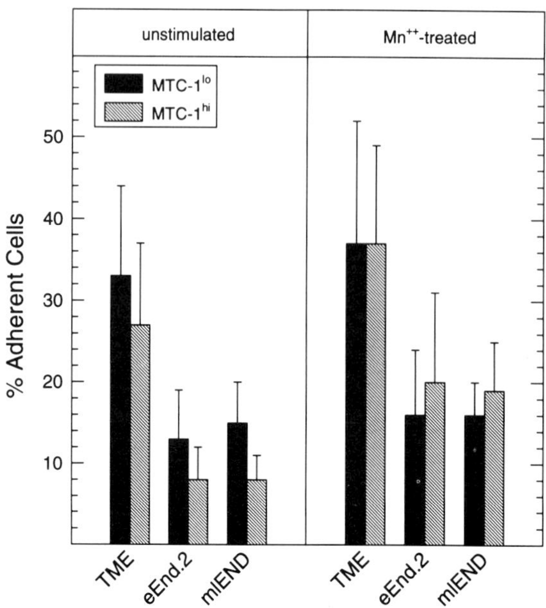


**FIG. 6.** Adhesion of $\alpha_{IEL}\beta_7$ expression variants to endothelioma cell lines. Integrin $\alpha_{IEL}\beta_7$ expression variants were selected by repeated cell sorting of the T-cell hybridoma MTC-1 using MAb M290. Confluent monolayers of endothelioma cells were incubated for 30 min at 37°C with fluorescence-labeled MTC-1$^{lo}$ or MTC-1$^{hi}$ variant sublines. Cells were used untreated or were preincubated for 15 min at 37°C with 1 m*M* $MnCl_2$. Unbound cells were removed by centrifugation of plates in an inverted orientation and adherence was measured by fluorimetry. Results are expressed as percent of bound cells and represent the mean ± SE from four independent experiments.

thelial cells with TNF or integrin activation with $Mn^{2+}$ did not reveal detectable $\alpha_{IEL}\beta_7$-dependent adhesion. Preincubation of MTC-1$^{hi}$ cells with MAb M290 also did not affect endothelial cell binding. These data suggest that the $\alpha_{IEL}\beta_7$ integrin does not participate in endothelial cell interactions of lymphocytes migrating to mucosal sites. Instead, the $\alpha_{IEL}\beta_7$ receptor appears to mediate subsequent lymphocyte binding to epithelial cells (44,45).

## SUMMARY

Integrin receptors are important for regulating lymphocyte recirculation and recruitment to sites of inflammation. The murine B-cell lymphoma 38C13 expresses surface-associated integrin $\alpha_4$ subunits but no detectable mRNA transcripts for either $\beta_1$ or $\beta_7$ chains. Integrin $\beta_1$ or $\beta_7$ subunits were expressed in 38C13 cells, thus allowing a functional comparison of LPAM-1 ($\alpha_4\beta_7$) and VLA-4 ($\alpha_4\beta_1$) in an identical cellular environment. Whereas 38-$\beta_7$ transfectants strongly bound to mucosal HEVs in situ, 38-$\beta_1$

cells and 38-LXSN controls failed to adhere. Adhesion of 38-$\beta_7$ cells was completely inhibited by MAbs directed against the mucosal addressin MAdCAM-1. Subsequent analysis of MAdCAM-1 adhesion revealed distinct binding specificities of $\alpha_4\beta_1$ and $\alpha_4\beta_7$ integrins. Thus, 38-$\beta_7$ transfectants strongly bound to purified MAdCAM-1, whereas unstimulated 38-$\beta_1$ cells did not. However, treatment of 38-$\beta_1$ cells with $Mn^{2+}$ induced low-level binding to MAdCAM-1. MAdCAM-1 adhesion of 38-$\beta_7$ cells was constitutive and was not enhanced by $Mn^{2+}$ treatment. The results therefore establish the LPAM-1/MAdCAM-1 interaction as the functionally dominant adhesion pathway for regulating lymphocyte homing to mucosal sites.

Unlike their adhesion to MAdCAM-1, 38-$\beta_7$ and 38-$\beta_1$ transfectants did not differ in their binding capacity for VCAM-1 or fibronectin. In addition, LPAM-1–mediated adhesion to MAdCAM-1 or VCAM-1 transfected CHO cells with the same efficiency. LPAM-1 may therefore contribute significantly to cellular functions previously attributed to VLA-4. Additional functional analysis of the intraepithelial lymphocyte integrin $\alpha_{IEL}\beta_7$, which is structurally related to LPAM-1, did not reveal detectable binding activity for MAdCAM-1, VCAM-1, or fibronectin and was not involved in lymphocyte binding to several endothelioma lines. Instead, the $\alpha_{IEL}\beta_7$ integrin appears to mediate subsequent lymphocyte binding to epithelial cells (44,45).

## ACKNOWLEDGMENT

The authors are indepted to Drs. M. J. Briskin, C. Berlin, and E. C. Butcher for providing purified MAdCAM-1 and MAb MECA-367. We also thank Dr. D. Naor for providing the LB lymphoma, Dr. D. Vestweber for MAb 16HB5, Drs. R. Hallmann and A. Hamann for endothelioma lines, Dr. J. Clements for purified VCAM-1, Dr. P. J. Kilshaw for the MTC-1 cell line, and Drs. G. Lippford and C. Evans for critically reading the manuscript. This work was supported by the Gerhard Hess Program of the Deutsche Forschungsgemeinschaft.

## REFERENCES

1. Picker LJ, Butcher EC. Physiological and molecular mechanisms of lymphocyte homing. *Annu Rev Immunol* 1992;10:561–91.
2. Mackay CR. T cell memory: the connection between function, phenotype and migration pathways. *Immunol Today* 1991;12:189–92.
3. Butcher EC. Leukocyte-endothelial cell recognition: three (or more) steps to specificity and diversity. *Cell* 1991;67:1033–6.
4. Dustin ML, Springer TA. Role of lymphocyte adhesion receptors in transient interactions and cell locomotion. *Annu Rev Immunol* 1991;9:27–66.
5. Shimizu Y, Newman W, Tanaka Y, Shaw S. Lymphocyte interactions with endothelial cells. *Immunol Today* 1992;13:106–12.
6. Zimmerman GA, Prescott SM, McIntyre TM. Endothelial cell interactions with granulocytes: tethering and signaling molecules. *Immunol Today* 1992;13:93–100.

7. Springer TA. Traffic signals for lymphocyte recirculation and leukocyte emigration: the multistep paradigm. *Cell* 1994;76:301–14.
8. Gallatin WM, Weissman IL, Butcher EC. A cell-surface molecule involved in organ-specific homing of lymphocytes. *Nature* 1983;304:30–4.
9. Hamann A, Jablonski Westrich D, Jonas P, Thiele HG. Homing receptors examined: mouse LECAM-1 (MEL-14 antigen) is involved in lymphocyte migration into gut-associated lymphoid tissue. *Eur J Immunol* 1991;21:2925–9.
10. Lasky LA, Singer MS, Dowbenko D, et al. An endothelial ligand for L-selectin is a novel mucin-like molecule. *Cell* 1992;69:927–38.
11. Baumhueter S, Singer MS, Henzel W, et al. Binding of L-selectin to the vascular sialomucin CD34. *Science* 1993;262:436–8.
12. True DD, Singer MS, Lasky LA, Rosen SD. Requirement for sialic acid on the endothelial ligand of a lymphocyte homing receptor. *J Cell Biol* 1990;111:2757–64.
13. Berg EL, Robinson MK, Warnock RA, Butcher EC. The human peripheral lymph node vascular addressin is a ligand for LECAM-1, the peripheral lymph node homing receptor. *J Cell Biol* 1991;114:343–9.
14. Berg EL, Golstein LA, Jutila MA, et al. Homing receptors and vascular addressins: cell adhesion molecules that direct lymphocyte traffic. *Immunol Rev* 1989;108:5–18.
15. Streeter PR, Berg EL, Rouse BT, Bargatze RF, Butcher EC. A tissue-specific endothelial cell molecule involved in lymphocyte homing. *Nature* 1988;331:41–6.
16. Briskin MJ, McEvoy LM, Butcher EC. The mucosal vascular addressin, MAdCAM-1, displays homology to immunoglobulin and mucin-like adhesion receptors and to IgA1. *Nature* 1993;363:461–4.
17. Berlin C, Berg EL, Briskin MJ, et al. $\alpha_4\beta_7$ integrin mediates binding to the mucosal vascular addressin MAdCAM-1. *Cell* 1993;74:185–95.
18. Berg EL, McEvoy LM, Berlin C, Bargatze RF, Butcher EC. L-selectin-mediated lymphocyte rolling on MAdCAM-1. *Nature* 1993;366:695–8.
19. Holzmann B, McIntyre BW, Weissman IL. Identification of a murine Peyer's patch-specific lymphocyte homing receptor as an integrin molecule with an α chain homologous to human VLA-4 α. *Cell* 1989;56:37–46.
20. Holzmann B, Weissman IL. Peyer's patch-specific lymphocyte homing receptors consist of a VLA-4-like α chain associated with either of two integrin β chains, one of which is novel. *EMBO J* 1989;8:1735–41.
21. Issekutz TB. Inhibition of in vivo lymphocyte migration to inflammation and homing to lymphoid tissues by the TA-2 monoclonal antibody. A likely role for VLA-4 in vivo. *J Immunol* 1991;147:4178–84.
22. Hamann A, Jablonski-Westrich D, Holzmann B, Andrews D, Butcher EC. α4 integrins and lymphocyte homing to mucosal tissues in vivo. *J Immunol* 1994;152:3282–93.
23. Hu MC, Crowe DT, Weissman IL, Holzmann B. Cloning and expression of mouse integrin βp (β7): a functional role in Peyer's patch-specific lymphocyte homing. *Proc Natl Acad Sci USA* 1992;89:825–58.
24. Elices MJ, Osborn L, Takada Y, et al. VCAM-1 on activated endothelium interacts with the leukocyte integrin VLA-4 at a site distinct from the VLA-4/fibronectin binding site. *Cell* 1990;60:577–84.
25. Rice GE, Munro JM, Bevilacqua MP. Inducible cell adhesion molecule 110 (INCAM-110) is an endothelial receptor for lymphocytes. A CD11/CD18-independent adhesion mechanism. *J Exp Med* 1990;171:1369–74.
26. Guan JL, Hynes RO. Lymphoid cells recognize an alternatively spliced segment of fibronectin via the integrin receptor $\alpha_4\beta_1$. *Cell* 1990;60:53–61.
27. Mould AP, Wheldon LA, Komoriya A, Wayner EA, Yamada KM, Humphries MJ. Affinity chromatographic isolation of the melanoma adhesion receptor for the IIICS region of fibronectin and its identification as the integrin $\alpha_4\beta_1$. *J Biol Chem* 1990;265:4020–4.
28. Wayner EA, Garcia-Pardo A, Humphries MJ, McDonald JA, Carter WG. Identification and characterizaton of the T lymphocyte adhesion receptor for an alternative cell attachment domain (CS-1) in plasma fibronectin. *J Cell Biol* 1989;109:1321–30.
29. Weg VB, Williams TJ, Lobb RR, Nourshargh S. A monoclonal antibody recognizing very late activation antigen-4 inhibits eosinophil accumulation in vivo. *J Exp Med* 1993;177:561–6.

30. Davis LS, Oppenheimer-Marks N, Bednarczyk JL, McIntyre BW, Lipsky PE. Fibronectin promotes proliferation of naive and memory T cells by signaling through both the VLA-4 and VLA-5 integrin molecules. *J Immunol* 1990;145:785–93.
31. Miyake K, Weissman IL, Greenberger JS, Kincade PW. Evidence for a role of the integrin VLA-4 in lympho-hemopoiesis. *J Exp Med* 1991;173:599–607.
32. Nojima Y, Humphries MJ, Mould AP, et al. VLA-4 mediates CD3-dependent $CD4^+$ T cell activation via the CS1 alternatively spliced domain of fibronectin. *J Exp Med* 1990;172: 1185–92.
33. Williams DA, Rios M, Stephens C, Patel VP. Fibronectin and VLA-4 in haematopoietic stem cell-microenvironment interactions. *Nature* 1991;352:438–41.
34. Baron JL, Madri JA, Ruddle NH, Hashim G, Janeway CA Jr. Surface expression of α4 integrin by CD4 T cells is required for their entry into brain parenchyma. *J Exp Med* 1993;177:57–68.
35. Van Dinther-Janssen ACHM, Horst E, Koopman G, et al. The VLA-4/VCAM-1 pathway is involved in lymphocyte adhesion to endothelium in rheumatoid synovium. *J Immunol* 1991;147:4207–10.
36. Yednock TA, Cannon C, Fritz LC, Sanchez-Madrid F, Steinman L, Karin N. Prevention of experimental autoimmune encephalomyelitis by antibodies against α4β1 integrin. *Nature* 1992;356:63–6.
37. Rice GE, Bevilaqua MP. An inducible endothelial cell surface glycoprotein mediates melanoma adhesion. *Science* 1989;246:1303–6.
38. Taichman DB, Cybulsky MI, Djaffar I, et al. tumor cell surface α4β1 integrin mediates adhesion to vascular endothelium: demonstration of an interaction with the N-terminal domains of INCAM-110/VCAM-1. *Cell Regul* 1991;2:347–55.
39. Kilshaw PJ, Murant SJ. A new surface antigen on intraepithelial lymphocytes in the intestine. *Eur J Immunol* 1990;20:2201–7.
40. Parker CM, Cepek KL, Russell GJ, et al. A family of β7 integrins on human mucosal lymphocytes. *Proc Natl Acad Sci USA* 1992;89:1924–8.
41. Cerf-Bensussan N, Bègue B, Gagnon J, Meo T. The human intraepithelial lymphocyte marker HML-1 is an integrin consisting of a β7 subunit associated with a distinctive α chain. *Eur J Immunol* 1992;22:273–7.
42. Cerf-Bensussan N, Jarry A, Brousse N, Lisowska-Grospierre B, Guy-Grand D, Griscelli C. A monoclonal antibody (HML-1) defining a novel membrane molecule present on human intestinal lymphocytes. *Eur J Immunol* 1987;17:1279–85.
43. Kilshaw PJ. Expression of the mucousal T cell integrin αM290β7 by a major subpopulation of dendritic cells in mice. *Eur J Immunol* 1993;23:3365–8.
44. Roberts K, Kilshaw PJ. The mucosal T cell integrin αM290β7 recognizes a ligand on mucosal epithelial cell lines. *Eur J Immunol* 1993;23:1630–5.
45. Cepek KL, Parker CM, Madara JL, Brenner MB. Integrin $\alpha^E\beta7$ mediates adhesion of T lymphocytes to epithelial cells. *J Immunol* 1993;150:3459–70.
46. Sarnacki S, Bègue B, Buc H, Le Deist F, Cerf-Bensussan N. Enhancement of CD3-induced activation of human intestinal intraepithelial lymphocytes by stimulation of the β7-containing integrin defined by HML-1 monoclonal antibody. *Eur J Immunol* 1992;22:2887–92.
47. Strauch UG, Lifka A, Gosslar U, Kilshaw PJ, Clements J, Holzmann B. Distinct binding specificities of integrins α4β7 (LPAM-1), α4β1 (VLA-4) and αIELβ7. *Int Immunol* 1994; 6:263–75.
48. Ruegg C, Postigo AA, Sikorski EE, Butcher EC, Pytela R, Erle DJ. Role of integrin α4β7/α4βp in lymphocyte adherence to fibronectin and VCAM-1 and in homotypic cell clustering. *J Cell Biol* 1992;117:179–89.
49. Chan BM, Elices MJ, Murphy E, Hemler ME. Adhesion to vascular cell adhesion molecule 1 and fibronectin. Comparison of α4β1 (VLA-4) and α4β7 on the human molecule 1 and fibronectin. Comparison of α4β1 (VLA-4) and α4β7 on the human B cell line JY. *J Biol Chem* 1992;267:8366–70.
50. Postigo AA, Sanchez Mateos P, Lazarovits AI, Sanchez Madrid F, De Landazuri MO. α4β7 integrin mediates B cell binding to fibronectin and vascular cell adhesion molecule-1. *J Immunol* 1993;151:2471–83.
51. Hynes RO. Integrins: versatility, modulation, and signaling in cell adhesion. *Cell* 1992;69: 11–25.

*Topics in Molecular Medicine, Volume 1,*
edited by Wolfgang Siess, Reinhard Lorenz,
and Peter C. Weber. Raven Press, Ltd.,
New York © 1995.

# 18

# Splice Variants of CD44: Common Functions in Lymphocyte Activation and Lymphatic Spread of Tumor Cells

M. Zöller

*Department of Tumor Progression and Immune Defense, German Cancer Research Center, 69120 Heidelberg, Germany*

## WHAT IS KNOWN ABOUT CD44v?

CD44 comprises a family of glycoproteins that vary according to their glycosylation and their protein structure (1–14). The so-called hematopoetic or standard form of CD44 (CD44s) is composed of a short intracytoplasmatic tail, a transmembrane region, and two extracellular domains (2,13–15). This protein backbone is largely conserved in most variant isoforms of CD44 and can be expanded by the insertion of additional exons in the membrane proximal extracellular domain. There exist 10 additional exons (v1 to v10) (5,10,12–21). All of them, and any possible combinations of them, can be expressed by alternative splicing (13,16–22). Furthermore, the same cell, and especially tumor cells, can express more than one combination of variant exons (16,23–29). In addition to this variability and diversity, there exists a further peculiarity of the splicing mechanism of CD44. Alternative splicing is frequently observed when cells change their functional program. However, in most instances alternative splicing occurs only once, whereas the splicing of CD44 can change repeatedly and reversibly such that the same cell, depending on its activation status, can express CD44s and various forms of CD44v. Because the splicing of CD44 appears to be stringently regulated, it can be assumed that the different exons or combinations of exons exert defined and divergent functions.

What are the functions of the family of CD44 glycoproteins? CD44s is an adhesion molecule with binding domains for hyaluronate, glucosaminoglycans, and possibly for further elements of the extracellular matrix (26,30–46). In addition to binding to components of the extracellular matrix, the molecule is also involved in cell–cell interactions, especially in binding to high endothelial venules during lymphocyte homing (36,47–50). Furthermore, CD44 is involved in hematopoetic stem cell maturation, in lymphocyte activation, and in lymphocyte traffic (51–69). It is, however, not known to what extent these functions are exerted by CD44s or CD44v. Up to the present, only two functions have been clearly assigned to variant isoforms of CD44: lymphatic spread of tumor cells, and lymphocyte activation (16,27, 70,71). Even for these two functions of CD44v, it is unknown which variant exons or combinations of variant exons are required and what may be the functional principle.

Special interest in the variant isoforms of CD44 arose when it was discovered in a rat tumor model (72,73) that a metastasizing subline of a pancreatic adenocarcinoma expressed CD44v and that the metastasizing phenotype could be transfered to the nonmetastasizing subline by transfection with CD44v4–v7 cDNA (16). A search for variant exons or exon combinations that could confer the metastatic phenotype revealed that the combination of exons v6 and v7 may be sufficient (27). According to preliminary evidence, expression of additional exons does not interfere with tumor progression (J. Sleemann, personal communication). Since then, much work has been done to explore the relevance of CD44v expression to progression of human tumors, to define the ligand, and to unravel the functional principle. Screening of human tumor specimens revealed that some (74–89) but not all human tumors express CD44v (25,90–93), and there is evidence that for some tumor types expression of CD44v may be linked to tumor progression (74,85,86,87). The search for defined ligands of CD44 variant exons on elements of the extracellular matrix was less successful. Thus far, no ligand other than hyaluronic acid has been defined. It is still disputed whether variant exons tighten or relax the cell matrix interaction (W. Rudy, personal communication and unpublished data). In considering a possible cellular ligand, crosslinking studies pointed toward a homotypic interaction of CD44 splice variants on tumor cells (J. Sleemann, personal communication). However, the mechanism of function remained elusive.

## DEFINING THE STEP IN TUMOR PROGRESSION–LYMPHOCYTE ACTIVATION THAT REQUIRES CD44v

The cascade of events involved in tumor progression encompasses detachment from the primary tumor, penetration into blood or lymphatic vessels, transport into and extravasation out of blood or lymph vessels, embedding in foreign tissue, and outgrowth. The process of lymphocyte activation encompasses migration of lymphocytes via the lymphatic vessels to the draining lymph node, encounter with antigen, activation, and expansion. Effector functions are pursued after migration to the injured tissue. Because expression of CD44v is required for the lymphatic spread of tumor cells and for lymphocyte activation, it was important to determine first whether the molecule facilitates migration from the primary tumor inoculum and/or of peripheral lymphocytes to the draining lymph node. This was evaluated by injection of [$^{125}$I]UDR-labeled tumor cells into the footpad of rats after migration from the tumor cell depot and migration to the popliteal node, dependent on the concomitant application of the anti-CD44v monoclonal antibody (MAb) 1.1ASML, which recognizes an epitope on exon v6.

Migration of lymphocytes was examined by painting the skin with fluorescein isothiocyanate (FITC) and examining the numbers of FITC-labeled cells in the skin and the draining lymph node dependent on the concomitant application of MAb 1.1ASML (Fig. 1). Migration from the primary tumor inoculum and migration of tumor cells into the draining lymph node was not impaired by anti-CD44v. This accounted for a spontaneously metastasizing tumor line (ASML) (data not shown) as well as for an a priori nonmetastasizing line, into which the metastatic phenotype was transfered by transfection with CD44v cDNA (AS-14). The same observation was true for FITC-labeled lymphocytes in the skin. Neither their migration from the dermis nor their migration into the draining node was influenced by anti-CD44v. It should be noted, however, that migration from the periphery as well as migration to the draining lymph node was slightly impaired in the presence of anti-CD44s (Ox50). This accounted for tumor cells and lymphocytes (Fig. 1).

Although migration of tumor cells and lymphocytes into the draining lymph node could not be inhibited by anti-CD44v, expansion of tumor cells and of antigen-stimulated lymphocytes was severely impaired in the presence of 1.1ASML (Fig. 2). From application of radiolabeled tumor cells, we know that, at maximum, approximately 0.1% of tumor cells reach the draining node and that even of those tumor cells a considerable number are going to die. Nevertheless, immunohistology revealed that as early as 10 days after tumor cell application islets of tumor cells were visible in the draining node (data not shown). Evaluating the number of tumor cells in the draining lymph node by soft agar cloning confirmed expansion of the surviving tumor cells. However, in the majority of animals treated with the MAb 1.1ASML the number of recovered tumor cells decreased continuously and at 3 to 4 weeks no tumor cell colonies could be detected. In animals, in which some tumor cells survived in the draining node, the cells slowly started to proliferate at 3 to 4 weeks after tumor cell application (Fig. 2). When lymphocytes enter a draining lymph node after application of antigen, blast transformation takes place and individual, antigen-specific cells undergo 10 to 15 rounds of cell division. The clonal expansion can be monitored by evaluating under limiting dilution conditions the increase in the number of antigen-specific lymphocytes. When rats received a local depot of allogeneic cells, expansion of alloantigen-specific lymphocytes in the draining node was severely impaired in the presence of MAb 1.1ASML, which was injected intravenously at a dose of 200 μg twice per week. The requirement of CD44 expression during lymphocyte expansion could clearly be assigned to the variant exons,

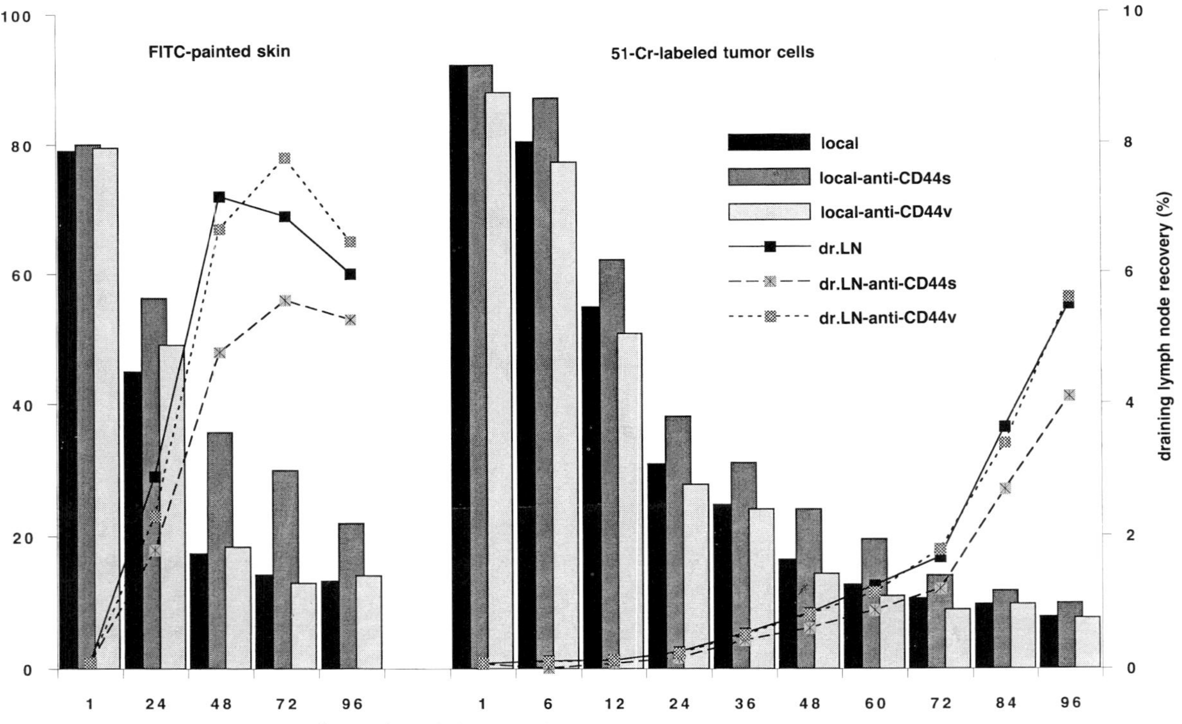


**FIG. 1.** Expression of CD44v is not required for migration in lymphatic vessels and for entering the draining lymph node. **Right:** [$^{125}$]IUDR-labeled AS-14 ($5 \times 10^6$) cells were injected into the footpad. Where indicated, rats concomitantly received an injection of 200 μg anti-CD44v. (1.1ASML) or anti-CD44s (Ox50) intravenously. Rats were killed at the indicated time points. Migration of tumor cells to the draining popliteal node were evaluated by determining the remaining radioactivity in foot and draining lymph node in a gamma counter. **Left:** Rats were painted with 2 ml PBS containing 1 mg/ml FITC and, where indicated, received 200 μg anti-CD44v or anti-CD44s concomitantly. Rats were sacrificed at the indicated time points. Lymphocytes were isolated from the dermis by digestion with trypsin and separation from keratinocytes by Percoll gradient centrifugation; the draining lymph node (dr.LN) was meshed. The number of FITC-labeled lymphocytes in the digest of the dermis and the draining lymph node was evaluated by fluorescence-activated cell scanning (FACS) analysis.

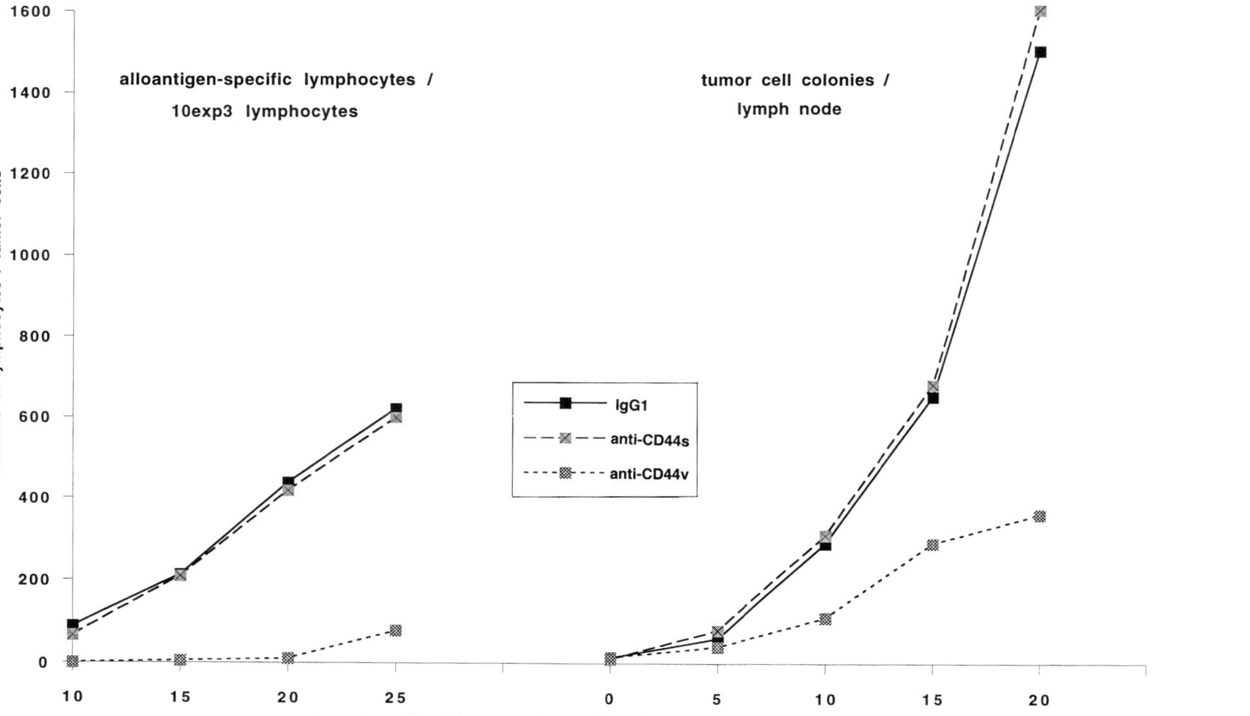


**FIG. 2.** Expression of CD44v is required for the growth of tumor cells and for clonal expansion of lymphocytes in the draining lymph node. **Right:** After intrafootpad application of $5 \times 10^5$ AS-14 cells, expansion of tumor cells in the draining lymph node was monitored by excising the node and cloning dispersed lymph node cells in soft agar at the indicated time points and counting the colonies after 8 days of culture. Although no colonies could be detected in the draining lymph node suspension of the majority of rats that received intravenous injections of 200 μg anti-CD44v twice per week, some rats developed metastases, although significantly delayed. The figure depicts examples of this latter group of animals. **Left:** BDX rats received a local depot of allogeneic lymphocytes from DA rats. Where indicated, rats were treated twice per week with intravenous injections of 200 μg anti-CD44v (1.1ASML) or anti-CD44s (Ox50), both antibodies of the IgG1 subclass. The number of alloantigen-specific helper T cells was monitored by determining the frequency of cells proliferating in response to irradiated DA lymphocytes under limiting dilution conditions. Proliferation was evaluated by [$^3$H]thymidine incorporation.

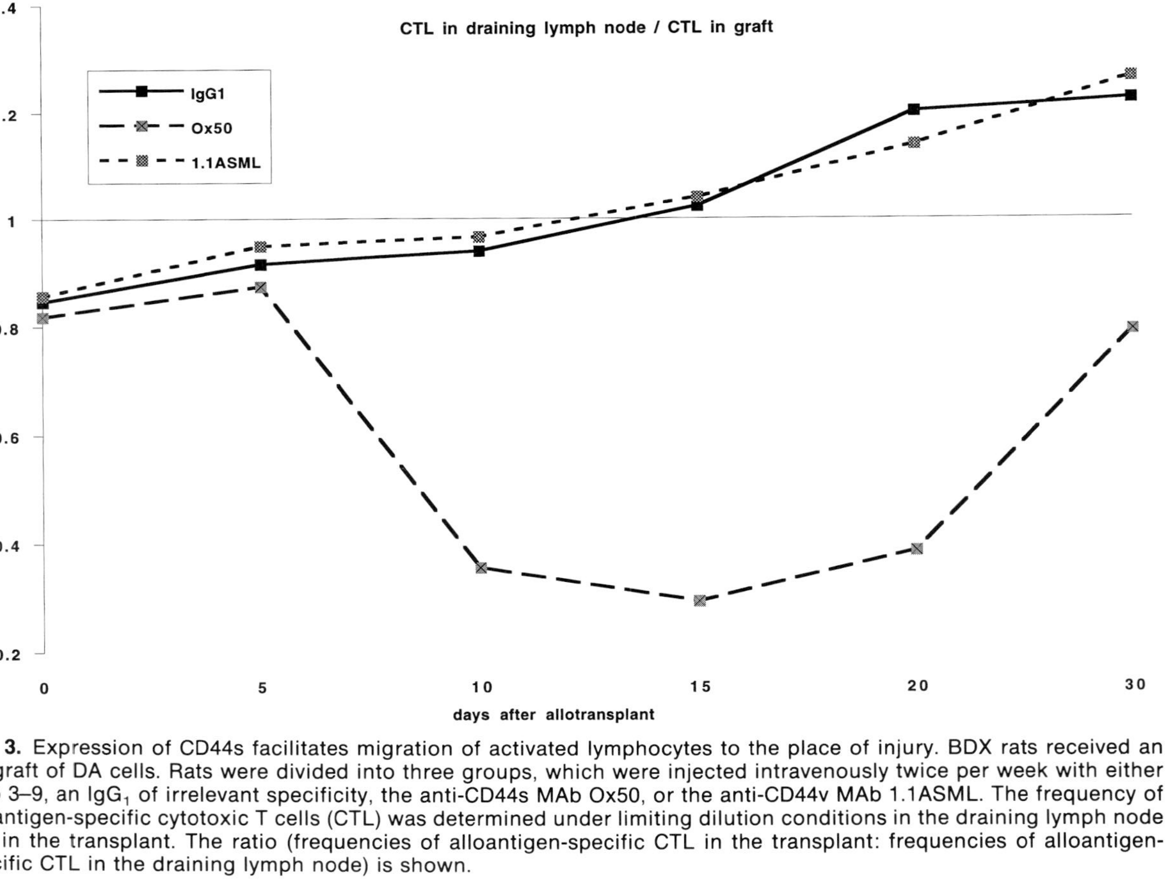


**FIG. 3.** Expression of CD44s facilitates migration of activated lymphocytes to the place of injury. BDX rats received an allograft of DA cells. Rats were divided into three groups, which were injected intravenously twice per week with either MAb 3–9, an $IgG_1$ of irrelevant specificity, the anti-CD44s MAb Ox50, or the anti-CD44v MAb 1.1ASML. The frequency of alloantigen-specific cytotoxic T cells (CTL) was determined under limiting dilution conditions in the draining lymph node and in the transplant. The ratio (frequencies of alloantigen-specific CTL in the transplant: frequencies of alloantigen-specific CTL in the draining lymph node) is shown.

as an antibody recognizing an epitope on standard domains of CD44 did not interfere with clonal expansion (Fig. 2).

The last steps during lymphocyte activation and tumor metastasis formation, i.e., the trafficking of activated lymphocytes to the place of injury and the settlement of metastasizing tumor cells in distant organs, appears to be, like the migration process, independent of the expression of CD44v. The settlement of secondary metastases is difficult to monitor experimentally. However, it was noted with ASML cells, whose settlement in the draining lymph node is significantly delayed during application of the MAb 1.1ASML, that formation of lung metastasis followed the same time schedule of embedding in lymph node and reaching the lung as in untreated rats. Furthermore, although again with a delay of 3 to 6 weeks, lung tissue finally was completely replaced by miliary metastases of ASML even under continuous treatment with anti-CD44v (data not shown). The migration of activated lymphocytes can be followed more precisely by determining the number of antigen-specific lymphocytes in the draining organ and at the place of injury. Evaluation of the ratio of antigen-specific cells in the allograft versus those in the draining lymph node (Fig. 3) revealed that this ratio was not influenced by the application of anti-CD44v. It was, however, significantly reduced under the influence of anti-CD44s (Fig. 3).

Comparative studies of lymphocyte activation and metastasis formation in the lymphatic system clearly revealed that expression of CD44v is required for expansion in the draining lymph node, whereas migration into and out of the lymph node are not influenced by anti-CD44v. In line with published evidence (69), the traffic out of the draining node appears to require or to be facilitated by domains in the standard part of the molecule.

## PARTICIPATION OF CD44v IN BINARY CELL–CELL INTERACTIONS BETWEEN ANTIGEN-PRESENTING CELLS AND TUMOR CELLS–LYMPHOCYTES

After T lymphocytes have entered the draining lymph node, they must encounter the nominal antigen presented in peptide form on specialized antigen-presenting cells (95,96). It is now well established that binding of the T-cell receptor to the antigen peptide–MHC complex on the antigen-presenting cell is not sufficient to initiate activation. There is a requirement also for strengthening the binding via bystander molecules. It has been shown for many adhesion molecules that ligand binding causes transduction of signals either in the ligated cell, the receptor-bearing cell, or in both (96–100), and that in the absence of these second signals T cells become anergic rather than activated (101,102). If variant isoforms of CD44 are involved or are essential for signal transduction between tumor cells–lymphocytes and antigen-presenting cells, metastasis formation should be precluded in the ab-

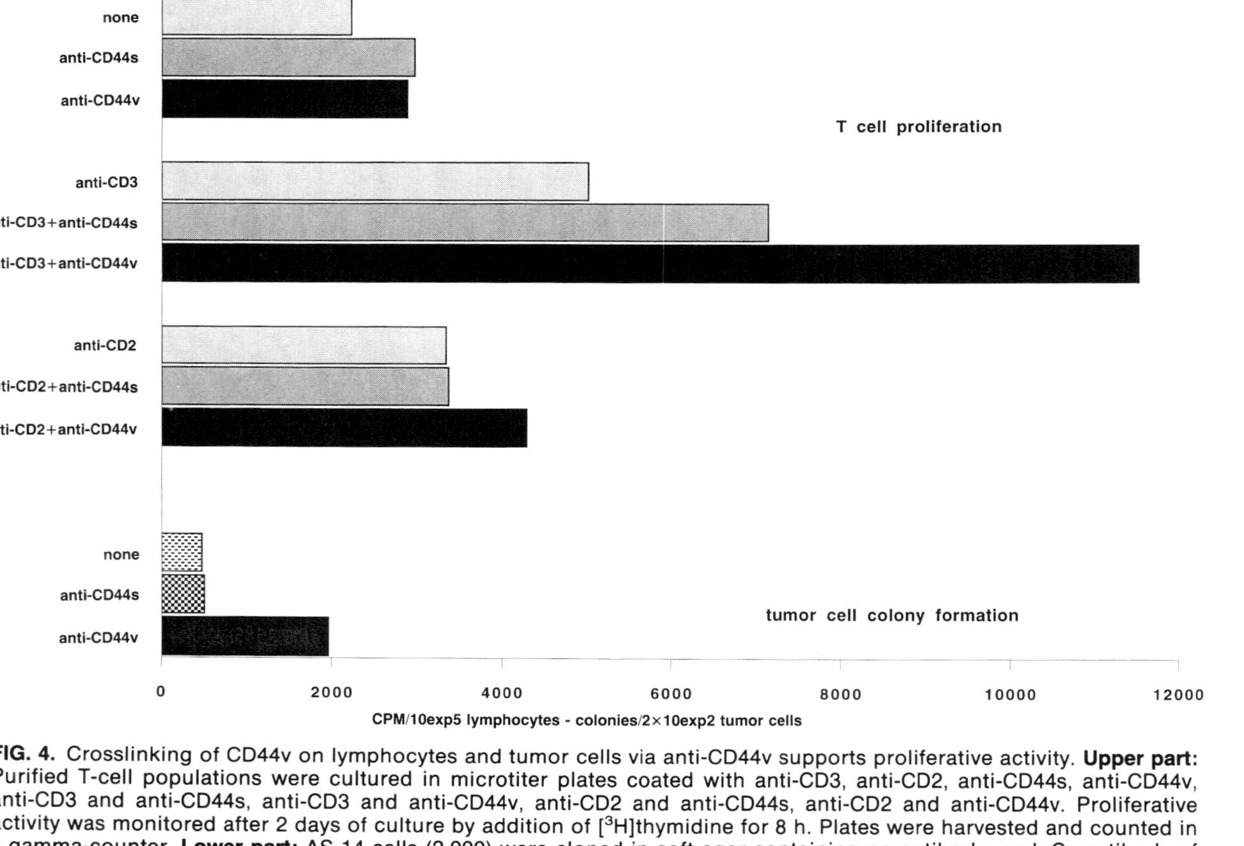


**FIG. 4.** Crosslinking of CD44v on lymphocytes and tumor cells via anti-CD44v supports proliferative activity. **Upper part:** Purified T-cell populations were cultured in microtiter plates coated with anti-CD3, anti-CD2, anti-CD44s, anti-CD44v, anti-CD3 and anti-CD44s, anti-CD3 and anti-CD44v, anti-CD2 and anti-CD44s, anti-CD2 and anti-CD44v. Proliferative activity was monitored after 2 days of culture by addition of [$^3$H]thymidine for 8 h. Plates were harvested and counted in a gamma-counter. **Lower part:** AS-14 cells (2,000) were cloned in soft agar containing no antibody, an $IgG_1$ antibody of irrelevant specificity, anti-CD44s (Ox50), or anti-CD44v (1.1ASML). The number of colonies after 7 days of culture is shown.

sence of antigen-presenting cells, as is known for lymphocyte activation. In addition, anti-CD44v should not be inhibitory where the requirement of antigen presentation can be circumvented, e.g., by cross-linking of the T-cell receptor. In fact, when isolated T cells were cultured on anti-CD2– or anti-CD3–coated plates, anti-CD44v did not display any inhibitory effect. Instead, the proliferative activity of lymphocytes was significantly augmented by addition of anti-CD44v (Fig. 4). A similar phenomenon was noted with tumor cell cultures. Addition of the MAb 1.1ASML had no impact on the proliferation of tumor cells (data not shown). However, when tumor cells were cloned in soft agar, admixture of MAb 1.1ASML significantly increased the efficiency of colony formation. No such effect was observed in the presence of anti-CD44s (Fig. 4).

These data not only prove that anti-CD44v is in itself neither cytotoxic nor cytostatic but indicate that, by ligand binding and/or homotypic aggregation of CD44v, signals may be transduced, initiating proliferation of tumor cells and lymphocytes. Therefore, we finally asked whether under physiologic conditions antigen-presenting cells in the draining lymph node may be the relevant substrate for metastasizing tumor cells and lymphocytes. In fact, although the described CD44v-positive rat tumor lines still metastasized in the nude/nude mouse via the lymphatic system, formation of lymph node metastases has never been observed in the SCID mouse. Interestingly, even after subcutaneous application of tumor cells, metastasis formation was frequently noted in the spleen of SCID mice, which contains some residual cells capable of antigen presentation (Table 1).

In line with this notion was the finding that purified T cells cultured with antigen-presenting cells did not proliferate in response to the nominal antigen in the presence of anti-CD44v, whereas anti-CD44s had no impact on

**TABLE 1.** *Lymphatic spread of tumor cells in nude/nude and SCID mice*

| | | | | Metastases | | |
|---|---|---|---|---|---|---|
| Tumor line[a] | Host[b] | Survival time (days) | Local growth $\phi$ (cm) | Lymph node | Lung | Other organs |
| AS | Rat | 30–45 | 2.5 | None | None | None |
| AS-14 | Rat | 35–48 | 2.2 | Draining | 4–20 | None |
| ASML | Rat | 43–65 | None | Systemic | Miliary | None |
| AS | nu/nu | 31–36 | 2.3 | None | None | None |
| AS-14 | nu/nu | 44–63 | 0.5 | Draining | 4–10 | None |
| ASML | nu/nu | 43–70 | None | Draining | 20–53 | None |
| AS | SCID | 17–20 | 2.8 | None | None | None |
| AS-14 | SCID | 15–27 | 2.0 | None | 0–2 | None |
| ASML | SCID | 35–90 | 0.9 | None | Miliary | Spleen |

[a] AS, CD44v-negative, nonmetastasizing line; AS-14, CD44v cDNA-transfected, metastasizing line; ASML, CD44v-positive, metastasizing line.

[b] Syngeneic BDX rats, nu/nu BALB/c mice, and SCID mice received $5 \times 10^5$ tumor cells subcutaneously.

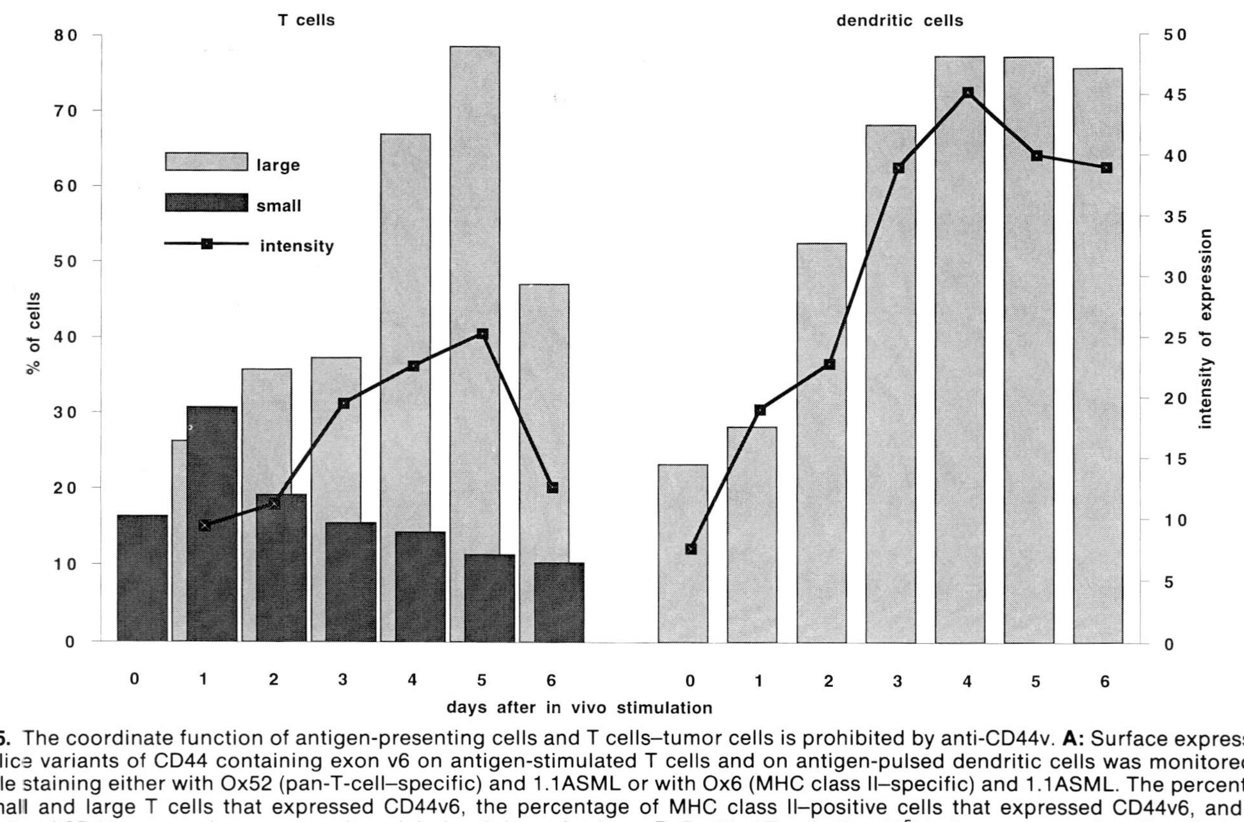


**FIG. 5.** The coordinate function of antigen-presenting cells and T cells–tumor cells is prohibited by anti-CD44v. **A:** Surface expression of splice variants of CD44 containing exon v6 on antigen-stimulated T cells and on antigen-pulsed dendritic cells was monitored by double staining either with Ox52 (pan-T-cell–specific) and 1.1ASML or with Ox6 (MHC class II–specific) and 1.1ASML. The percentage of small and large T cells that expressed CD44v6, the percentage of MHC class II–positive cells that expressed CD44v6, and the intensity of CD44v expression were monitored during 6 days of culture. **B:** Purified T cells ($2 \times 10^5$) in the presence of ovalbumin (upper part) and AS-14 tumor cells ($1 \times 10^4$) (lower part) were cultured with $1 \times 10^4$ or $1 \times 10^3$ dendritic cells and 10 μg/ml of an $IgG_1$ MAb of irrelevant specificity, or of anti-CD44s (Ox50), or of anti-CD44v (1.1ASML). Proliferative activity was monitored after 3 days of culture by addition of [$^3$H]thymidine for 12 h. Plates were harvested and counted in a gamma-counter.

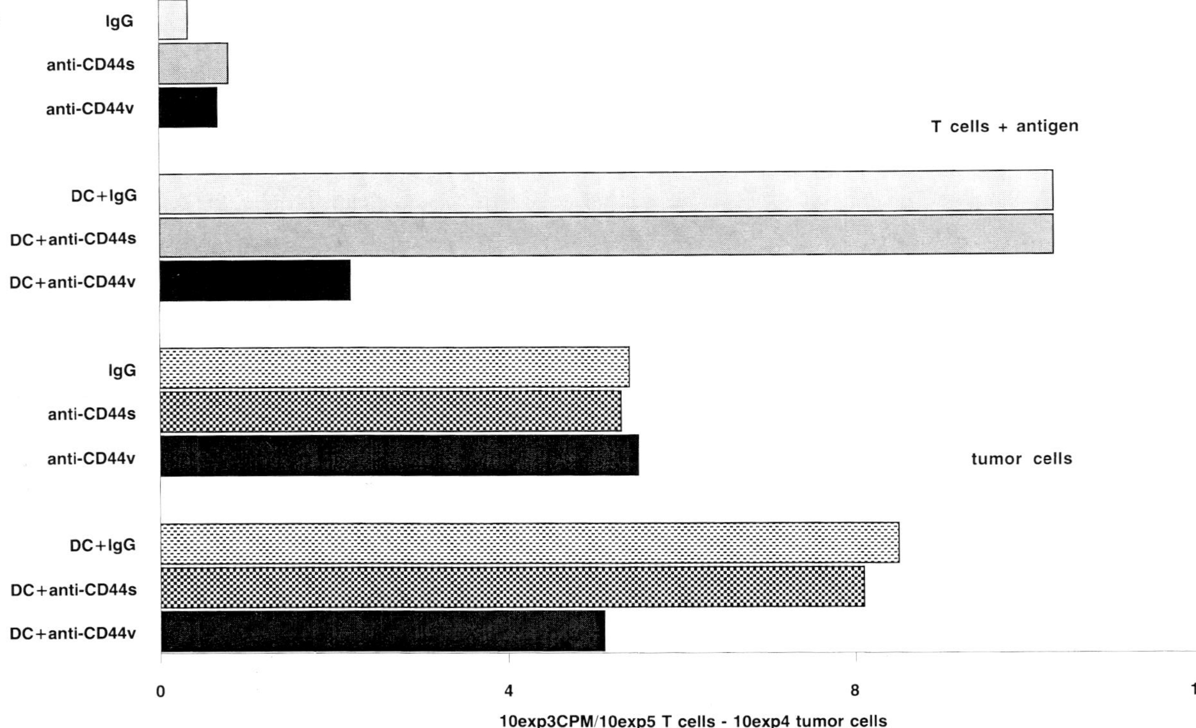
B
IgG
anti-CD44s
anti-CD44v
T cells + antigen
DC+IgG
DC+anti-CD44s
DC+anti-CD44v
IgG
anti-CD44s
anti-CD44v
tumor cells
DC+IgG
DC+anti-CD44s
DC+anti-CD44v
0
4
8
12
10exp3CPM/10exp5 T cells - 10exp4 tumor cells

T-cell proliferation. After depletion of macrophages by adherence to plastic twice for 1 h each, dendritic cells were isolated by Percoll gradient centrifugation to collect the low-density population. This population expressed MHC class II antigens (data not shown), and expression of CD44v was initiated by encounter with antigen. The kinetics of expression of CD44v on dendritic cells resembled those of T lymphocytes during antigenic stimulation, the only difference being a slightly accelerated appearance of CD44v mRNA (data not shown) and protein on antigen-presenting cells (Fig. 5A). As shown in Fig. 5B, dendritic cells supported the proliferation of purified T cells in response to nominal antigen. However, the proliferative response was strongly reduced in the presence of anti-CD44v, whereas anti-CD44s had no effect. Neither anti-CD44v, anti-CD44s, nor an isotype-matched control MAb of irrelevant specificity displayed any activity when added in solution to the purified T-cell population. The same observation accounted for the proliferation of tumor cells in the presence of anti-CD44s or anti-CD44v. When dendritic cells were added to the admixture of tumor cells and antibodies, tumor cell proliferation was increased in the presence of anti-CD44s or of an MAb of irrelevant specificity but not in the presence of anti-CD44v. This finding indicates that antigen-presenting cells assist even the autonomous growth of tumor cells. However, the amount of support will be inhibited by anti-CD44v (Fig. 5).

It is well established that it is the antigen-presenting cell that provides the required growth factors for the clonal expansion of lymphocytes after encounter with nominal antigen (103–106). Hence, it is tempting to speculate that by the same pathway the isolated tumor cell may be supported after ligation via CD44v. This is not thus far proven, nor is it known whether binding between tumor cells–lymphocytes and antigen-presenting cells via CD44v represents a homotypic or a heterotypic interaction. Nevertheless, comparative analysis of lymphocyte activation and metastasis formation in lymphatic tissue provides convincing evidence that the pathophysiologic process of tumor progression follows a physiologic program, which at a central point involves signal transduction after binary cell interactions initiated by CD44v ligand binding. Further unraveling this lymphocyte activation–metastasis formation pathway may well provide general schemes for therapeutic intervention.

## SUMMARY

Variant isoforms of the adhesion molecule CD44 (CD44v) have been described to initiate lymphatic spread of locally growing rat tumor cells. This unexpected finding was shown to be of general validity: all rat tumors that spontaneously metastasize via the lymphatic system express variant isoforms of CD44, tumor cells of different histology that express variant isoforms after transfection with CD44v-cDNA metastasize, and metastasis formation can be inhibited by anti-CD44v.

Taking into account that tumor progression comprises a cascade of biologic processes, it was speculated that tumor cells may adopt a physiologic program. In fact, variant isoforms of CD44 are abundantly expressed during organogenesis and expression is conserved throughout life on stem cells in skin, gut, and bone marrow. In addition, expression of CD44v is transiently observed during lymphocyte activation. Based on the particularly tight link between metastasis formation and the process of lymphocyte activation, experiments were designed to elucidate the functional mechanism of variant isoforms of CD44. It was shown that variant isoforms are involved in binary cell–cell interactions, the ligand of CD44v-positive tumor cells and lymphocyte probably being an antigen-presenting cell located in the draining lymph node. After ligand binding via CD44v, antigen-presenting cells provide support for proliferation to both lymphocytes and isolated tumor cells. CD44v appears not to be involved in homing and trafficking of lymphocytes, nor in migration of metastasizing tumor cells. These functions, however, are modulated by domains of the standard form of CD44 (CD44s).

In conclusion, CD44v probably functions by means of signal transduction. The initial trigger appears to be the activation of an antigen-presenting cell by CD44v ligand binding. This could well explain the requirement of CD44v on lymphocytes and tumor cells for settlement and expansion in lymphatic tissue.

## ACKNOWLEDGMENT

This investigation was supported by the Deutsche Forschungsgemeinschaft, grant No Zo5-1, the Mildred Scheel Stiftung, the Tumorzentrum Heidelberg/Mannheim, and the Sandoz Stiftung für Therapeutische Forschung.

## REFERENCES

1. Jalkanen S, Bargatze RF, de los Toyos J, Butcher EC. A lymphoid cell surface protein involved in endothelial cell recognition and lymphocyte homing in man. *Eur J Immunol* 1986;16:1195–1202.
2. Goldstein LA, Zhou DFH, Picker LJ, et al. A human lymphocyte homing receptor, the Hermes antigen, is related to cartilage proteoglycan core and link proteins. *Cell* 1989;56: 1063–5.
3. Zhou DFH, Ding JF, Picker LF, Bargatze RF, Butcher EC, Goeddel DV. Molecular cloning and expression of Pgp-1—the mouse homolog of the human H-CAM (Hermes) lymphocyte homing receptor. *J Immunol* 1989;143:3390–5.
4. Quackenbush EJ, Vera S, Greaves A, Letarte M. Confirmation by peptide sequence and co-expression on various cell types of the identity of CD44 and P85 glycoprotein. *Mol Immunol* 1990;27:947–55.
5. Brown TA, Bouchard T, St John T, Wayner E, Carter WG. Human keratinocytes express

a new CD44 core protein (CD44E) as a heparan-sulfate intrinsic membrane proteoglycan with additional exons. *J Cell Biol* 1991;113:207–21.
6. Lokeshwar VB, Bourguignon LYW. Post-translational protein modification and expression of ankyrin-binding site(s) in GP85 (Pgp-1/CD44) and its biosynthetic precursors during T-lymphoma membrane biosynthesis. *J Biol Chem* 1991;266:17983–9.
7. Camp RL, Kraus TA, Pure E. Variations in the cytoskeletal interaction and posttranslational modification of the CD44 homing receptor in macrophages. *J Cell Biol* 1991;115: 1283–92.
8. Omary MB, Trowbridge IS, Letarte M, Kagnoff MF, Isacke CM. Structural heterogeneity of human Pgp-1 and its relationship with p85. *Immunogenetics* 1988;27:460–4.
9. Kansas GS, Wood GS, Daily MO. A family of cell-surface glycoproteins defined by a putative anti-endothelial cell receptor antibody in man. *J Immunol* 1989;142:3050–7.
10. Goldstein LA, Butcher EC. Identification of mRNA that encodes an alternative form of H-CAM (CD44) in lymphoid and nonlymphoid tissues. *Immunogenetics* 1990;32:389–97.
11. Stamenkovic I, Amiot M, Pesando JM, Seed B. A lymphocyte molecule implicated in lymph node homing is a member of the cartilage link protein family. *Cell* 1989;56:1057–62.
12. Screaton GR, Bell MV, Jackson DG, Cornelis FB, Gerth U, Bell JI. Genomic structure of DNA encoding the lymphocyte homing receptor CD44 reveals at least 12 alternatively spliced exons. *Proc Natl Acad Sci USA* 1992;89:12160–4.
13. Idzerda RL, Carter WG, Nottenburg C, Wayner EA, Gallatin WM, St. John T. Isolation and DNA sequence of a cDNA clone encoding a lymphocyte adhesion receptor for high endothelium. *Proc Natl Acad Sci USA* 1989;86:4659–63.
14. Stamenkovic L, Aruffo A, Amiot M, Seed B. The hematopoietic and epithelial forms of CD44 are distinct polypeptides with different adhesion potentials for hyaluronate-bearing cells. *EMBO J* 1991;10:343–8.
15. Dougherty GJ, Lansdorp PM, Cooper DL, Humphries RK. Molecular cloning of CD44R1 and CD44R2, two novel isoforms of the human CD44 lymphocyte "homing" receptor expressed by hemopoietic cells. *J Exp Med* 1991;174:1–5.
16. Günthert U, Hofmann M, Rudy W, et al. A new variant of glycoprotein CD44 confers metastatic potential to rat carcinoma cells. *Cell* 1991;65:13–24.
17. Jackson DG, Buckley J, Bell JI. Multiple variants of the human lymphocyte homing receptor CD44 generated by insertions at a single site in the extracellular domain. *J Biol Chem* 1992;267:4732–9.
18. Screaton GR, Bell MV, Bell JI, Jackson DG. The identification of a new alternative exon with highly restricted tissue expression in transcripts encoding the mouse Pgp-1 (CD44) homing receptor. Comparison of all 10 variable exons between mouse, human, and rat. *J Biol Chem* 1993;268:12235–8.
19. Hofmann M, Rudy W, Zöller M, et al. CD44 splice variants confer metastatic behavior in rats: homologous sequences are expressed in human tumor cell lines. *Cancer Res* 1991; 51:5292–7.
20. Haggerty JG, Bretton RH, Milstone LM. Identification and characterization of a cell surface proteoglycan on keratinocytes. *J Invest Dermatol* 1992;99:374–80.
21. Tölg C, Hofmann M, Herrlich P, Ponta H. Splicing choice from ten variant exons establishes CD44 variability. *Nucleic Acids Res* 1993;21:1225–9.
22. Hughes EN, Colombatti A, August JT. Murine cell surface glycoproteins. *J Biol Chem* 1981;256:1014–21.
23. Shtivelman E, Bishop JM. Expression of CD44 is repressed in neuroblastoma cells. *Mol Cell Biol* 1991;11:5446–53.
24. Birch M, Mitchell S, Hart IR. Isolation and characterization of human melanoma cell variants expressing high and low levels of CD44. *Cancer Res* 1991;51:6660–7.
25. Faasen AE, Schrager JA, Klein DJ, Oegema TR, Couchman JR, McCarthy J. A cell surface chondroitin sulfate proteoglycan, immunologically related to CD44, is involved in type I collagen-mediated melanoma cell motility and invasion. *J Cell Biol* 1992;116:521–31.
26. Gjerset RA, Arya J, Volkman S, Haas M. Association of induction of a fully tumorigenic phenotype in murine radiation-induced T-lymphoma cells with loss of differentiation antigens, gain of CD44, and alterations in p53 protein levels. *Mol Carcinogen* 1992;5:190–8.
27. Rudy W, Hofmann M, Schwartz-Albiez R, et al. The two major CD44 proteins expressed

on a metastatic rat tumor cell line are derived from different splice variants: each one individually suffices to confer metastatic behavior. *Cancer Res* 1993;53:1262–8.

28. East JA, Hart IR. CD44 and its role in tumour progression and metastasis. *Eur J Cancer* 1993;14:1921–2.
29. Carter WG, Wayner EA. Characterization of the class III collagen receptor, a phosphorylated, transmembrane glycoprotein expressed in nucleated human cells. *J Biol Chem* 1988;263:4193–201.
30. Kalomiris EL, Bourguignon LYW. Mouse T lymphoma cells contain a transmembrane glycoprotein (GP85) that binds ankyrin. *J Cell Biol* 1988;106:319–27.
31. Wayner E, Carter W. Identification of multiple cell adhesion receptors for collagen and fibronectin in human fibrosarcoma cells possessing unique a and common b submits. *J Cell Biol* 1987;105:1873–84.
32. Horst E, Meijer CJ, Duijvestijn AM, Hartwig N, Van der Harten HJ, Pals ST. The ontogeny of human lymphocyte recirculation: high endothelial cell antigen (HECA-452) and CD44 homing receptor expression in the development of the immune system. *Eur J Immunol* 1990;20:1483–9.
33. Aruffo A, Stamenkovic I, Melnick M, Underhill CB, Seed B. CD44 is the principal cell surface receptor for hyaluronate. *Cell* 1990;61:1303–13.
34. Miyake K, Medina KL, Hayashi SI, Ono S, Hamaoka T, Kincade PW. Monoclonal antibodies to Pgp-1/CD44 block lympho-hemopoiesis in long-term bone marrow cultures. *J Exp Med* 1990;171:477–88.
35. Belitsos PC, Hildreth JEK, August JT. Homotypic cell aggregation induced by anti-CD44 (Pgp-1) monoclonal antibodies and related to CD44 (Pgp-1) expression. *J Immunol* 1990; 144:1661–70.
36. Miyake K, Underhill CB, Lesley J, Kincade PW. Hyaluronate can function as a cell adhesion molecule and CD44 participates in hyaluronate recognition. *J Exp Med* 1990; 172:69–75.
37. St John T, Meyer J, Idzerda R, Gallatin WM. Expression of CD44 confers a new adhesive phenotype on transfected cells. *Cell* 1990;60:45–52.
38. Culty M, Nguyen HA, Underhill CB. The hyaluronan receptor (CD44) participates in the uptake and degradation of hyaluronan. *J Cell Biol* 1992;116:1055–62.
39. Thomas L, Byers HR, Vink J, Stamlenkovic I. CD44H regulates tumor cell migration on hyaluronate-coated substrate. *J Cell Biol* 1992;118:971–7.
40. Lesley J, Hyman R. CD44 can be activated to function as an hyaluronic acid receptor in normal murine T cells. *Eur J Immunol* 1992;22:2719–23.
41. Jalkanen S, Jalkanen M. Lymphocyte CD44 binds the COOH-terminal heparan-binding domain of fibronectin. *J Cell Biol* 1992;116:817–25.
42. Thomas L, Etoh T, Stamenkovic I, Mihm MC Jr, Byers HR. Migration of human melanoma cells on hyaluronate is related to CD44 expression. *J Invest Dermatol* 1993;100: 115–20.
43. Underhill C, Nguyen HA, Shizari M, Culty M. CD44 positive macrophages take up hyaluronan during lung development. *Dev Biol* 1993;155:324–36.
44. Knudson W, Bartnik E, Knudson GB. Assembly of pericellular matrices by COS-7 cells transfected with CD44 lymphocyte-homing receptor genes. *Proc Natl Acad Sci USA* 1993;90:4003–7.
45. Peach RJ, Hollenbaugh D, Stamenkovic I, Aruffo A. Identification of hyaluronic acid binding sites in the extracellular domain of CD44. *J Cell Biol* 1993;122:257–64.
46. Neame SJ, Isacke CM. The cytoplasmic tail of CD44 is required for basolateral localization in epithelial MDCK cells but does not mediate association with the detergent-insoluble cytoskeleton of fibroblasts. *J Cell Biol* 1993;121:1299–1310.
47. Berg EL, Goldstein LA, Jutila MS, et al. Homing receptors and vascular addressins: cell adhesion molecules that direct lymphocyte traffic. *Immunol Rev* 1989;108:5–18.
48. Picker LJ, Nakache M, Butcher EC. Monoclonal antibodies to human lymphocyte homing receptors define a novel class of adhesion molecules on diverse cell types. *J Cell Biol* 1989;109:927–37.
49. Jalkanen S, Jalkanen M, Bargatze R, Tammi M, Butcher EC. Biochemical properties of glycoproteins involved in lymphocyte recognition of high endothelial venules in man. *J Immunol* 1988;141:1615–23.

50. Gallatin WM, Wayner EA, Hoffman PA, St. John T, Butcher EC, Carter WG. Structural homology between lymphocyte receptors for high endothelium and class III extracellular matrix receptor. *Proc Natl Acad Sci USA* 1989;86:4654–8.
51. Haynes BF, Liao HX, Patton KL. The transmembrane hyaluronate receptor (CD44): multiple functions, multiple forms. *Cancer Cells* 1991;3:347–50.
52. Huet S, Groux H, Caillou B, Valentin H, Prieur AM, Bernard A. CD44 contributes to T cell activation. *J Immunol* 1989;143:798–801.
53. Shimizu Y, van Seventer GA, Siraganian R, Wahl L, Shaw S. Dual role of the CD44 molecule in T-cell adhesion and activation. *J Immunol* 1989;143:2457–63.
54. Hale LP, Haynes BF. Bromelian treatment of human T cells removes CD44, CD45RA, E2/MIC2, CD6, CD7, CD8, and Leu 8/LAM1 surface molecules and markedly enhances CD2-mediated T cell activation. *J Immunol* 1992;149:3809–16.
55. Culty M, Miyake K, Kincade PW, Silorski E, Butcher EC, Underhill C. The hyaluronate receptor is a member of the CD44 (H-CAM) family of cell surface glycoproteins. *J Cell Biol* 1990;111:2765–74.
56. Denning SM, Le PT, Singer KH, Haynes BF. Antibodies against the CD44 p80, lymphocyte homing receptor molecule augment human peripheral blood T cell activation. *J Immunol* 1990;144:7–15.
57. Murakami S, Miyake K, June CH, Kincade PW, Hodes RJ. Il-5 induces a Pgp-1 (CD44) bright B cell subpopulation that is highly enriched in proliferative and Ig secretory activity and binds to hyaluronate. *J Immunol* 1990;145:3618–27.
58. Rothman BL, Blue ML, Kelley KA, Wunderlich D, Mierz DV, Aune TM. Human T cell activation by OKT3 is inhibited by a monoclonal antibody to CD44. *J Immunol* 1991;147: 2493–9.
59. Seth A, Gote L, Nagarkatti M, Nagarkatti PS. T-cell-receptor-independent activation of cytolytic activity of cytotoxic T lymphocytes mediated through CD44 and gp90MEL-14. *Proc Natl Acad Sci USA* 1991;88:7877–81.
60. Westphal JR, de Waal RMW. The role of adhesion molecules in endothelial cell accessory function. *Mol Biol Rep* 1992;17:47–59.
61. Böhrer C, Berlin C, Jablonski-Westrich D, Holzmann B, Thiele HG, Hamann A. Lymphocyte activation and regulation of three adhesion molecules with supposed function in homing: LECAM-1 (MEL-14 antigen), LPAM-½ (4-integrin) and CD44 (Pgp-1). *Scand J Immunol* 1992;35:107–20.
62. Pierres A, Lipcey C, Mawas C, Olive D. A unique CD44 monoclonal antibody identifies a new T cell activation pathway. *Eur J Immunol* 1992;22:413–7.
63. Chong ASF, Boussy IA, Graf LH, Scuder P. Stimulation of IFN-g, TNF-a and TNF-b secretion in IL-2-activated T cells: costimulatory roles for LFA-1, LFA-2, CD44 and CD45 molecules. *Cell Immunol* 1992;144:69–79.
64. Toyama-Sorimachi N, Miyake K, Miyasaka M. Activation of CD44 induces ICAM-1/LFA-1-independent, $Ca^{2+}$-, $Mg^{2+}$-independent adhesion pathway in lymphocyte-endothelial cell interaction. *Eur J Immunol* 1993;23:439–46.
65. Tan PH, Santos EB, Rossbach HC, Sandmaier BM. Enhancement of natural killer activity by an antibody to CD44. *J Immunol* 1993;150:812–20.
66. Yang H, Binns RM. CD44 is involved in porcine natural cytotoxicity. *Cell Immunol* 1993;149:227–36.
67. Galandrini R, Albi N, Tripodi G, et al. Antibodies to CD44 trigger effector functions of human T cell clones. *J Immunol* 1993;150:4225–35.
68. Funaro A, Spagnoli GC, Momo M, Knapp W, Malavasi F. Stimulation of T cells via CD44 requires LFA-1 interactions and IL-2 production. *Hum Immunol* 1994;40:267–78.
69. Camp RL, Scheynius A, Johansson C, Pure E. CD44 is necessary for optimal contact allergic responses but is not required for normal leukocyte extravasation. *J Exp Med* 1993;178:497–507.
70. Seiter S, Arch R, Reber S, et al. Prevention of tumor metastasis formation by anti-variant CD44. *J Exp Med* 1993;177:443–55.
71. Arch R, Wirth K, Hofmann M, et al. Participation in normal immune responses of a metastasis-inducing splice variant of CD44. *Science* 1992;257:682–5.
72. Matzku S, Komitowski D, Miltenberger M, Zöller M. Characterization of BSp73, a spon-

taneous rat tumor and its in vivo selected variants showing different metastasizing capacities. *Invasion Metast* 1983;3:109–23.
73. Matzku S, Wenzel A, Liu S, Zöller M. Antigenic differences between metastatic and nonmetastatic BSp73 rat tumor variants characterized by monoclonal antibodies. *Cancer Res* 1989;49:1294–9.
74. Koopman G, Heider KH, Horst E, et al. Activated human lymphocytes and aggressive non-Hodgkin's lymphomas express a homologue of the rat metastasis-associated variant of CD44. *J Exp Med* 1993;177:897–904.
75. Kuppner MC, Van Meir E, Gauthier T, Hamou MF, De Tribolet N. Differential expression of the CD44 molecule in human brain tumours. *Int J Cancer* 1992;50:572–7.
76. Li H, Hamou MF, de Tribolet N, et al. Variant CD44 are expressed in human brain metastases but not in glioblastomas. *Cancer Res* 1993;53:5345–9.
77. Pilkington GJ, Akinwunmi J, Ognjenovic N, Rogers JP. Differential binding of anti-CD44 on human gliomas in vitro. *Neuroreport* 1993;4:259–62.
78. Cannistra SA, Kansas GS, Niloff J, DeFranzo B, Kim Y, Ottensmeier C. Binding of ovarian cancer cells to peritoneal mesothelium in vitro is partly mediated by CD44H. *Cancer Res* 1993;53:3830–8.
79. Heider KH, Hofman M, Horst E, et al. A human homologue of the rat metastasis-associated variant of CD44 is expressed in colorectal carcinomas and adenomatous polyps. *J Cell Biol* 1993;120:227–33.
80. Kim JC, Ishii S, Ford R, Thomas P, Steele G Jr, Jessup JM. Epitopes for homotypic binding of human carcinoembryonic antigen also participate in adhesion to extracellular matrix proteins. *Proceedings of the American Association for Cancer Research* 1993;34: 197.
81. Jothy S, McClure D, LeDuy L, Günthert U. Selective expression of CD44 messenger RNA splice variants in human colon carcinoma. *Proceedings of the American Association for Cancer Research* 1993;34:553.
82. Tanabe KK, Ellis LM, Saya H. Expression of CD44R1 adhesion molecule in colon carcinomas and metastases. *Lancet* 1993;341:725–6.
83. Mayer B, Jauch KW, Günthert U, et al. De-novo expression of CD44 and survival in gastric cancer. *Lancet* 1993;342:1019–22.
84. Heider KH, Dämmrich J, Skroch-Angel P, et al. Different expression of CD44 splice variants in intestinal and diffuse type human gastric carcinomas and normal gastric mucosa. *Cancer Res* 1993;53:4197–203.
85. Wielenga VJM, Heider KH, Offerhaus JA, et al. Expression of CD44 variant proteins in human colorectal cancer is related to tumor progression. *Cancer Res* 1993;53:4754–6.
86. Kaufmann M, Heider KH, Sinn HP, von Minckwitz G, Ponta H, Herrlich P. Surface expression of distinct CD44 variant exon epitopes on primary breast cancer strongly correlates with poor survival. *N Engl J Med* [*in press*].
87. Liu AY. Expression of CD44 in prostate cancer cells. *Cancer Lett* 1994;76:63–9.
88. Guo Y, Ma J, Wang J, et al. Inhibition of human melanoma growth and metastasis in vivo by anti-CD44 monoclonal antibody. *Cancer Res* 1994;54:1561–5.
89. Dall P, Heider KH, Hekele A, et al. Surface expression of distinct CD44 variant epitopes on cervical carcinoma. *Cancer Res* 1994;54:3337–41.
90. Shtivelman E, Bishop JM. Expression of CD44 is repressed in neuroblastoma cells. *Mol Cell Biol* 1991;11:5446–53.
91. Salmi M, Gron-Virta K, Sointu P, Grenman R, Kalimo H, Jalkanen S. Regulated expression of exon v6 containing isoforms of CD44 in man: downregulation during malignant transformation of tumors of squamocellular origin. *J Cell Biol* 1993;122:431–42.
92. Penneys NS, Kaiser M. Cylindroma expresses immunohistochemical markers linking it to eccrine coil. *J Cutan Pathol* 1993;20:40–3.
93. Kee BL, Dadi HK, Tran-Paterson R, Quackenbush EJ, Andrulis IL, Letarte M. CD10 and CD44 genes of leukemic cells and malignant cell lines show no evidence of transformation-related alterations. *J Cell Physiol* 1991;148:414–20.
94. Unanue ER, Allen PM. The basis for the immunoregulatory role of macrophages and other accessory cells. *Science* 1987;236:551–7.
95. Allen PM. Antigen processing at the molecular level. *Immunol Today* 1987;8:270–3.

96. Altman A, Mustelin T, Coggeshall KM. T lymphocyte activation: a biological model of signal transduction. *Crit Rev Immunol* 1990;10:347–91.
97. Janeway CA. The T cell receptor as a multicomponent signalling machine. CD4/CD8 coreceptors and CD45 in T cell activation. *Annu Rev Immunol* 1992;10:645–74.
98. Littman DR. Role of cell-to-cell interactions in T lymphocyte development and activation. *Curr Opin Cell Biol* 1989;1:920–8.
99. Springer TA. Adhesion receptors of the immune system. *Nature* 1990;346:425–33.
100. Wegener AMK, Letourneur F, Hoeveler A, Brocker T, Luton F, Malissen B. The T cell receptor/CD3 complex is composed of at least two autonomous transduction modules. *Cell* 1992;68:83–95.
101. Nossal GJV. Immunological tolerance and the collaboration between antigen and lymphokines in lymphocyte signalling. *Science* 1989;245:145–53.
102. Ullman KS, Northrop JP, Verweij CL, Crabtree GR. Transmission of signals from the T lymphocyte antigen receptor to genes responsible for cell proliferation and immune function: the missing link. *Annu Rev Immunol* 1990;8:421–52.
103. Balkwill FR, Burke F. The cytokine network. *Immunol Today* 1989;10:299–304.
104. Steinman RM. The dendritic cell system and its role in immunogenicity. *Annu Rev Immunol* 1991;9:271–96.
105. Gruber MF, Webb DSA, Gerrard TL. Stimulation of human monocytes via CD45, CD44 and LFA-3 triggers macrophage-colony-stimulating factor production: synergism with lipopolysaccharide and IL-1β. *J Immunol* 1992;148:1113–8.
106. Olweus J, Lund-Johanson F, Horejsi V. CD53, a protein with four membrane spanning domains, mediates signal transduction in human monocytes and B cells. *J Immunol* 1993; 151:707–16.

*Topics in Molecular Medicine, Volume 1,*
edited by Wolfgang Siess, Reinhard Lorenz,
and Peter C. Weber. Raven Press, Ltd.,
New York © 1995.

# 19

# The Melanoma-Associated Glycoprotein MUC18: A Putative Endothelial Cell Adhesion Molecule

Judith P. Johnson

*Institute for Immunology, 80336 Munich, Germany*

Cell adhesion molecules play a critical role in the development and maintenance of normal tissue structure, in the regulation of cell migration, and in a myriad of intercellular interactions (1,2). Changes in the adhesive interactions of cells are characteristic of malignant tumors and are implicated in the development of metastatic disease as tumor cells leave the primary tumor mass, migrate through the tissues, traverse the vascular system, and establish foci of growth in new environments (3,4). The loss, inactivation, or downregulation of normally expressed cell adhesion molecules, such as E-cadherin and DCC, occurs early in the development of most human carcinomas (5,6) and is believed to weaken interactions among the tumor cells, freeing them from regulatory influences and encouraging their separation from the tumor mass. In addition to the loss of normal adhesive interactions, metastasizing tumor cells also undertake a variety of new interactions and, consistent with this, express new cell adhesion molecules. Many solid tumors express ligands for inducible endothelial adhesion molecules that can be shown to mediate tumor cell–endothelial cell adhesion in vitro. These include modified Lewis blood group antigens, which can serve as ligands for P- and E-selectin and which are frequently expressed by gastrointestinal tumors (7), and the integrin VLA-4, a ligand for VCAM-1 normally expressed on activated T lymphocytes but also found on advanced melanomas (3). The molecule described here, MUC18, may also fall into this category. MUC18 was identified during a search for molecules that are upregulated during the development of cutaneous melanoma and its progression to metastatic disease (8). Although its function has not yet been established, structural characteristics suggest that MUC18 may mediate cell adhesion, and its expression pattern suggests that it may be involved in leukocyte–endothelial cell interactions.

## MUC18 HAS THE STRUCTURAL CHARACTERISTICS OF A CELL ADHESION MOLECULE

MUC18 was originally defined by a monoclonal antibody (MAb) that reacted with the majority of metastatic melanomas but only rarely with benign melanocytic tumors (8). Studies using MAbs directed against different epitopes confirmed that the expression of MUC18 is upregulated in malignant lesions and that it appears to increase with tumor progression (9,10).

MUC18 is a cell surface glycoprotein with an apparent $M_r$ of 113 kDa on melanoma cells. Isolation of the MUC18 encoding cDNA from a melanoma expression library revealed that it is a novel member of the immunoglobulin (Ig) superfamily (Fig. 1) (11). MUC18 consists of five Ig-like domains; the two N-terminal domains are V type and the remaining domains are C2 type. MUC18 has the characteristics of a typical cell surface glycoprotein, with eight potential glycosylation sites, a transmembrane region, and a cytoplasmic tail of 63 amino acids. Sequence comparison with available databases using the FASTA and BLAST programs (Table 1) showed that MUC18 is most closely related to SC1/BEN/DM-GRASP, a developmentally regulated homophilic cell adhesion molecule identified in the chicken (12–14). In addition to an overall sequence identity of 28%, SC1 shares with MUC18 the Ig domain pattern of V-V-C2-C2-C2. However, the expression patterns of the two molecules appear to be distinct, as SC1 is broadly expressed in adults both in the nervous system and in the hematopoietic system (15), whereas MUC18 appears limited to smooth-muscle cells and blood vessels

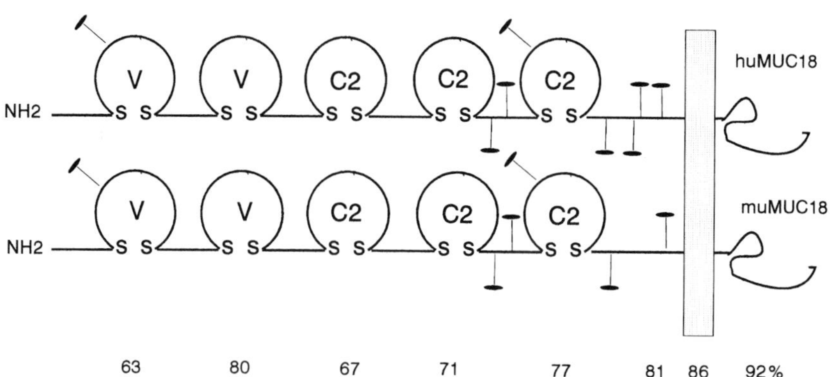


**FIG. 1.** The predicted structure of the human and murine MUC18 molecules. Potential glycosylation sites are indicated. The numbers refer to the percent of amino acid identity observed between the human and mouse molecules for each domain.

**TABLE 1.** *Proteins with sequence similarity to human MUC18*

| Molecule | % Identity | Overlap[a] |
|---|---|---|
| SC1/DM-GRASP/BEN | 28 | 493 |
| DCC | 23 | 374 |
| CEA | 19 | 368 |
| NCAM | 20 | 335 |
| VCAM-1 | 18 | 335 |
| Fasciclin II | 16 | 229 |
| F3 | 25 | 119 |

[a] Overlap in number of amino acids.

(see below). Among the other molecules related to MUC18 are DCC, a putative tumor suppressor gene identified in human colorectal carcinomas (6), the inducible endothelial–leukocyte adhesion molecule VCAM-1 (16), members of the CEA (carcinoembryonic antigen) family (17), and cell adhesion molecules that play an important role in the development of the nervous system. These include NCAM and F3 in mammals as well as fasciclin isolated from insects (18–20).

MUC18 has two structural characteristics that suggest a putative function as a cell adhesion molecule. First, each of the MUC18-related molecules listed in Table 1 has been shown to mediate intercellular adhesion through either homophilic or heterophilic interactions. Second, MUC18, as isolated from melanoma cells, carries a carbohydrate modification known as HNK-1 or CD57, which is expressed on many cell adhesion molecules of the nervous system (21). This sulfated glucuronic acid moiety has been shown to contribute to cell adhesion in several systems (22,23).

To examine the putative adhesive activity of this molecule, MUC18-negative human and mouse cell lines were transfected with the full-length MUC18 cDNA cloned into the inducible eukaryotic expression vector pMAMneo. An example of one of these cells is shown in Fig. 2. As can be seen, the murine L cells transfected with MUC18 cDNA in a sense orientation express this glycoprotein at levels comparable to that expressed by human melanoma cells. Despite this, the MUC18-expressing L cells failed to show an increase in aggregation when measured either in a suspension or in a solid-phase assay. L cells transfected with a cDNA-encoding CEA, which can function as a homophilic cell adhesion molecule (24), clearly demonstrated an increased rate of aggregation in the same experiments. Because cell-specific post-translational modifications, such as glycosylation or the presence of additional molecules, can affect the adhesive activity of a cell adhesion molecule (25,26), MUC18-mediated homotypic adhesion may be detectable if the appropriate cell is used. On the other hand, many cell adhesion molecules of the Ig superfamily interact with heterophilic ligands and mediate primarily heterotypic cell adhesion (27). Therefore, the MUC18 ligand may be expressed on another cell type.

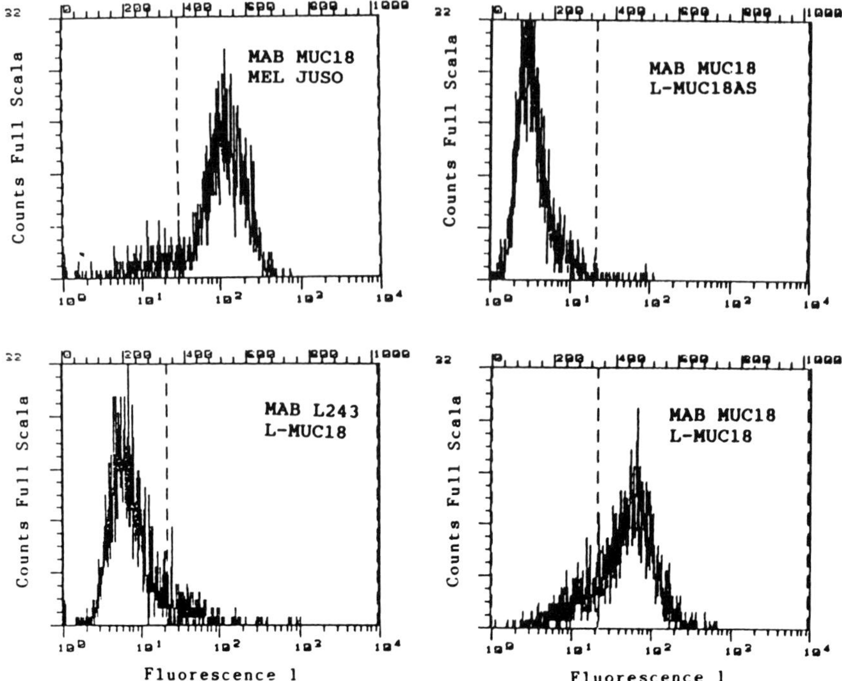


**FIG. 2.** Expression of MUC18 by cDNA transfectants. FACs profiles showing the reactivity of antibodies to MUC18 with human melanoma cells (Mel JuSo), mouse L cells expressing antisense MUC18 cDNA (L-MUC18AS), and mouse L cells expressing sense MUC18 cDNA (L-MUC18). As negative control, the reactivity of L-MUC18 is also shown with MAb L243 directed against human HLA-DR.

## IN NORMAL ADULT TISSUES MUC18 IS EXPRESSED ON BLOOD VESSELS

An insight into the possible location of a putative MUC18 ligand, as well as into the normal function of the molecule, might be obtained from an analysis of the MUC18 expression pattern in normal tissues. This was examined with both Northern blot analysis and immunohistochemistry (28,29). Analysis of cell lines indicated that the expression of MUC18 mRNA and its surface expression are well correlated and are essentially limited to cell lines derived from tumors of neuroectodermal lineage. In contrast, examination of mRNA isolated from normal human tissues indicated that the 3.3-kb MUC18 mRNA is ubiquitous, being expressed in virtually all tissues examined. The highest levels, relative to the amounts of GAPDH mRNA, were seen in placenta and lung, and the lowest in skeletal muscle, kidney, and liver (29). Immunohistochemical analysis indicated that this reflects the expression of MUC18 in blood vessels, particularly by those in vascular smooth muscle.

Other smooth-muscle cells also express MUC18, e.g., those in the gastrointestinal tract, dermis, and lung. However, cardiac and skeletal muscle were not reactive with anti-MUC18 antibodies. Probing the Northern blots with a cDNA probe directed against smooth muscle α-actin showed that, indeed, the tissues with high levels of MUC18 mRNA also expressed high levels of α-actin mRNA. All of these observations indicate that the normal expression of MUC18 in vivo is in smooth-muscle cells. In contrast to the in vivo observations, short-term smooth-muscle cell cultures derived either from arteries or from restenosic lesions did not show any MUC18 protein expression (unpublished observations).

Although MUC18 expression could generally be localized to vascular smooth muscle, endothelial cells also occasionally appeared to be stained. To examine this more closely, endothelial cells derived from human umbil-

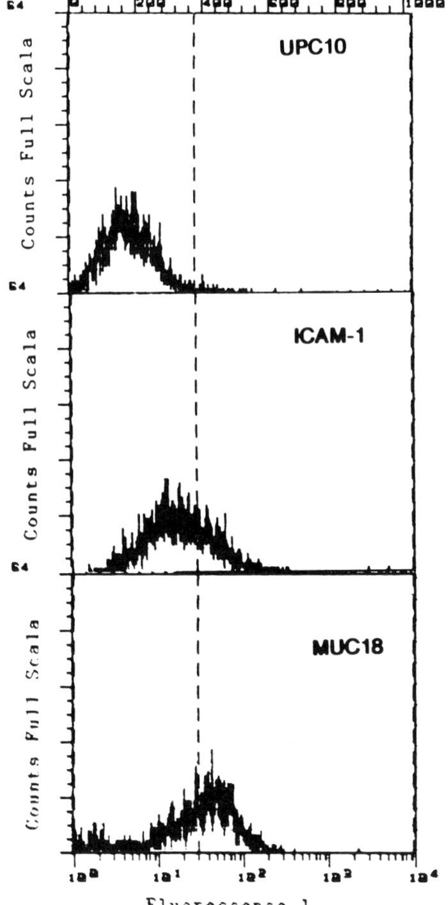


**FIG. 3.** Expression of MUC18 on endothelial cells. FACs profiles of cultured human umbilical cord endothelial cells incubated with Upc10, an isotype control, anti–ICAM-1 (P3.58BA-14), and anti-MUC18.

ical cord (HUVECs) were grown in culture and tested for their expression of surface MUC18. As shown in Fig. 3, immunofluorescence studies indicate that cultured HUVECs express high levels of MUC18 comparable to that observed for the inducible cell adhesion molecule ICAM-1 after exposure to TNFα. Although the expression of MUC18 on HUVECs was unaffected by exposure to IFNγ or TNFα, it nevertheless appears to be associated with activation of the cells. Immunohistochemical analysis of tissue sections of umbilical cord indicated that whereas MUC18 expression was clearly present on the vascular smooth muscle, it could not be detected on the endothelia. This suggests that the expression of MUC18 on the endothelial cells could be induced by a growth factor present in the culture medium.

## REGULATION OF MUC18 EXPRESSION

To gain an understanding of how the expression of this molecule might be regulated, the human MUC18 gene was isolated and the potential regulatory regions examined (29). MUC18 is a single-copy gene stretching over a distance of approximately 12 kb and consisting of at least 16 exons. Like NCAM and in contrast to most members of the Ig supergene family, each of the Ig-like domains is encoded by multiple exons. However, MUC18 is unique because it is the first example reported in which an Ig domain is encoded by three exons. Analysis of the murine homologue indicates that this intron–exon organization is maintained in mouse MUC18 (30).

Examination of the nucleotide sequence 5′ to the start of the MUC18 gene revealed that it lacks classical TATA and CAAT boxes. However, this region is GC-rich and shows homology to an initiator sequence that has also been shown to mediate transcription initiation (31). Primer extension analysis indicated that MUC18 has only a single transcriptional start and that this is located in the middle of the predicted initiator element (Fig. 4).

Putative binding sites for a number of transcriptional factors were found in the 5′ region of the MUC18 gene (29,32). MUC18 is undetectable on epidermal melanocytes in vivo but is strongly expressed by the majority of melanocytic tumors, and may therefore be a gene normally expressed during a particular stage in melanocyte development. However, elements that have

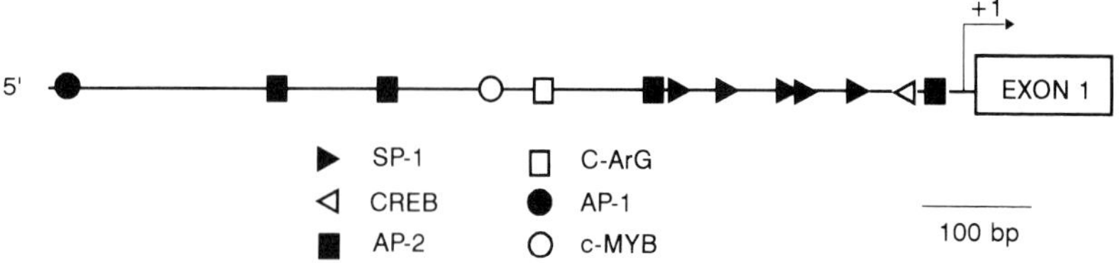


**FIG. 4.** Location of putative transcription factor binding sites in the 5′ region of the human MUC18 gene. The single transcription start site is indicated by +1.

been identified as being necessary for melanocyte specific expression (33) were not found in this region. In contrast, two elements that have been associated with the ability of smooth-muscle cells to express smooth muscle α-actin, the CArG box and c-myb binding element, were found (Fig. 4) (34), suggesting that MUC18 is indeed a smooth-muscle cell gene.

Inspection of the sequence of the 5′ upstream region also suggests that MUC18 is a gene whose expression can be regulated by exogenous signals. Putative consensus sequences for cAMP regulatory proteins, as well as for the transcription factors AP-1 and AP-2, are present in this region.

To examine whether MUC18 might be regulated in vitro by cAMP or by phorbol ester (which can activate AP-1 and AP-2), a variety of cell lines were exposed to these agents and their effects on MUC18 expression examined. No effects were observed on MUC18-negative cells of epithelial, hematopoietic, or neuroectodermal origin. However, two MUC18-expressing glioma cell lines showed an increase in cell surface expression after exposure to agents that are able to increase cAMP levels. In contrast to the effects of cAMP elevation, exposure to phorbol ester resulted in a loss or downregulation of surface expression in several cells. Both of these effects were seen only 24 to 48 h after treatment, suggesting that the changes observed on the cell surface may reflect transcriptional regulation. In contrast to the observations in MUC18-positive glioma cells, no alteration in MUC18 surface expression could be observed in melanoma cells after exposure to cAMP or to phorbol ester. These preliminary observations provide direct evidence that MUC18 expression, at least in some cells, can be regulated by exogenous agents.

## POTENTIAL ROLE OF MUC18 IN TUMOR GROWTH AND DEVELOPMENT

The correlation between MUC18 expression and tumor progression initially observed in malignant melanoma raises the question of whether MUC18 plays a role in the growth or metastasis of certain tumors. Two approaches have been used in attempts to investigate this. The first involves an examination of the role of MUC18 in the growth and metastatic pattern of human melanoma cells in immune-compromised mice. The second involves the development of a homologous model system. A large number of human melanoma cell lines have been characterized in terms of their growth patterns and metastatic ability in nude mice. In collaboration with M. Bar-Eli and colleagues, a panel of nine such cells has been investigated (35). Five of the cell lines metastasized in nude mice and all were found to express high levels of MUC18 mRNA and protein. Of the four cell lines that did not metastasize, two had no detectable MUC18 mRNA or protein, one had detectable mRNA but no protein, and one had large amounts of both mRNA and protein. These observations are consistent with the hypothesis that the expression of MUC18 protein is necessary for metastasis of human melanoma cells in the nude mouse. However, MUC18 expression alone is not sufficient, as one high-expressing cell failed to metastasize. Studies are currently under way to determine whether transfection of MUC18 cDNA into the MUC18-negative, nonmetastasizing tumor cells alters in any way the behavior of the tumor cells.

The major disadvantage of this model is that it is a heterologous system. For the molecule to function, the human MUC18 must be able to interact with the murine counterpart of its ligand. Moreover, although cross-species interactions among molecules are frequently observed, this is not always the case (36). To be able to study MUC18 in a homologous system, the murine MUC18 homologue was isolated and characterized (30). Murine MUC18 is similar to its human counterpart in overall structure (five Ig domains, V-V-C2-C2-C2 pattern) (Fig. 1) and in amino acid sequence (74% identity over the entire molecule). Interestingly, the least sequence similarity (63%) is found in the most N-terminal domain. Because the N-terminal domain is frequently important in ligand interactions for members of the Ig superfamily (37,38), this could mean that effective interactions between MUC18 and its murine ligand may not occur.

The expression pattern of murine MUC18 is also very similar to that of its human counterpart. Northern blot analysis and in situ hybridization indicated that murine MUC18 is expressed in normal blood vessels and also in transformed melanocyte cell lines and murine melanomas (30). Therefore, on the basis of both structure and expression pattern, murine MUC18 appears similar to the human gene. The mouse therefore appears to be an appropriate model system in which to investigate both the normal function of MUC18 and its role in tumor growth and metastasis.

## PERSPECTIVES

Despite the fact that a ligand has not yet been identified, MUC18 has many characteristics suggesting that it may function to mediate cell adhesion. The fact that MUC18 is expressed on endothelial cells both in vivo and in vitro, together with the observation that its expression can be regulated by cAMP modulation and by exposure to phorbol esters, raises the possibility that MUC18 is an inducible endothelial cell adhesion molecule. Many endothelial cell adhesion molecules have been identified in recent years, and each appears to mediate temporally distinct and cell type-specific events in leukocyte migration (39,40). This suggests that the search for the MUC18 ligand should be focused on leukocytes. The original discovery of MUC18 as a melanoma-associated molecule is also not inconsistent with a potential role as an endothelial cell adhesion molecule, as these molecules and their ligands are increasingly found to be expressed by tumor cells (41,42). Tumor cells, at least in vitro, can clearly use endothelial cell ligands to attach to and to migrate through the endothelium (43,44). In in vivo models, tumor cell–leukocyte interactions also enhance tumor survival and metastasis formation (45), an effect that has been partially attributed to the production of growth factors, angiogenic factors, and proteolytic enzymes by the leukocytes, but also partially to an indirect interaction with the vascular endothelium (46). The expression of MUC18 by human melanoma cells may be a further example of a more general phenomenon: that nonhematopoietic tumor cells metastasize in part by usurping the functions of the molecules that serve to direct and mediate normal leukocyte trafficking.

## ACKNOWLEDGMENT

This work was supported in part by a grant from the Deutsche Krebshilfe, Mildred Scheel Stiftung, Bonn, Germany.

## REFERENCES

1. Ruitshauser U, Jessel TM. Cell adhesion molecules in vertebrate neural development. *Physiol Rev* 1988;68:819–57.
2. Edelman GM, Crossin KL. Cell adhesion molecules: implications for a molecular histology. *Annu Rev Biochem* 1991;60:155–90.
3. Albeda SM. Role of integrins and other cell adhesion molecules in tumor progression and metastasis. *Lab Invest* 1993;68:4–17.
4. Johnson JP. Cell adhesion molecules of the immunoglobulin supergene family and their role in malignant transformation and progression to metastatic disease. *Cancer Metastasis Rev* 1991;10:11–22.
5. Mayer B, Johnson JP, Leitl F, et al. E-cadherin expression in primary and metastatic gastric cancer: downregulation correlated with cellular dedifferentiation and glandular disintegration. *Cancer Res* 1993;53:1690–5.
6. Fearon ER, Cho KR, Nigro JM, et al. Identification of a chromosome 18q gene that is altered in colorectal cancers. *Science* 1990;247:49–56.
7. Itzkowitz SH, Yuan M, Fukushi Y, et al. Lewis-x and sialylated Lewis-x related antigen expression in human malignant and nonmalignant colonic tissues. *Cancer Res* 1986;46:2627–32.
8. Lehmann JM, Holzmann B, Breitbart EW, Schmiegelow P, Riethmüller G, Johnson JP. Discrimination between benign and malignant cells of melanocytic lineage by two novel antigens, a glycoprotein with a molecular weight of 113,000 and a protein with a molecular weight of 76,000. *Cancer Res* 1987;47:841–5.
9. Holzmann B, Bröcker EB, Lehmann JM, et al. Tumor progression in human melanoma: five stages defined by their antigenic phenotypes. *Int J Cancer* 1987;39:466–71.
10. Shih IE, Elder DE, Speicher D, Johnson JP, Herlyn M. Isolation and functional characterization of the A32 melanoma-associated antigen. *Cancer Res* 1994;54:2514–20.
11. Lehmann JM, Riethmüller G, Johnson JP. MUC18, a marker of tumor progression in human melanoma shows sequence similarity to the neural cell adhesion molecules of the immunoglobulin superfamily. *Proc Natl Acad Sci USA* 1989;86:9891–5.
12. Tanaka H, Matsui T, Agata A, et al. Molecular cloning and expression of a novel adhesion molecule, SC1. *Neuron* 1991;7:535–45.
13. Pourquie O, Corbel C, La Caer JP, Rossier J, LeDouarin NM. BEN, a surface molecule of

the immunoglobulin superfamily expressed in a variety of developing systems. *Proc Natl Acad Sci USA* 1992;89:5261–5.

14. Burns FR, von Kannen S, Guy L, Raper JA, Kamholz J, Chang S. DM-GRASP, a novel immunoglobulin superfamily axonal surface protein that supports neurite extension. *Neuron* 1991;7:209–20.
15. Pourquie O, Coltry M, Thomas JL, LeDouarin NM. A widely distributed antigen developmentally regulated in the nervous system. *Development* 1990;109:743–52.
16. Osborn L, Hession C, Tizard R, et al. Direct expression cloning of vascular cell adhesion molecule 1, a cytokine-induced endothelial protein that binds to lymphocytes. *Cell* 1989; 59:1203–11.
17. Thompson J, Zimmermann W. The carcinoembryonic antigen (CEA) gene family: structure, expression and evolution. *Tumor Biol* 1988;9:63–83.
18. Gennarini G, Cibelli G, Rougon G, Mattei MG. The mouse neuronal cell surface protein F3: a phosphatidylinositol-anchored member of the immunoglobulin superfamily related to chicken contactin. *J Cell Biol* 1989;109:775–88.
19. Harrelson AL, Goodman CS. Growth cone guidance in insects: fasciclin II is a member of the immunoglobulin superfamily. *Science* 1988;242:700–8.
20. Lander AD. Understanding the molecules of neural cell contacts: emerging patterns of structure and function. *Trends Neurosci* 1989;12:189–95.
21. Kruse J, Mailhammer R, Wernecke H, et al. Neural cell adhesion molecules and myelin-associated glycoprotein share a common carbohydrate moiety recognized by monoclonal antibodies L2 and HNK-1. *Nature* 1984;311:153–5.
22. Hoffman S, Edelman GM. A proteoglycan with HNK-1 antigenic determinants is a neuron-associated ligand for cytotactin. *Proc Natl Acad Sci USA* 1987;84:2523–7.
23. Künemund V, Jungalwala FB, Ficher G, Chou DKH, Keilhauer G, Schachner M. The L2/HNK-1 carbohydrate of neural cell adhesion molecules is involved in cell interactions. *J Cell Biol* 1988;106:213–23.
24. Oikaea S, Inuzuka C, Kuroki M, Matsuok Y, Kosati G, Nakazato H. Cell adhesion activity of nonspecific cross reactive antigen (NCA) and carcinoembryonic antigen (CEA) expressed on CHO cell surfaces: homophilic and heterophilic adhesion. *Biochem Biophys Res Commun* 1989;164:39–45.
25. Diamond MS, Staunton DE, Marlin SD, Springer TA. Binding of the integrin Mac-1 (CD11b/CD18) to the third immunoglobulin-like domain of ICAM-1 (CD54) and its regulation by glycosylation. *Cell* 1991;65:961–71.
26. Kadmon G, Kowitz A, Altevogt P, Schachner M. Functional cooperation between the neural adhesion molecules L1 and N-CAM is carbohydrate dependent. *J Cell Biol* 1990; 110:209–18.
27. Springer TA. Adhesion receptors of the immune system. *Nature* 1990;346:425–34.
28. Sers C, Riethmüller G, Johnson JP. MUC18, a melanoma-progression associated molecule—a potential role in tumor vascularization and hematogenous spread? *Cancer Res* 1994;54:5689–94.
29. Sers C, Kirsch K, Rothbächer U, Riethmüller G, Johnson JP. Genomic organization of the melanoma-associated glycoprotein MUC18: implications for the evolution of the immunoglobulin domains. *Proc Natl Acad Sci USA* 1993;90:8514–8.
30. Rothbächer U, Sers C, Riethmüller G, Johnson JP. Characterization of the human melanoma metastasis associated molecule MUC18 in the mouse: unique gene structure and expression in murine melanoma. (Submitted for publication.)
31. Roy AL, Meisterernst M, Pognonec P, Roeder RG. Cooperative interaction of an initiator-binding transcription initiation factor and the helix-loop-helix activator USF. *Nature* 1991; 354:245–8.
32. Johnson JP, Rothbächer U, Sers C. The progression associated antigen MUC18: a unique member of the immunoglobulin supergene family. *Melanoma Res* 1993;3:337–40.
33. Jackson I, Chambers DM, Budd PS, Johnson R. The tyrosine-related protein-1 gene has a structure and promoter sequence very different from tyrosinase. *Nucleic Acids Res* 1991; 19:3799–804.
34. Blank RS, McQuinn TC, Yin KC, et al. Elements of the smooth muscle a-actin promoter required in cis for transcriptional activation in smooth muscle. *J Biol Chem* 1992;267:984–9.
35. Luca M, Hunt B, Bucana CD, Johnson JP, Fidler IJ, Bar-Eli M. Direct correlation between

MUC18 expression and metastatic potential of human melanoma cells. *Melanoma Res* 1993;3:35–41.
36. Johnston SC, Hibbs ML, Dustin ML, Springer TA. On the species specificity of the interaction of LFA-1 with intercellular adhesion molecules. *J Immunol* 1990;145:1181–7.
37. Staunton DE, Dustin ML, Erickson HP, Springer TA. The arrangement of the immunoglobulin-like domains of ICAM-1 and the binding sites for LFA-1 and rhinovirus. *Cell* 1990;61:243–54.
38. Hahn WC, Menu E, Bothwell ALM, Sims PJ, Bierer BE. Overlapping but nonidentical binding sites on CD2 for Cd58 and a second ligand CD59. *Science* 1992;256:1805–7.
39. Osborn L. Leukocyte adhesion to endothelium in inflammation. *Cell* 1990;62:3–6.
40. Butcher EC. Leukocyte-endothelial cell recognition: three (or more) steps to specificity and diversity. *Cell* 1991;67:1033–6.
41. Johnson JP, Stade BG, Holzmann B, Schwäble W, Riethmüller G. De novo expression of cell adhesion molecule ICAM-1 in melanoma and increased risk of metastasis. *Proc Natl Acad Sci USA* 1989;86:641–4.
42. Jonhiic N, Martin-Padura I, Pollicino T, et al. Regulated expression of vascular cell adhesion molecule-1 in human malignant melanoma. *Am J Pathol* 1992;141:1323–30.
43. Rice GE, Bevilacqua MO. An inducible endothelial cell surface glycoprotein mediates melanoma adhesion. *Science* 1989;246:1303–6.
44. Takada A, Ohmori K, Yoneda T, et al. Contribution of carbohydrate antigens sialyl Lewis a and sialyl Lewis x to adhesion of human cancer cells to vascular endothelium. *Cancer Res* 1993;53:354–61.
45. Aeed PA, Nakajima M, Welch DR. The role of polymorphonuclear leukocytes (PMN) on the growth and metastatic potential of 1376NF mammary adenocarcinoma cells. *Int J Cancer* 1988;42:748–59.
46. Starkey JR, Liggitt HD, Jones W, Hosick HL. Influence of migratory blood cells on the attachment of tumor cells to vascular endothelium. *Int J Cancer* 1984;34:535–43.

*Topics in Molecular Medicine, Volume 1*,
edited by Wolfgang Siess, Reinhard Lorenz,
and Peter C. Weber. Raven Press, Ltd.,
New York © 1995.

# 20

# Prostacyclin Analogues in Experimental Metastasis

M. R. Schneider, M. Schirner, R. B. Lichtner, and H. Graf

*Research Laboratories, Schering AG, 13342 Berlin, Germany*

The majority of cancer patients die because of their metastases rather than from the primary tumor. The process of tumor dissemination, called the metastastic cascade, can be subdivided into several steps: detachment from the primary tumor and intravasation, transport through blood and lymphatic vessels, interaction with components of the blood system, adherence to the endothelial lining, extravasation into a distant organ, and finally growth and angiogenesis (1,2).

Many efforts have been undertaken to develop compounds that specifically interfere with certain steps of the metastatic cascade, thereby reducing the number and/or growth of metastases. However, no such drug is available at present for clinical use (3).

A class of compounds with potential in antimetastatic therapy are prostacyclin and its stable analogues (Fig. 1). It was first shown by Honn and co-workers (4) that prostacyclin can reduce the number of lung colonies of intravenously injected melanoma cells. These authors also demonstrated that prostacyclin inhibits tumor cell-induced platelet aggregation, phorbol ester- or 12-(S)-HETE–stimulated adhesion of tumor cells to endothelial cells, subendothelial cell matrix or fibronectin, and diminished endothelial cell retraction (5). More recently, comparable data were obtained in a series of tumor lines and by various investigators (2).

Prostacyclin is one of the most potent natural inhibitors of platelet aggregation. It causes dilatation of blood vessels and is a strong inhibitor of platelet and leukocyte adherence to the endothelial lining. Furthermore, it is "cytoprotective" in different damage models (6). However, prostacyclin is metabolically very unstable, with a half-life of seconds to minutes, and it can be administered only intravenously (6). Therefore, many efforts have been undertaken to obtain more stable prostacyclin analogues. One of these analogues is cicaprost (Fig. 1), which is metabolically stable, with a half-life of 4

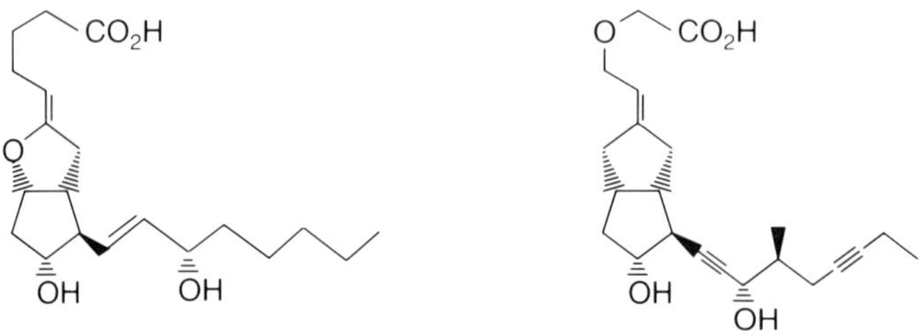

**FIG. 1.** Chemical structures of prostacyclin (left) and cicaprost (right).

to 6 h and is almost completely bioavailable after oral administration (7,8). It is not only the most potent prostacyclin analogue thus far but is also the most specific one, i.e., it binds only to the prostacyclin receptor (9). Using cicaprost as well as other stable analogues, such as iloprost or TEI8153, data comparable to those described above for prostacyclin itself were obtained in tumor metastasis models (2,10).

## EFFECT OF CICAPROST IN THE ARTIFICIAL METASTASIS ASSAY

It was our intention to intensely characterize the antimetastatic potential of cicaprost. In the artificial metastasis assay, a large number of tumor cells are injected i.v. so that no detachment from the primary tumor and intravasation are necessary. Treatment of animals is usually performed only once, shortly before tumor cell injection. Cicaprost strongly reduced the number of lung colonies in the B16 B/6 and the B16 F1 melanoma cell lines, as well as the number and incidence of liver nodules in the M 5076 mouse sarcoma (2,10–12). However, in some other tumor lines, such as the murine B16a melanoma, sarcoma 180, and rat MTLn3 mammary carcinoma and RUCA endometrial carcinoma, no effect in this artificial metastasis assay was found when cicaprost was given only once. This means that cicaprost affects the lodgment step of tumor cells in some but not all tumor lines.

## EFFECT OF CICAPROST ON SPONTANEOUSLY METASTASIZING TUMORS

Because the artificial metastasis assay is not relevant to the clinical situation, our major interest was to determine the effect of this drug in spontaneously metastasizing tumors reflecting all stages of the metastatic cascade. For this purpose, a variety of tumor lines and assay systems were used (Fig. 2). In the first type of assay, daily oral treatment was started shortly before s.c. or orthotopic tumor implantation and was continued until the end of the

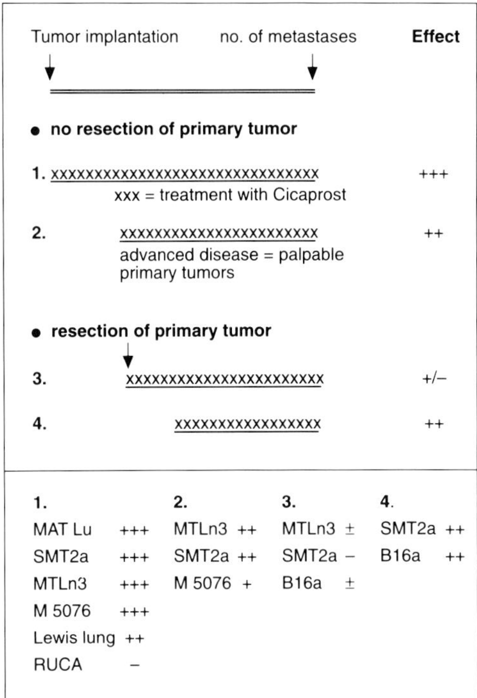


**FIG. 2.** Antimetastatic effect of cicaprost in spontaneously metastasizing tumors.

experiment. The primary tumor was not removed, and therefore metastatic cells were released during the entire period of the experiment (2,10).

In the R 3327 MAT Lu prostate carcinoma of the rat, the 13762 NF MTLn3 and SMT 2a mammary carcinomas of the rat, and the M 5076 sarcoma of the mouse, cicaprost caused a dose-dependent inhibition of the median number of lung or liver metastases. The degree of inhibition was 80% or higher (2,10,12–14). In the Lewis lung carcinoma, a 50% reduction in metastasis number was achieved (2). No effect was seen in the RUCA endometrial carcinoma of the rat (Fig. 2). It is important to mention that in all of these tumor lines cicaprost had no effect on primary tumor growth (2,10). Therefore, a merely cytotoxic effect can be excluded. However, in the two mammary carcinomas investigated, a strong inhibition of lymph node weights as a measure of lymph node metastases was found (13,14).

Because in the MTLn3 mammary carcinoma line no effect in the artificial metastasis assay but a very pronounced activity in the orthotopic implantation model was determined, it became clear that cicaprost might act not only on steps in the metastatic cascade such as invasion and tumor cell–blood cell interaction or adhesion but also on steps of tumor cell release or after tumor cells had left the circulation. Therefore, the efficacy of cicaprost was tested in an advanced tumor stage with treatment beginning when already palpable primary tumors and, in some models, also palpable lymph nodes were present (Fig. 2). In the SMT 2a mammary carcinoma, therapy was started at day 10 and continued until day 32. Cicaprost at an oral dose of 0.1 mg/kg significantly reduced the median number of metastases to about 20% of the control value (Fig. 3) (15,16). A similar effect was determined in the MTLn3 mammary carcinoma, using treatment from day 11 to day 21 after tumor cell implantation (15).

To prove that cicaprost has an effect on metastasis when the metastatic cells have already left the circulation, it is necessary to remove the primary

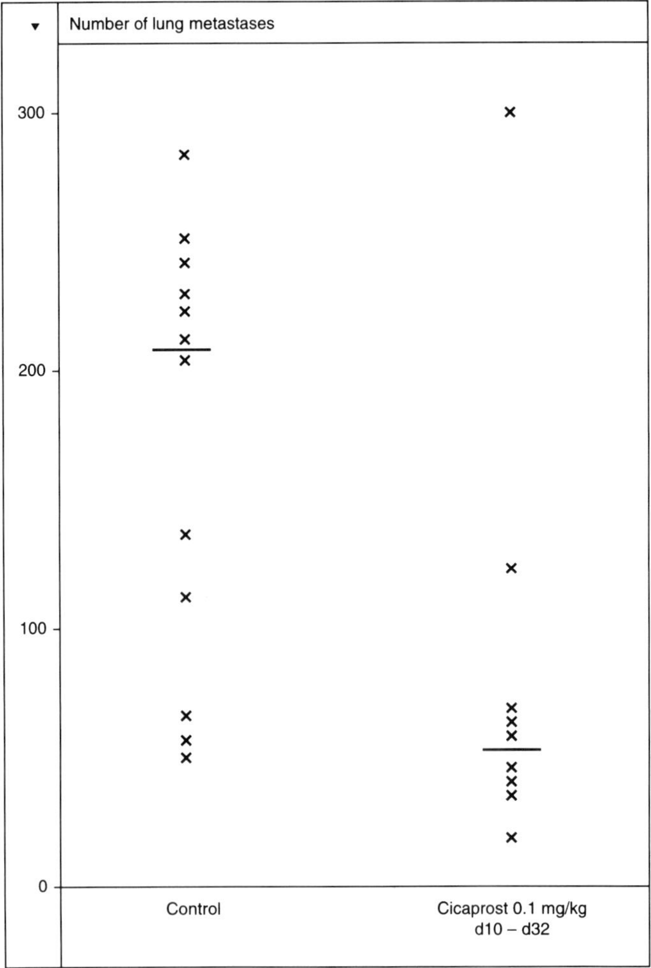


**FIG. 3.** Effect of cicaprost (0.1 mg/kg p.o. daily from day 10 until end of experiment, day 32) on number of lung metastases of Wistar–Furth rats bearing s.c. implanted SMT 2a mammary carcinomas.

tumor at a time point after the first metastatic tumor cells have settled in the target tissue and to start treatment thereafter (Fig. 2). When treatment was started at the day of tumor resection and continued until the end of the experiment, no profound effect was observed. This may be due to interference of cicaprost with the stress reaction caused by surgery, and requires further clarification. However, when therapy was started about 10 days after resection of the primary tumor and continued until the end of the experiment, a significant inhibition of metastases numbers was achieved in the B16a melanoma and the SMT 2a mammary carcinoma (Figs. 2 and 4) (17). The presence of micrometastases at the time of surgery and at the start of

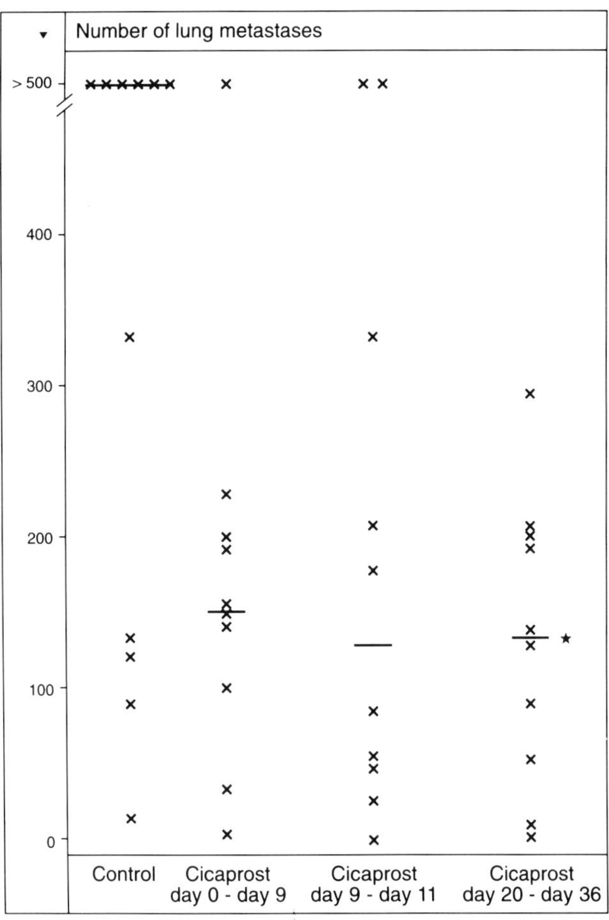


**FIG. 4.** Effect of cicaprost (0.1 mg/kg p.o.) on number of lung metastases of Wistar–Furth rats bearing s.c. implanted SMT 2a mammary carcinomas. The primary tumor was removed at day 10 after implantation. Cicaprost treatment was performed in three schedules: preoperative (days 0–9), perioperative (days 9–11), and postoperative (days 20–36).

treatment was proven histologically. In the SMT 2a tumor line, these results were reproduced in a second experiment. Treatment from day 20 to day 36 caused a significant inhibition of lung metastasis number (Fig. 4). In this assay, in addition, a preoperative (day 0 to 9) and a perioperative (day 9 to 11) treatment schedule was included. Both schedules led to an inhibition of the median number of metastases, although this was not statistically significant. Nevertheless, this experiment shows that cicaprost may, at least in this tumor model, have multiple modes of action.

The in vivo data on the antimetastatic action of cicaprost summarized here indicate that this stable prostacyclin analogue has pronounced inhibitory effects on metastases in a variety of spontaneously metastasizing tumor lines. This effect was evident not only when treatment was started immediately with tumor implantation but also in advanced tumor stages and even after the primary tumor was resected. This implies a mode of action in addition to or even instead of the mechanisms described for prostacyclin analogues that concentrated on effects during circulation of tumor cells.

## STUDIES ON THE MODE OF ACTION OF CICAPROST

Prostacyclin and its analogues act via specific membrane receptors. Ligand binding leads to the activation of G-proteins, which interact with the effector system adenylate cyclase. cAMP thus formed, then activates protein kinase A, which in turn leads to the phosphorylation of a pattern of proteins, e.g., from the signal transduction system and the transcription machinery, resulting in effects on gene transcription (18). We investigated whether the presence of functional prostacyclin receptors in tumor cells correlates with the antimetastatic effects of cicaprost in the respective tumor lines. For this purpose, cells were incubated with cicaprost and the increase in cAMP was taken as a surrogate marker for a functional, specific prostacyclin receptor (18,19). Table 1 shows that some but not all of the tumor lines investigated expressed prostacyclin receptors. In the MTLn3 mammary carcinoma line, an increase in cAMP was observed, which correlated with a strong effect of cicaprost in the spontaneously metastasizing tumor system. In the RUCA endometrial carcinoma line, however, no receptor was expressed and cicaprost had no antimetastatic effect in vivo (Fig. 2).

In the R3327 MAT Lu prostate carcinoma, cicaprost had a pronounced effect on metastases number (Fig. 2) (13). We have established two sublines in tissue culture, MAT Lu S1 and S2. The S1 line showed a functional prostacyclin receptor but the S2 line did not. The S1 line was cloned further. Among these clones, S1/H3 and S1/H1 were characterized. In the S1/H3 line an increase in cAMP was found, but not in the S1/H1 line.

Because transendothelial migration is an important step in the metastatic cascade during intra- and extravasation of tumor cells, we measured the

**TABLE 1.** *Prostacyclin receptor-mediated increase in cAMP in tumor cells: correlation with antimetastatic effects*

| Cell line | Increase in cAMP | Antimetastatic effect/ response in vitro |
|---|---|---|
| MTLn3 | + | +/+ |
| RUCA | − | −/ND |
| MAT Lu S1 | + | ND/+ |
| MAT Lu S2 | − | ND/− |
| S1/H3 | + | ND/+ |
| S1/H1 | − | ND/− |
| B16a | + | +/+ |
| B16 F1 | + | +/+ |
| B16 F10 | − | −/− |
| SK Mel 28 | + | ND/+ |
| Hs 939 | − | ND/− |

ND, not determined.

influence of cicaprost on the transendothelial migration of these tumor cell clones. Bovine pulmonary endothelial cells were cultured for 8 days on collagen-coated porous filters. Tumor cells were seeded on top of endothelial monolayers and the number of cells that migrated through the endothelial cell monolayer in response to thrombin (1 U/ml) in the presence or absence of cicaprost ($5 \times 10^{-8}$ *M*) after 5 h was determined (20). As shown in Table 1, transendothelial migration of tumor cells was inhibited by cicaprost only if the respective cells expressed a functional prostacyclin receptor.

To further corroborate this correlation of a functional receptor with antimetastatic effects, we have used several melanoma lines. Again, some lines expressed functional prostacyclin receptors, whereas others did not (Table 1). In the B16a (amelanotic) mouse melanoma, cicaprost had a strong effect on metastasis number in vivo, in accordance with the correlation stated above (see Fig. 2) (17).

Moreover, as we observed an increase in melanotic cells, we determined both photometrically and morphometrically the influence of cicaprost on melanin content. In a typical experiment (Fig. 5), cells were incubated with increasing concentrations of cicaprost for 9 days. Melanin was extracted with 10% DMSO–1 N NaOH and the melanin content was measured photometrically at 492 nm versus a standard curve. A significant concentration-dependent increase in melanin by cicaprost was observed. This was also found with shorter and longer incubation times. Interestingly, in the B16a line, neither forskolin, nocloprost, retinoic acid, or calcitriol caused an increase in melanin under identical conditions. After cells became confluent, cicaprost reproducibly reduced the cell number to about 70% of the control. When B16a cells were cultured in a three-dimensional collagen gel, cicaprost ($10^{-9}$ to $10^{-7}$ *M*) concentration-dependently increased the number of melanotic cells. In addition, the cellular phenotype changed from epitheloid or

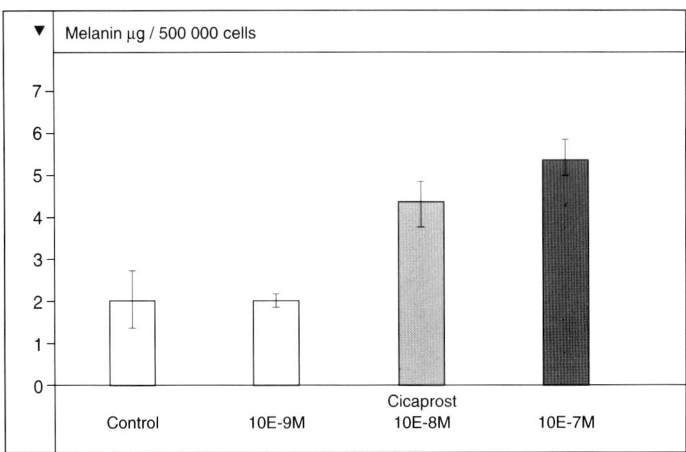


**FIG. 5.** Effect of cicaprost on melanin content in B16a melanoma cells after 9 days of culture. Melanin was measured photometrically. Mean ± SE of three dishes determined in quadruplicate.

rounded cells to very flat cells with many protrusions resembling dendrite-like processes.

Using the photometric method of melanin determination, an enhancement of melanin content was also found in the B16 F1 and the human SK Mel 28 line but not in the B16 F10 and the human Hs 939 line. Again, these findings perfectly correlate with the presence or absence of functional prostacyclin receptors (Table 1). To evaluate whether this increase in melanotic cells, which can be taken as an alteration of the differentiation status of the melanoma cell, has a consequence in vivo, we pretreated B16 F1 or B16 F10 cells in vitro with $1 \times 10^{-8}$ *M* of cicaprost for 30 min. The cells were washed and 250,000 cells were injected into the tail veins of mice. Fourteen days later the number of metastatic lung colonies was counted. Cicaprost significantly decreased the number of lung colonies in the B16 F1 line, whereas in the B16 F10 line no significant effect was observed (Fig. 6). These data substantiate the correlation proposed. The activity of cicaprost on lung colonizing potential was further determined in the B16 F1 line using a single pretreatment of mice with 1.0 mg/kg cicaprost 30 min before i.v. tumor cell inoculation. Cicaprost reduced the number of lung colonies significantly to about 25% of the control value.

To further analyze the mode of action of cicaprost in the MTLn3 mammary carcinoma line, which has functional prostacyclin receptors and in which cicaprost had a pronounced effect on spontaneous lung and lymph node metastases (14), a series of in vitro investigations was performed. No effect of cicaprost was found on adhesion of tumor cells to various matrix

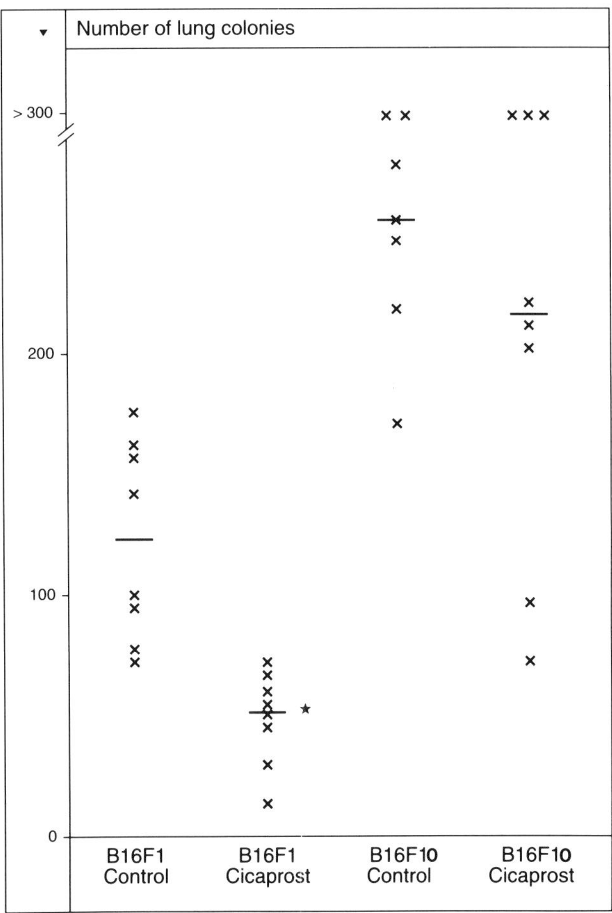


**FIG. 6.** Effect of in vitro pretreatment of B16 F1 and B16 F10 melanoma cells with cicaprost $1 \times 10^{-8}$ *M* for 30 min on lung colonizing potential of i.v. injected cells (250,000 cells/mouse).

proteins (Table 2). In accordance with in vivo data that cicaprost has no effect on primary tumor growth, no antiproliferative activity in vitro in cells growing in monolayer culture or in three-dimensional culture in matrigel or collagen gel could be demonstrated. However, significant changes in the expression of various cytokeratins and of vimentin (21) by incubation of MTLn3 cells grown in matrigel with cicaprost were found. Moreover, EGF-stimulated transendothelial migration of MTLn3 cells was abolished by cicaprost (Table 2) (20).

Taken together, the data on the mode of action of prostacyclin and stable prostacyclin analogues in metastasis now available do not yet provide a

**TABLE 2.** *Effects of cicaprost on MTLn3 cells*

| Parameter | Experimental conditions | Effect |
|---|---|---|
| cAMP | 2D, plastic | ↑ |
| Adhesion | Matrix proteins | → |
| Growth | 3D, matrigel | → |
| | 3D, collagen | → |
| Transendothelial migration | BPEC, +EGF | ↓ |
| Cytokeratin, vimentin expression | Matrigel, FACS | Changes |

definitive picture. Figure 7 shows a summary of the activities of such compounds in the metastatic cascade known thus far. In the past, the predominant activities of prostacyclin analogues were seen in the inhibition of tumor cell–platelet interaction, of the adhesion of tumor cells to matrix components, and of endothelial cell retraction (5). This was shown for a number of tumor cell lines and may be of importance in these lines. However, there are other tumor lines, e.g., the MTLn3 line, in which cicaprost is strongly active in inhibiting metastases of spontaneously metastasizing tumors but has no effect on adhesion in vitro or in the artificial metastasis assay. In such cell lines, other modes of action appear to be of major importance.

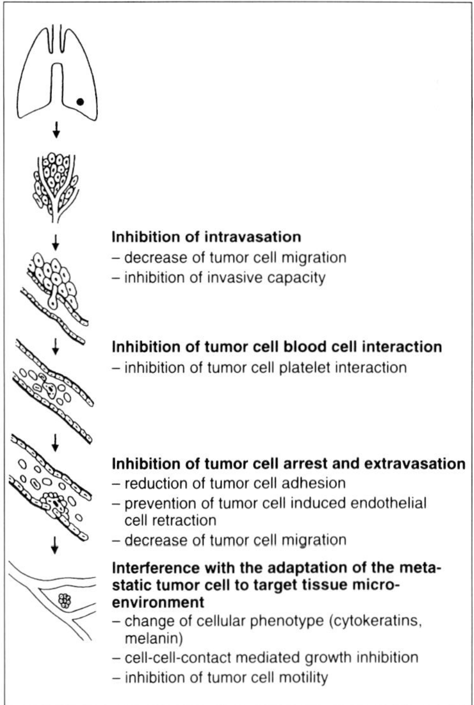


**FIG. 7.** Possible sites of the antimetastatic action of cicaprost.

A further proof that the mode of action of these compounds goes beyond events within the circulation is deduced from in vivo data from assays in which the primary tumor was removed and cicaprost treatment was started several days thereafter. In these assays, there appears to be an interference with the adaptation of the metastatic tumor cell to target tissue environment. Thus far, we have found changes in the cellular phenotype with regard to alterations in cytokeratin expression and induction of melanin. Of great importance is the direct correlation of the presence of a functional prostacyclin receptor in the tumor cell with antimetastatic effects of cicaprost in the various cell lines. More investigations are needed to further clarify the complex mode of action of these compounds in metastasis.

## CONCLUSIONS

Prostacyclin and some stable prostacyclin analogues have been shown to interfere with steps of the metastatic cascade, such as tumor cell–blood cell interaction and tumor cell adhesion, and to inhibit lung colony formation in the artificial metastasis assay. More recent data using the metabolically stable and orally active analogue cicaprost revealed that this drug strongly inhibits metastasis in a variety of spontaneously metastasizing tumor lines when treatment is initiated at the time of tumor implantation or at an advanced tumor stage. Because antimetastatic effects were also found when the primary tumor was removed before the start of treatment, effects of such compounds on metastatic cells already localized in the target tissue are strongly suggested. Thus far, effects of cicaprost on transendothelial migration of tumor cells, on alterations in cytokeratin expression, and on an increase in melanin content have been shown. This indicates that this drug may change the cellular phenotype and thus may interfere with the adaptation of the tumor cell to the target tissue environment.

## REFERENCES


1. Goldfarb RH, Brunson KW. Overview of current understanding of tumor spread. In: Goldfarb RH, ed. *Fundamental aspects of cancer.* Dordrecht: Kluwer, 1989:28–32.
2. Schneider MR, Schirner M. Antimetastatic prostacyclin analogs. *Drugs Future* 1993;18: 29–48.
3. Honn KV, Onoda JM, Menter DG, et al. Possible strategies for antimetastatic therapy. *Prog Clin Biol Res* 1986;212:217–49.
4. Honn KV, Cicone B, Skoff A. Prostacyclin: a potent antimetastatic agent. *Science* 1981; 212:1270–2.
5. Honn KV, Tang DG, Chen YQ. Platelets and cancer metastasis: more than an epiphenomenon. *Semin Thromb Hemost* 1992;18:392–415.
6. Moncada S. Biological importance of prostacyclin. *Br J Pharmacol* 1982;76:3–31.
7. Skuballa W, Schillinger E, Stürzebecher CS, Vorbrüggen H. Synthesis of a new chemically and metabolically stable prostacyclin analogue. *J Med Chem* 1986;29:313–5.
8. Stürzebecher CS, Haberey M, Müller B, et al. Pharmacological profile of a novel carba-

cyclin derivative with metabolic stability and oral activity in the rat. *Prostaglandins* 1986; 31:95–109.

9. Thierauch K-H, Dinter H, Stock G. Prostaglandins and their receptors: I. Pharmacologic receptor description, metabolism and drug use. *J Hypertens* 1993;11:1315–8.
10. Schirner M, Schneider MR. Antimetastatic potential of the stable prostacyclin analogue cicaprost. In: Rubanyi GM, Vane J, eds. *Prostacyclin: new perspectives for basic research and novel therapeutic indication.* Amsterdam: Excerpta Medica; 1992:247–75.
11. Costantini V, Giampietri A, Allegrucci M, Agnelli G, Nenci GG, Fioretti MC. Mechanisms of the antimetastatic activity of stable prostacyclin analogues: modulation of host immunocompetence. In: Samuelsson B, Pamwell PW, Paoletti R, Folco G, Granström E, eds. *Advances in prostaglandin, thromboxane, and leukotriene research.* New York: Raven Press, 1991:917–20.
12. Schirner M, Schneider MR. Cicaprost inhibits metastases of animal tumors. *Prostaglandins* 1992;42:451–61.
13. Schirner M, Schneider MR. The stable prostacyclin analogue cicaprost inhibits metastasis of tumors of R3327 MAT Lu prostate carcinoma and SMT 2A mammary carcinoma. *J Cancer Res Clin Oncol* 1992;118:497–501.
14. Schirner M, Lichtner RB, Schneider MR. The stable prostacyclin analogue cicaprost inhibits metastasis to lungs and lymph nodes in the 13762NF MTLn3 rat mammary carcinoma. *Clin Exp Metast* 1994;12:24–30.
15. Schirner M, Lichtner RB, Graf H, Schneider MR. Efficacy of cicaprost on metastasis in advanced tumor disease. Third International conference on eicosanoids and other bioactive lipids in cancer, inflammation and radiation injury, Washington, DC, Oct. 13–16, 1993.
16. Schneider MR. Prostacyclin and its analogues: antimetastatic effects and mechanisms of action. *J Cancer Detect Prev* 1994;13:349–64.
17. Honn KV, Schirner M, Schneider MR. Prostacyclin, thromboxane $A_2$, and tumor metastasis. *Cancer Metastasis Rev* [*in press*].
18. Thierauch K-H, Dinter H, Stock G. Prostaglandins and their receptors: II. Receptor structure and signal transduction. *J Hypertens* 1994;12:1–5.
19. Gorman RR, Bunting S, Miller OV. Modulation of human platelet adenylate cyclase by prostacyclin (PGX). *Prostaglandins* 1977;13:377–86.
20. Graf H. Endothelial control of cell migration and proliferation. *Eur Heart J* 1993;14(suppl 1):183–6.
21. Lichtner RB, Jilian JA, Glasser SR, Nicolson GL. Characterization of cytokeratins expressed in metastatic rat mammary adenocarcinoma cells. *Cancer Res* 1989;49:104–11.

*Topics in Molecular Medicine, Volume 1*, edited by Wolfgang Siess, Reinhard Lorenz, and Peter C. Weber. Raven Press, Ltd., New York © 1995.

# 21

# Interventional Approaches In Vivo for Selectin-Dependent Inflammatory Injury

Peter A. Ward

*Department of Pathology, University of Michigan Medical School, Ann Arbor, Michigan 48109-0602*

Selectins, which are now considered prerequisite for the earliest events in the inflammatory response, are believed to mediate the "rolling" of leukocytes along the endothelial surface (reviewed in 1–3). The rolling is believed to be due to intermittent adhesive interactions between leukocytes and activated endothelial cells. Because of low-affinity adhesive contact, leukocytes are only transiently delayed in their rapid movement within the vascular lumen. Thereafter, other adhesive events take place, including leukocytes $\beta_2$ integrins (CD11/CD18 heterodimers) interactions with endothelial ICAM-1, resulting in high-affinity binding that leads to flattening of leukocytes on the endothelial surface and virtual cessation of movement (1). Recently, a third adhesion process has been described, featuring the role of endothelial PECAM-1 with unknown counter-receptors on neutrophils (4,5).

The requisite role of selectins and $\beta_2$ integrins for maintenance of an intact inflammatory response comes from two types of families of patients with genetic deficiencies of adhesion molecules. Leukocyte adhesion deficiency (LAD) I is due to defective CD18 and in its full clinical expression is a lethal mutation. LAD II is associated with defective formation of sialyl Lewis$^x$, a complex oligosaccharide which is an important ligand for selectins (reviewed in 6). Because other ligands for selectins are present, LAD II does not appear to cause the degree of clinical problems seen in LAD I.

## IN VIVO STRATEGIES FOR DEFINING SELECTIN REQUIREMENTS

There are several strategies for determining the role of selectins in in vivo inflammatory responses (summarized in Table 1). The most definitive approach is the use of monoclonal antibodies (MAbs) that block the ligand-binding site of a given selectin molecule. Although this appears to be an ideal

**TABLE 1.** *Interventions for definitions of selectin requirements*

| |
|---|
| Blocking antibodies |
| Soluble counter-receptors |
| Sialyl Lewis$^a$ |
| Sialyl Lewis$^x$ |
| Other factors |
| Selectin–Ig chimeric proteins |

approach, there are some serious drawbacks, as will be discussed below. The second way in which selectin requirements can be defined in vivo is by the use of synthetic oligosaccharides, such as sialyl Lewis$^x$. It should be pointed out that these compounds, in general, do not necessarily define the precise nature of the selectin(s) involved because the oligosaccharides are often reactive with all selectins. Nevertheless, if sialyl Lewis$^x$ is highly protective in vivo in suppressing neutrophil accumulation whereas the non-fucosylated form (sialyl-*N*-acetyllactosamine) is not, this provides strong evidence for the involvement of selectin molecules (7,8). There are some suggestions that chemical derivatization of reactive oligosaccharides will confer on these molecules selectin-specificity (9), but it remains to be determined if this will be confirmed in vivo. The third approach for in vivo definition of selectin molecules is the use of selectin–Ig chimeric proteins. These molecules have been developed so that the very short $t_{1/2}$ value of recombinant selectin molecules after in vivo infusion can be overcome by the presence of the CH2 and CH3 domains of the Fc region of human IgG, which confers an extended half-life in vivo to the selectin chimeric molecules (10).

## PROBLEMS WITH IN VIVO BLOCKADE OF SELECTINS

As would be expected, all selectin blocking interventions have demonstrated drawbacks (Table 2). The most important issue with the MAbs is to be certain that they are able to block the ligand-binding site of the selectin molecule. This can be determined by a variety of in vitro assays that feature neutrophil binding to activated endothelial monolayers or to selectins present in solid phase. A major problem with anti–L-selectin MAbs is their tendency to induce neutropenia after i.v. infusion (11). For this reason, we

**TABLE 2.** *Complications in the use of selectin-blocking factors*

| |
|---|
| Unpredictable reliability of MAbs |
| Problems with Fc region of MAbs |
| Short $t_{1/2}$ for oligosaccharides and $F(ab')_2$ MAbs |
| Expense of synthesis for preparations |
| Low affinities of binding, especially for selectin–Ig chimeras |

have been forced to use $F(ab')_2$ forms of anti–L-selectin MAb, which effectively block L-selectin–mediated neutrophil accumulation in vivo. Another instance in which the intact MAb cannot be used is the case of anti–E-selectin (CL-3), which, when used in the intact form, not only fails to block E-selectin–dependent injury in the lung but has a tendency to accentuate the injury (12). This is due to the presence of the Fc region of the MAb, which appears to substitute for the blocked E-selectin molecule and now imposes an Fc receptor-mediated binding of neutrophils to the MAb-coated endothelial surface. The use of $F(ab')_2$ versions of this MAb overcomes these difficulties. It is apparent that the use of $F(ab')_2$ forms of MAb carries with it the serious problem that $F(ab')_2$ molecules are very rapidly excreted by the kidney, giving the $F(ab')_2$ a $t_{1/2}$ of less than 15 min. Obviously, this creates substantial problems if these preparations are to be used in vivo.

The use of oligosaccharides, such as sialyl Lewis$^x$, that compete with the naturally occurring ligand for selectin binding in vivo has provided impressive results in in vivo models of inflammatory lung injury (7,8). However, because of the short $t_{1/2}$ of these compounds (10 to 15 min), this again poses a serious problem for in vivo usage, especially in view of the current expense of synthesizing these compounds. However, new synthetic approaches as well as development of derivatized forms of these compounds to increase the $t_{1/2}$ value may alleviate these difficulties.

The chief problem with the use of selectin–Ig chimeric proteins, aside from the expense of preparing these compounds, appears to be their problem of low affinity (10). They are considerably less effective in blocking selectin-dependent inflammatory reactions than are MAbs (13). Because of the presence of a portion of the Fc region of human IgG, the selectin–Ig chimeric proteins lend themselves to probing of their binding to counter-receptors present in sections of tissues.

## IN VIVO DEFINITION OF SELECTIN REQUIREMENTS IN LUNG INFLAMMATION

The use of animal models has permitted identification of selectins that are required for neutrophil recruitment and development of injury, as measured by increased permeability and hemorrhage (Table 3). Neutrophil accumulation is quantitated by measurement of myeloperoxidase (MPO) in lung tissue extracts (12). These approaches have been applied to four models of inflam-

**TABLE 3.** *Inflammatory injury models*

| |
|---|
| IgG immune complex-induced lung injury |
| IgA immune complex-induced lung injury |
| Lung injury after systemic activation of complement |
| Ischemia–reperfusion injury of extremities and secondary lung injury |

matory injury (Table 3). In all of these, lung injury develops after intrapulmonary (intra-alveolar) deposition of IgG or IgA immune complexes (14,15), after systemic activation of complement in response to i.v. infusion of cobra venom factor (CVF) (16) and after ischemia (4 h) and reperfusion (4 h) events involving the hind limbs of rats (17). The requirements for selectins in these models of injury have been defined by blocking MAbs or by the use of selectin–Ig chimeric proteins. The results of these studies are summarized in Table 4.

In the IgG immune complex model of acute lung injury, which requires recruitment of neutrophils in a complement-dependent manner, blocking of selectins demonstrates requirements for L- and E-selectins but not for P-selectin (11–13). Blocking of the relevant selectins in this model of lung injury is associated with a parallel reduction in increase of MPO levels, consistent with the conclusion that L- and E-selectins are required for neutrophil recruitment in this inflammatory model. The pattern of a requirement for E- but not for P-selectin is not unexpected, in view of the fact that neutrophil accumulation peaks 3 to 4 h after initiation of immune complex deposition. This time course is parallel to the known time requirement for in vitro endothelial expression of E-selectin (A. A. Vaporciyan, P. A. Ward, unpublished observations). It should also be pointed out that in this model of injury, additional adhesion-promoting molecules are also required for neutrophil recruitment, including ICAM-1 and the $\beta_2$ integrin, CD11a/CD18 (LFA-1), but not CD11b/CD18 (Mac-1) (15). PECAM-1 is also required (15). Therefore, although E- and L-selectins are required in this model of lung injury, they are not sufficient for the neutrophil recruitment process.

Inflammatory lung injury resulting from intrapulmonary deposition of IgA immune complexes is associated with severe injury to alveolar and vascular endothelial cells, but the process leading to injury does not involve the participation of neutrophils (15). It appears that the injury process can be related to an oxidant(s) (from oxygen and L-arginine) generated by lung macrophages. On the basis of this information, it is not surprising that the process leading to cell injury in this model is independent of a requirement

**TABLE 4.** *Selectin requirements in lung models*

| Model | Selectin requirement[a] | | |
|---|---|---|---|
| | L- | E- | P- |
| IgG immune complex lung injury | 52 | 33 | <1 |
| IgA Immune complex lung injury | <1 | <1 | <1 |
| CVF-induced lung injury | 54 | <1 | 51 |
| Ischemia–reperfusion injury | | | |
| Skeletal muscle | 49 | 13 | 5 |
| Lung | 82 | 79 | 4 |

[a] As determined by reductions in permeability or hemorrhage, using MAbs or selectin–Ig chimeric proteins.

for any of the selectins (Table 4). To make the story complete, in this model of lung injury there are requirements for ICAM-1 and for both LFA-1 and Mac-1 (15). The requirements for ICAM-1 and $\beta_2$ integrins appear to be inconsistent with the lack of need for recruited neutrophils in this model. However, this seeming discrepancy might be related to the fact that alveolar macrophages are anchored to alveolar epithelial type II cells, which contain ICAM-1 and might permit the macrophage to anchor to the epithelial cells via LFA-1 and/or Mac-1 contained on the macrophage. Alternatively, it is possible that macrophage activation requires the presence of either or both of the $\beta_2$ integrins. However, this would not explain the ICAM-1 requirement.

Infusion of CVF results in intense intravascular activation of complement, causing neutrophils to become stimulated within the vascular compartment (16). These cells self-aggregate but also become attached to the pulmonary vascular endothelium, where endothelial cell damage rapidly develops. All of these events occur within a 30-min period. When assessed for selectin requirements, this model maps for L- and P-selectins but not for E-selectin (Table 4) (7,13,17,18). It appears likely that P-selectin upregulation occurs rapidly as a result of complement activation and generation of C5a. C5a would then cause mast cells and basophils to release histamine, which is known to cause upregulation of endothelial P-selectin. Alternatively, C5b-9, if generated on the surfaces of endothelial cells in vivo after CVF infusion, might also cause upregulation of endothelial or P-selectin, as has been shown to occur in vitro (19). It should be pointed out that CVF-induced lung injury also requires ICAM-1, LFA-1, and Mac-1, but not PECAM-1.

The last model of tissue injury in which selectin requirements have been defined in rats involves events that occur after 4 h of ischemia and 4 h of reperfusion of both hind limbs. In this model, injury occurs in the vasculature of crural skeletal muscle as well as in the lung. In both cases, the injury is complement-dependent and neutrophil-dependent. In the case of vascular injury in the hind limbs, it is apparent that L-selectin is required; however, neither E- nor P-selectin is required under the conditions employed (17). There are, however, also requirements for ICAM-1, LFA-1, and Mac-1 for full development of vascular injury (20). In this model of ischemia–reperfusion, lung damage also develops and is complement- and neutrophil-dependent. In contrast to the selectin requirements in the skeletal muscle vasculature (see above), lung vascular injury is highly dependent on both L- and E-selectins but not on P-selectin (Table 4) (21). In the lung, ICAM-1, LFA-1, and Mac-1 are concomitantly required for neutrophil accumulation and tissue injury. What is rather surprising in this ischemia–reperfusion model is the difference in the requirement for selectins between the two vascular beds (skeletal muscle and lung microvasculature). The discrepant patterns for selectin requirements may be attributed to two possibilities. There may be intrinsic differences in the ability of the two vascular beds to express E- and

P-selectins, with the skeletal muscle vasculature being unable to express either E- or P-selectin, in contrast to the lung vasculature in which E-selectin is readily expressed. Alternatively, it is possible that because the vasculature of the skeletal muscle has been the direct target of ischemic injury, it is therefore less susceptible to reversibility of leakage by the presence of either anti–E- or anti–P-selectin. If it is true that vascular beds vary in their ability to express E- and P-selectins, this has important implications for an understanding of inflammatory responses. This possibility is supported by data from an experimental model of nephrotoxic nephritis (22). In this condition, IgG immune complexes are deposited along the glomerular vascular basement membrane. This results in complement activation, neutrophil adherence, and focal damage to glomerular endothelial and epithelial cells. Under these conditions, E-selectin plays no demonstrable role, although ICAM-1, LFA-1, and Mac-1 are essential for the full expression of glomerular damage. These findings suggest that differences exist between organs with respect to the participation of vascular adhesion molecules, which indicates that care must be exercised in extrapolating data from one organ system to another.

## CONCLUSIONS

Selectins play important roles in inflammatory lung injury in which recruitment of blood neutrophils is a prerequisite. An important linkage exists between cytokine expression (TNF$\alpha$, and IL-1) and upregulation of endothelial adhesion molecules (ICAM-1, E-selectin). Inflammatory reactions in lung have variable requirements for selectins (P- and E-), depending on the circumstances and the time course of the response. The requirements for selectins can be defined by the use of blocking MAbs or by the use of selectin–Ig chimeric proteins or appropriate fucosylated and sialylated oligosaccharides. Selectin engagement in these inflammatory reactions is necessary but not sufficient for the transmigration of neutrophils. Additional adhesion factors, such as endothelial ICAM-1 and neutrophil $\beta_2$ integrins, are necessary for the ultimate transmigration of neutrophils. There are suggestions that selectin requirements for neutrophil-dependent inflammatory injury also vary depending on the organ involved. A more complete understanding of tissue injury may provide new approaches to blockade of the inflammatory response.

## REFERENCES

1. Lobb RR. Integrin-immunoglobulin superfamily interactions in endothelial-leukocyte adhesion. In: Harlan JM, Lin DY, eds. *Adhesion—its role in inflammatory disease*. New York: WH Freeman, 1992:1–18.
2. Bevilacqua MP, Nelson RM. Selectins. *J Clin Invest* 1993;91:379–87.

3. Lasky LA. Selectins: interpreters of cell-specific carbohydrate information during inflammation. *Science* 1992;258:964–9.
4. Muller WA, Weigl SA, Deng X, Phillips DM. PECAM-1 is required for transendothelial migration of leukocytes. *J Exp Med* 1993;178:449–60.
5. Vaporciyan AA, DeLisser HM, Yan HC, et al. Involvement of platelet-endothelial cell adhesion molecule-1 in neutrophil recruitment in vivo. *Science* 1993;262:1580–2.
6. Harlan JM. Leukocyte adhesion deficiency syndrome: insights into the molecular basis of leukocyte emigration. *Clin Immunol Immunopathol* 1993;67:S16–24.
7. Mulligan MS, Paulson JC, De Frees S, Zheng ZL, Lowe JB, Ward PA. Protective effects of sialylated oligosaccharides in immune complex-induced acute lung injury. *J Exp Med* 1993;178:623–31.
8. Mulligan MS, Paulson JC, De Frees S, Zheng ZL, Lowe JB, Ward PA. Protective effects of oligosaccharides in P-selectin-dependent lung injury. *Nature* 1993;364:149–51.
9. Feizi T. Blood group related oligosaccharides are ligands in cell adhesion. *Biochem Soc Trans* 1992;20:274–8.
10. Watson SR, Fennie C, Lasky LA. Neutrophil influx into an inflammatory site inhibited by a soluble homing receptor-IgG chimaera. *Nature* 1991;349:164–7.
11. Mulligan MS, Miyasaka M, Tamatani T, Jones ML, Ward PA. Requirements for L-selectin in neutrophil-mediated lung injury in rats. *J Immunol* 1994;152:832–40.
12. Mulligan MS, Varani J, Dame MK, et al. Role of endothelial-leukocyte adhesion molecule 1 (ELAM-1) in neutrophil-mediated lung injury in rats. *J Clin Invest* 1991;88:1396–1406.
13. Mulligan MS, Watson SR, Fennie C, Ward PA. Protective effects of selectin chimeras in neutrophil-mediated lung injury. *J Immunol* 1993;151:6410–17.
14. Mulligan MS, Vaporciyan AA, Miyasaka M, Tamatani T, Ward PA. Tumor necrosis factor alpha regulates *in vivo* intrapulmonary expression of ICAM-1. *Am J Pathol* 1993;142:1739–49.
15. Mulligan MS, Wilson GP, Todd RF, et al. Role of β1, β2 integrins and ICAM-1 in lung injury following deposition of IgG and IgA immune complexes. *J Immunol* 1993;150:2407–17.
16. Till GO, Johnson KJ, Kunkel R, Ward PA. Intravascular activation of complement and acute lung injury. Dependency on neutrophils and toxic oxygen metabolites. *J Clin Invest* 1982;69:1126–35.
17. Seekamp A, Warren JS, Remick DG, Till GO, Ward PA. Requirements for tumor necrosis factor-α and interleukin-1 in limb ischemia/reperfusion injury and associated lung injury. *Am J Pathol* 1993;143:453–63.
18. Mulligan MS, Smith CW, Anderson DC, et al. Role of leukocyte adhesion molecules in complement-induced lung injury. *J Immunol* 1993;150:2401–6.
19. Hattori R, Hamilton KK, McEver RP, Sims PJ. Complement proteins C5b-C9 induce secretion of high molecular weight multimers of endothelial von Willebrand factor and translocation of granule membrane protein GMP-140 to the cell surface. *J Biol Chem* 1989;264:9053–60.
20. Seekamp A, Mulligan MS, Till GO, et al. Role of β2 integrins and ICAM-1 in lung injury following ischemia-reperfusion of rat hind limbs. *Am J Pathol* 1993;143:464–72.
21. Seekamp A, Till GO, Paulson JC, et al. Role of selectins in local and remote tissue injury following ischemia and reperfusion. *Am J Pathol* 1994;144:592–8.
22. Mulligan MS, Johnson KJ, Todd RF, et al. Requirements for leukocyte adhesion molecules in nephrotoxic nephritis. *J Clin Invest* 1993;91:577–87.

*Topics in Molecular Medicine, Volume 1,*
edited by Wolfgang Siess, Reinhard Lorenz,
and Peter C. Weber. Raven Press, Ltd.,
New York © 1995.

# 22

# Expression of Adhesion Molecules in Pulmonary Inflammation and Asthma

Robert H. Gundel, Dean J. Lindell, and Peter D. Harris

*Pharmacology Section, Preclinical Research, Miles, Inc., Biotechnology, West Haven, Connecticut 06516*

Bronchial asthma is a disease of complex etiology which affects 5% to 8% of the world population (1). Asthma is characterized by episodes of reversible airway obstruction resulting in chest tightness, dyspnea, wheezy breathing, and cough. Acute asthmatic episodes are the result of the action of mast cell-derived mediators on airway smooth muscle and, in the mild asthmatic, are well controlled by treatment with inhaled bronchodilators (2). In moderate-to-severe asthmatics, however, chronic symptoms of asthma such as persistent airway obstruction and bronchial hyperresponsiveness predominate, leading to more frequent and more severe episodes which require much more aggressive treatment and often are refractory to presently available therapies (3,4). Recent studies indicate that the severity and chronicity of asthma symptoms are a manifestation of airway inflammation involving many different cells, with eosinophils and T lymphocytes predominating (5). Airway inflammation results from a potentially large number of stimuli that lead to events at both the cellular and the tissue level involving cellular communication facilitated by the upregulation and cell surface expression of adhesion molecules. Cell adhesion directs the initial adherence of inflammatory cells to vascular endothelium and subsequently facilitates their movement out of the vascular space to sites of inflammation. In addition, adhesion molecules can regulate the state of activation of inflammatory cells and are responsible for the retention of inflammatory cells at effector cell sites (6).

The regulation and expression of adhesion molecules have been the subject of profound research efforts in recent years. Activation of resident antigen-presenting cells in the lungs is most likely the primary process that leads to the cascade of adhesion molecule regulation events in allergic asthma. These events are complex and intricate, and have exposed a new paradigm in our understanding of the disease and possible avenues for novel therapeutic interventions. At the center of this issue are the potential modulatory role of various cytokines in adhesion molecule cell surface expression, the ability to elucidate potential targets for inhibition of the inflammatory process, and the development of new and more effective therapeutic treatments. This chapter reviews work from our laboratory investigating the role of cytokines in the regulation of adhesion molecule expression and the role of these adhesion molecules in the development of pulmonary inflammation in a primate model of extrinsic asthma.

## OVERVIEW OF ADHESION MOLECULES

Cell–cell adhesion and communication have important roles in the normal functioning of the immune system (7). For example, leukocyte–leukocyte adherence is required for the initiation of an inflammatory response, in that specific immunologic processes, such as antibody production or the generation of antigen-specific T cells, involve attachment of lymphocytes to antigen-presenting cells (8). Leukocyte–endothelial cell adhesion is required for the movement of leukocytes out of the vascular space into tissue. The adherence of leukocytes to vascular endothelium represents a pivotal step in the movement of leukocytes from the vascular space to specific sites of tissue inflammation, as demonstrated in a population of leukocyte adhesion-deficient (LAD) patients. These patients are unable to recruit leukocytes and mount an inflammatory response owing to lack of expression of normal numbers of CD18 ($\beta_2$ integrin) adhesion receptors on their cell surfaces (9–11). Consequently, LAD patients suffer from severe, recurrent, and often life-threatening bacterial infections. Finally, leukocyte–target cell adhesion enhances the effector cell capacity of leukocytes by allowing a localized release of toxic substances onto target cells, thus facilitating cytotoxicity (12). Therefore, adhesion among various cells of the immune system promotes normal immune system function and represents one of the primary mechanisms of host defense.

On the basis of their structural homology, cell adhesion molecules can be grouped into three gene families: the immunoglobulins, the integrins, and the selectins. In addition, the adhesion molecules can be classified on a functional basis, separating those involved in mucosal cell–cell adhesion, from those that function to bind luminal cells with the substratum and those that are expressed on the cells lining the luminal surface, which function in cell recruitment, activation, and transendothelial migration. The function of adhesion molecules in these capacities has been reviewed in detail elsewhere, and the reader is referred to these publications for further information (8,13–15).

## REGULATION OF ENDOTHELIAL–EPITHELIAL ADHESION MOLECULES

The exposure of cultured endothelial or epithelial cells to various inflammatory mediators (e.g., TNF) induces the expression of several adhesion molecules on the cell surface. The inducible expression of adhesion molecules on the cell surface is responsible for margination, diapedesis, and retention of leukocytes at specific sites of tissue inflammation. The kinetics of the induced expression differ among adhesion molecules but can be roughly categorized into those that are rapidly expressed and those that show a lag time and have delayed expression on the cell surface after stimulation.

Adhesion molecules that are rapidly expressed on the endothelial cell surface include two members of the selectin family, P-selectin and E-selectin. P-Selectin is not constitutively expressed but is rapidly expressed (within 5 min) after stimulation. This rapid upregulation of expression is facilitated by the fact that P-selectin is pre-formed and is stored in cytoplasmic Weibel–Palade bodies which, on stimulation, fuse with the cytoplasmic membrane and extrude P-selectin onto the cell surface (16). In contrast, E-selectin has a short lag phase (approximately 4 h) before it is expressed on the cell surface. This lag phase results from the requirement of new protein synthesis, because E-selectin is not pre-formed and stored in cytoplasmic granules as is the case with P-selectin (17–19). Peak cell surface expression of E-selectin occurs by 6 to 8 h after stimulation and returns to basal levels by 12 to 18 h. Because of the relatively rapid expression of these adhesion molecules on endothelial cell surfaces, it has been proposed that they are important in acute inflammation and in the initial adherence of leukocytes at specific sites of tissue inflammation.

Other adhesion molecules, such as ICAM-1 and VCAM-1, require more time to reach maximal cell surface expression (up to 12 or 18 h). Because of this delayed expression, it is tempting to categorize these adhesion molecules as being more important in chronic inflammation, a view that is supported by their extremely long duration of expression (up to 72 h) after cell stimulation (20,21). Therefore, in contrast to the selectins, members of the immunoglobulin supergene family, such as ICAM-1 and VCAM-1, may have a more prominent role in the maintenance of chronic inflammation.

## ROLE OF ADHESION MOLECULES IN PULMONARY INFLAMMATION AND AIRWAY DYSFUNCTION

To investigate the role of adhesion molecules in the development of pulmonary inflammation and subsequent airway dysfunction, we developed a nonhuman primate model of extrinsic asthma. The animals utilized for these studies are adult male cynomolgus monkeys (*Macaca fasicularis*) that have a naturally occurring, IgE-mediated respiratory hypersensitivity to *Ascaris suum* protein extract. The details of the development of this model, as well as the specific methodology utilized to study changes in airway function and airway cellular composition, are described in previous publications (22–24).

In our first series of studies, we reported that multiple antigen challenges led to the development of a profound airway eosinophilia that was chronologically related to damage or frank denudation of airway epithelium and a 10- to 100-fold increase in airway responsiveness to inhaled methacholine (20,22,23). Damage to the airways was assessed in these studies by whole-lung histology as well as by analysis of the presence of creola bodies in bronchoalveolar lavage (BAL) fluid, and a significant correlation was demonstrated between the degree of damage and the magnitude of the increase in airway responsiveness. Furthermore, analysis of immunohistologic staining of lung tissue demonstrated that ICAM-1 and E-selectin were upregulated on pulmonary vascular endothelium, whereas only ICAM-1 was expressed on airway epithelium (25). Interestingly, the expression of ICAM-1 on airway epithelium was most marked on the basolateral surface and basement membrane zone and was associated with a large accumulation of LFA-1–positive leukocytes. This provided associative evidence suggesting that upregulation of the expression of adhesion molecules on vascular endothelial cells and airway epithelial cells contributes to the pulmonary inflammation and airway dysfunction after multiple antigen inhalations.

To further examine the role of adhesion molecules in antigen-induced pulmonary inflammation, we examined a number of adhesion-blocking murine monoclonal antibodies (MAbs) in the primate model. In our first study, animals were treated intravenously with the anti–ICAM-1 MAb R6.5 before and during the course of multiple antigen challenges over a 10-day study period (25). Treatment with anti–ICAM-1 significantly inhibited the influx of eosinophils into the airways and the associated increase in airway responsiveness, suggesting that ICAM-1 has a pivotal role in antigen-induced pulmonary inflammation and airway dysfunction (Fig. 1). Because we had previously shown that, like ICAM-1, E-selectin is also expressed on vascular endothelial cells after antigen challenge, the anti–E-selectin MAb CL2 was tested in the model (25). After the same dosage regimen as in the anti–ICAM-1 treatment study, the results demonstrated that CL2 treatment had no effect on the influx of eosinophils into the airways or the increase in airway responsiveness to inhaled methacholine (Fig. 2). Therefore, although it is expressed quite prominently on vascular endothelium, E-selectin does not play a major role in the recruitment of eosinophils into the airways in the primate model. In our next study, we demonstrated that treatment with an anti–Mac-1 MAb (LM-2) had no effect on the influx of eosinophils into the airways but, surprisingly, totally blocked the increase in airway responsiveness to inhaled methacholine (Fig. 3). Analysis of EPO activity in BAL fluid

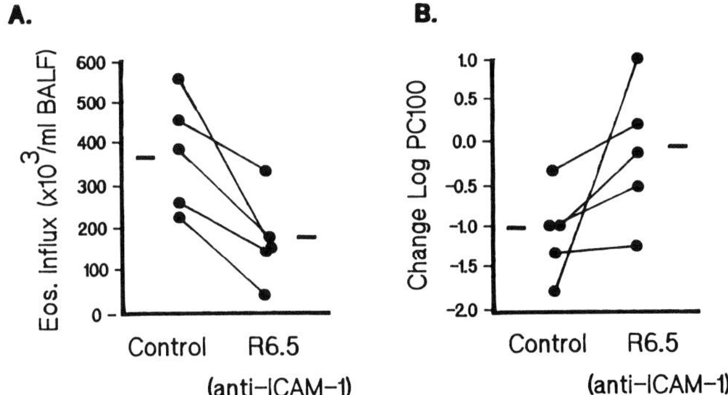


**FIG. 1. A,B:** Effects of intravenous administration of anti–ICAM-1 MAb (R6.5) on antigen-induced eosinophil influx into the airways and increases in airway responsiveness to inhaled methacholine. R6.5 treatment significantly reduced the influx of eosinophils and increases in airway responsiveness induced by repeated antigen challenge. (Modified from ref. 25.)

indicated that LM-2 treatment significantly decreased EPO activity compared to control studies, suggesting that although the eosinophils had actively infiltrated into the lungs they were not "activated" and degranulating to release their toxic cytoplasmic granule contents onto the airway epithelium. On the basis of our results to date, we hypothesized that the increase in airway responsiveness might be the result of an epithelial ICAM-1-mediated event. Therefore, we designed a study in which R6.5 (anti–ICAM-1) was administered by inhalation to selectively target the airway epithelium. The results of this study demonstrated that inhaled R6.5 significantly inhibited the antigen-induced airway epithelial damage and the airway hyperresponsiveness, supporting the hypothesis of the importance of epithelial cell ICAM-1 (Fig. 4).

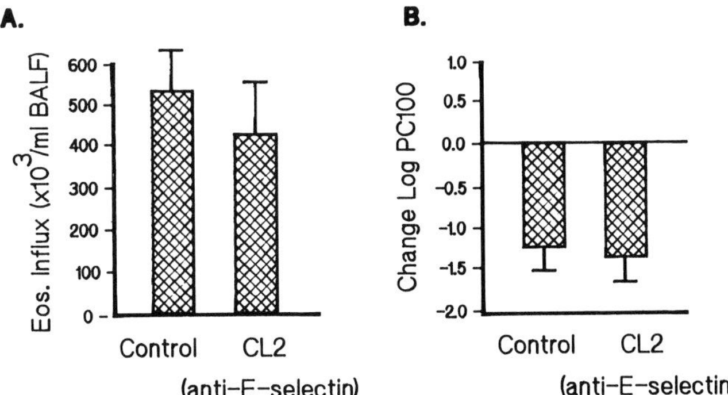


**FIG. 2. A,B:** Effects of intravenous administration of anti–E-selectin MAb (CL2) on the antigen-induced airway eosinophilia and airway hyperresponsiveness to inhaled methacholine. CL2 treatment had no inhibitory effect on the antigen-induced eosinophil influx into the airways or on the increase in airway responsiveness to methacholine.

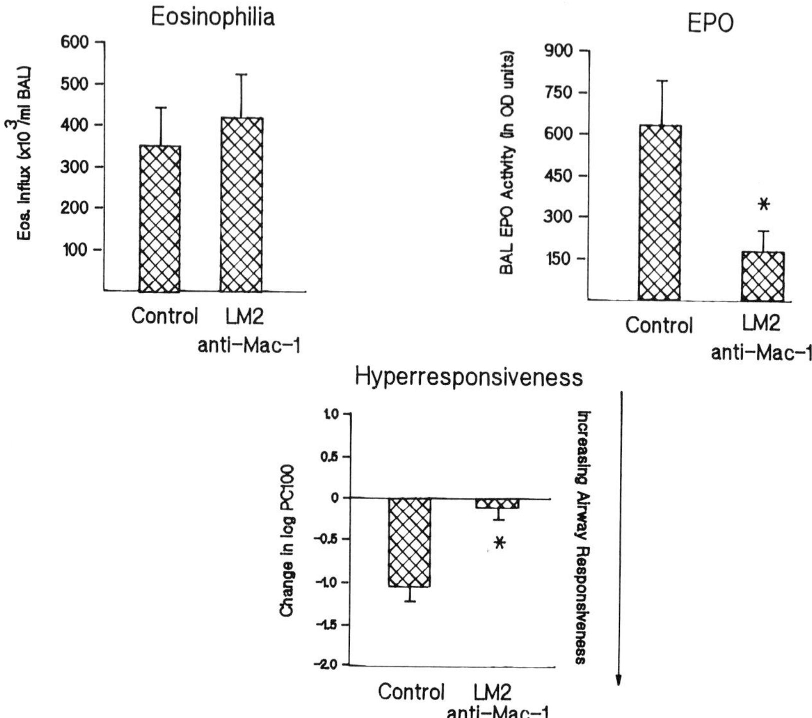


**FIG. 3.** Effects of intravenous administration of anti-Mac-1 MAb (LM2) on antigen-induced airway eosinophilia, eosinophil peroxidase activity in BAL fluid, and airway hyperresponsiveness to inhaled methacholine. LM2 administration had no effect on eosinophil infiltration into the airways. However, the level of eosinophil peroxidase in BAL fluid and the increase in airway responsiveness to inhaled methacholine were significantly decreased.

We have demonstrated that both ICAM-1 and Mac-1 contribute to the airway hyperresponsiveness induced by repeated antigen inhalation in primates. On the basis of these results, we postulate that eosinophil infiltration into the airways (binding to vascular endothelium and diapedesis) is partially mediated by LFA-1 (CD11a/CD18)–ICAM-1 interactions and that the activation of eosinophils once in the airway tissue, and the altered airway function associated with the development of airway eosinophilia, are mediated by an Mac-1–ICAM-1–dependent mechanism. Furthermore, the Mac-1–ICAM-1 binding of eosinophils to the basolateral surface of the airway epithelium results in a further activation and a subsequent localized degranulation. This localized degranulation leads to a high concentration of toxic granule-derived products (e.g., MBP) directly on the airway epithelium, which may be causal in the degradation and ultimate breakdown of the adhesion between the luminal cells and the substratum. This could result in a dysfunction of the mucosal lining caused by loss of epithelial cell integrity,

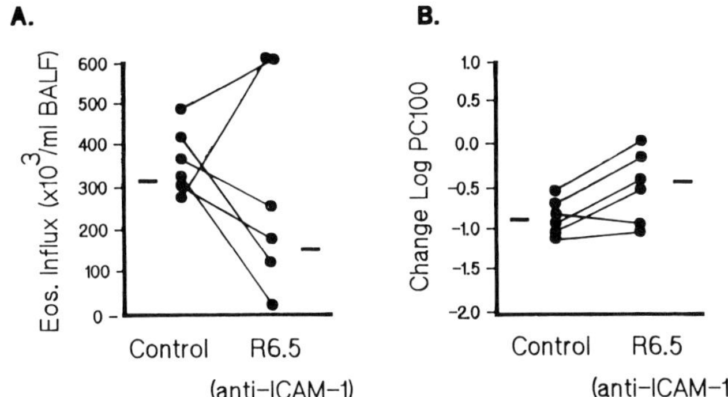


**FIG. 4. A,B:** Effects of inhaled anti–ICAM-1 (R6.5) MAb on antigen-induced airway eosinophilia and airway hyperresponsiveness. Inhaled R6.5 significantly reduced both the airway eosinophilia and the increase in airway responsiveness to inhaled methacholine.

leading to an increase in airway responsiveness via a number of possible mechanisms.

## INTERLEUKIN-4–INDUCED LEUKOCYTE TRAFFICKING

Interleukin-4 (IL-4) is a member of the IL-4 lymphokine family that includes IL-3, IL-4, IL-5, and GM-CSF. IL-4 was originally described as a B-cell growth factor and is produced mainly by T lymphocytes and mast cells (26). However, further study has demonstrated that IL-4 is a pleiotropic cytokine that has a wide variety of effects on various cells of the lymphoid system, including increased expression of MHC class II antigen, CD23, and Ig isotype class switching that promotes the production of IgE and $IgG_1$ antibodies (27,28). Recent studies have demonstrated that IL-4 selectively induces the expression of VCAM-1 on vascular endothelium, while having little or no effect on the expression of ICAM-1 or E-selectin (21,29). Other studies have demonstrated that IL-4 selectively promotes adherence of T lymphocytes and eosinophils, but not neutrophils, to cultured vascular endothelium via the interaction of VCAM-1 and the $\beta_1$ Integrin VLA-4 (CD49d/CD29) (21,29). These findings are potentially very important with regard to the development of allergic inflammation, which is characterized by a strong T-cell and eosinophil component, and they provide a possible mechanism for the selective recruitment of these cells into the tissue. Although the role of IL-4 in the upregulation of VCAM-1 expression on endothelial cells and in T-cell and eosinophil adhesion has been well characterized in vitro, the importance of this mechanism of selective recruitment in vivo is unknown. Recently, we have been investigating the role of IL-4 in the expression of adhesion molecules and leukocyte trafficking in cynomolgus monkeys.

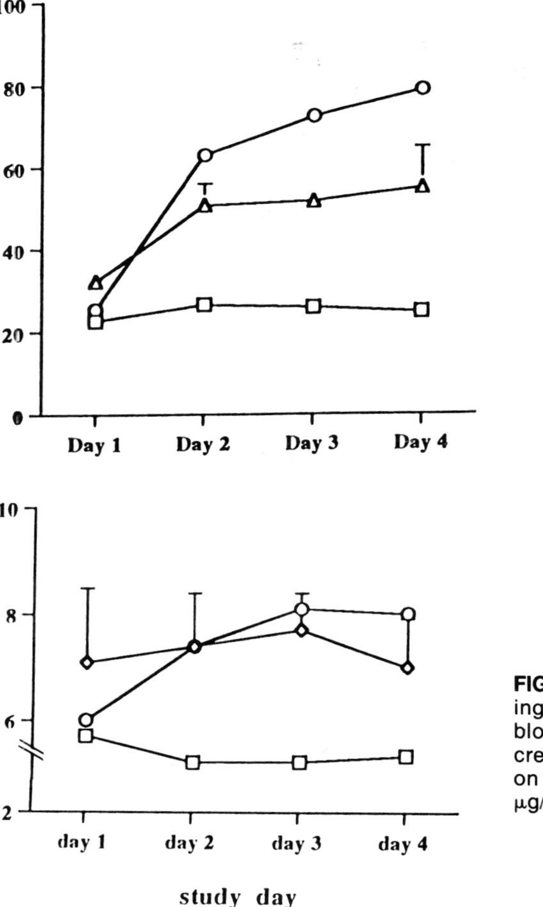


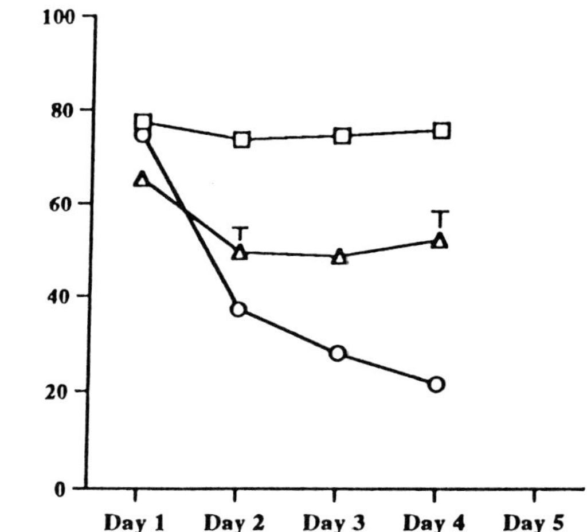


**FIG. 5.** Effects of IL-4 administration on the percentage of circulating neutrophils, lymphocytes, and total leukocytes in peripheral blood. IL-4 administration caused a significant dose-related increase in neutrophils, a decrease in lymphocytes, and had no effect on the total number of leukocytes in peripheral blood. –□–, IL-4 (2.5 μg/kg b.i.d.); –△–, IL-4 (25 μg/kg b.i.d.); –○–, IL-4 (250 μg/kg b.i.d.).

A series of initial studies was performed in cynomolgus monkeys to examine the effects of IL-4 on the expression of adhesion molecules in cutaneous tissues. Intradermal injection of IL-4 caused upregulation of VCAM-1 expression on small vessels, but had little or no effect on ICAM-1 expression. The presence of VCAM-1 was detected by 8 h after injection, persisted for up to 24 h, and was associated with mononuclear cell infiltration into the subcutaneous tissue.

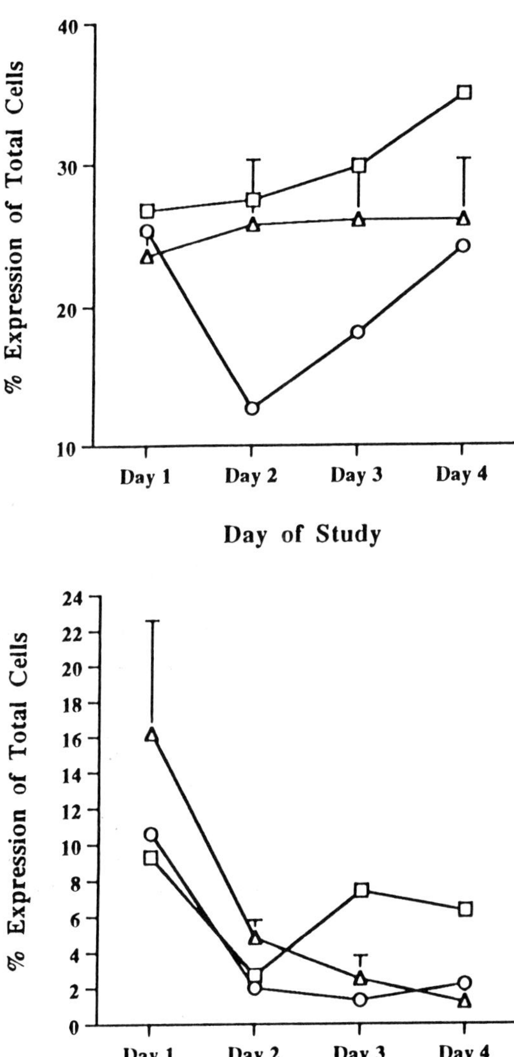


**FIG. 6.** Effects of IL-4 administration on peripheral blood mononuclear cell surface markers. IL-4 caused a transient decrease in $CD4^+$ **(top)** and a sustained decrease in $CD16^+$ cells **(bottom).** –□–, IL-4 (2.5 μg/kg b.i.d.); –△–, IL-4 (25 μg/kg b.i.d.); –○–, IL-4 (250 μg/kg b.i.d.).

In another series of experiments, IL-4 was administered subcutaneously twice a day over a period of 4 days. Circulating leukocytes, PBMC surface markers, and plasma cytokine levels were monitored before the first IL-4 injection and then on a daily basis over the entire 4 days. IL-4 administration led to a dramatic increase in the percentage of circulating neutrophils and a decrease in the percentage of circulating lymphocytes, but had little or no effect on the total number of circulating leukocytes (Fig. 5). These IL-4–induced effects were evident after 1 day of treatment and persisted for the duration of the study. Analysis of cell surface markers revealed that IL-4 induces a transient decrease in $CD4^+$ cells and a sustained decrease in $CD16^+$ cells (Fig. 6). Other cell surface markers, such as CD8, VLA-4, HLA-DR, and CD23, were unaffected during the period of IL-4 administration. These preliminary results with exogenously administered IL-4 suggest that this cytokine is important in the early recruitment of lymphocytes to sites of tissue inflammation by the selective upregulation of VCAM-1 on vascular endothelial cells, thus promoting lymphocyte trafficking out of the vascular space. This series of events may have profound implications for the initiation and subsequent potentiation of allergic inflammation.

## CONCLUSIONS

Cell–cell interactions and communication are facilitated by the expression of cell surface adhesion molecules, and they represent a critical part of the normal functioning of the immune system. The initiation of specific immunologic processes, including the interactions of T lymphocytes with antigen-presenting cells and the stimulation of lymphocyte proliferation and cytokine production, are also critically linked to cell adhesion.

Cell adhesion is pivotal in the process of leukocyte margination and subsequent diapedesis to sites of inflammation. Margination is initiated and directed by adhesion molecules present on or shed from leukocyte cell surfaces, endothelium, and extravascular tissue components under the influence of chemoattractants and cytokines. Adhesion molecules are responsible for the focal adherence of leukocytes, enhancement of their effector cell functions, and their retention in tissue inflammation. The studies described provide evidence to suggest the important role of adhesion molecules in each of these capacities. Furthermore, these studies and others suggest that inhibitors of cellular adhesion may represent a novel form of therapy for acute and chronic inflammatory diseases.

## ACKNOWLEDGMENT

The authors acknowledge the technical assistance of C. A. Torcellini and C. C. Clarke with the primate asthma model, and Maura Macaro and Keith Kelley for the PBMC surface marker data.

## REFERENCES

1. Burney PGJ. Current questions in the epidemiology of asthma. In: Holgate ST, Austen KF, Lichtenstein LM, Key AB, eds. Asthma: physiology, immunopharmacology and treatment. Boston: Academic, 1993:3–16.
2. Nelson HS. Adrenergic therapy of bronchial asthma. *J Allergy Clin Immunol* 1986;77:771–85.
3. O'Byrne PM, Dolovich J, Hargreave FE. Late asthmatic responses. *Am Rev Respir Dis* 1987;136:740–8.
4. Sybert A, Weiss EB. The late asthmatic response. In: Weiss EB, Segal MS, Stein M, eds. *Bronchial asthma: mechanisms and therapeutics*. Boston: Little Brown, 1985:808–40.
5. Larsen GL. The pulmonary late-phase response. *Hosp Pract* 1987;2:155–69.
6. Wright SD, Lo SK, Detmers PA. Specific and regulation of CD-18 dependent adhesions. In: Springer TA, Anderson DC, Rosenthal AS, Rothlein RR, eds. *Leukocyte adhesion molecules: structure, function and regulation,* New York: Springer-Verlag, 1989:190–207.
7. Kohl S, Springer TA, Schmalstieg FC, Loom SL, Anderson DC. Defective natural killer cytotoxicity and polymorponuclear leukocyte antibody-dependent cellular toxicity in patients with LFA-1/OKM-1 deficiency. *J Immunol* 1984;133:2972–8.
8. Springer TA. Adhesion receptors on the immune system. *Nature* 1990;346:425–34.
9. Harlan JN. Leukocyte-endothelial cell interactions. *Blood* 1985;65:513–25.
10. Anderson DC, Schmalstieg FC, Finegold MJ, et al. The severe and moderate phenotypes of heritable Mac-1, LFA-1 deficiency: their quantitative definition and relation to leukocyte dysfunction and clinical features. *J Infect Dis* 1985;152:668–89.
11. Anderson DC, Springer TA. Leukocyte adhesion deficiency: an inherited defect in the Mac-1, LFA-1 and P15095 glycoproteins. *Annu Rev Med* 1987;38:175–94.
12. Martz E, Gromkowski SH. Lymphocyte function-associated antigens: regulation of lymphocyte adhesions in vitro and in vivo. In: Henkart P, Martz E, eds. *Mechanisms of cell-mediated cytotoxicity*. New York: Plenum Press, 1985:1–17.
13. Gundel RH, Wegner CD, Letts LG. Leukocyte-endothelial adhesion. In: Holgate ST, Busse W, eds. *Asthma and rhinitus*. Boston: Blackwell Scientific, 1994:752–63.
14. Albelda SM. Endothelial and epithelial cell adhesion molecules. *Am J Respir Cell Mol Biol* 1991;4:195–203.
15. Larson RS, Springer TA. Structure and function of leukocyte integrins. *Immunol Rev* 1990;114:181–206.
16. Johnson GI, Cook RG, McEver RP. Cloning of GMP-140, a granule membrane protein of platelets and endothelium: sequence similarity to proteins involved in cell adhesion and inflammation. *Cell* 1989;56:1033–44.
17. Pober JS, Gimbrone MA, Lapierre LA, et al. Overlapping patterns of activation of human endothelial cells by interleukin-1, tumor necrosis factor and immune interferon. *J Immunol* 1986;137:1893–6.
18. Munro JM, Pober JS, Cotran RS. Tumor necrosis factor and interferon-gamma induced distinct patterns of endothelial activation and associated leukocyte accumulation in skin of Papio anubis. *Am J Pathol* 1989;135:121–33.
19. Pober JS, Cotran RS. Cytokines and endothelial cell biology. *Physiol Rev* 1990;70:427–43.
20. Wegner CD, Rothlein RR, Gundel RH. Adhesion molecules in the pathogenesis of asthma. *Agents Actions* 1991;34:529–44.
21. Schleimer RP, Sterbinsky SA, Kaiser J, et al. IL-4 induces adherence of human eosinophils and basophils but not neutrophils to endothelium. Associated with expression of VCAM-1. *J Immunol* 1992;148:1086–92.
22. Gundel RH, Gerritsen ME, Wegner CD. Antigen-coated Sepharose beads induce airway eosinophilia and airway hyperresponsiveness in cynomolgus monkeys. *Am Rev Respir Dis* 1989;140:629–33.
23. Gundel RH, Gerritsen ME, Gleich GJ, Wegner CD. Repeated antigen inhalation results in a prolonged airway eosinophilia and airway hyperresponsiveness in primates. *J Appl Physiol* 1990;68:779–86.
24. Gundel RH, Wegner CD, Torcellini CA, Letts LG. Antigen induced acute and late-phase responses in primates. *Am Rev Respir Dis* 1992;146:369–73.
25. Wegner CD, Gundel RH, Reilly P, Hayner N, Letts LG, Rothlein RR. Intercellular adhesion molecule-1 (ICAM-1) in the pathogenesis of asthma. *Science* 1990;247:456–9.

26. Howard M, Farrar J, Hilfiker M, et al. Identification of a T cell derived B cell growth factor distinct from interleukin-2. *J Exp Med* 1982;155:914–8.
27. Paul WE, Ohara J. B-cell stimulatory factor-1/interleukin-4. *Annu Rev Immunol* 1987;5: 429–46.
28. Lebman CA, Lebman A, Coffman RL. Interleukin 4 causes isotype switching to IgE in T-cell stimulated clonal B cell cultures. *J Exp Med* 1988;168:853–6.
29. Oppenheimer-Marks N, Davis LS, Tompkins Bogue C, Ramberg J, Lipsky PE. Differential utilization of ICAM-1 and VCAM-1 during the adhesion and transendothelial migration of human T lymphocytes. *J Immunol* 1991;147:2913–7.

*Topics in Molecular Medicine, Volume 1,*
edited by Wolfgang Siess, Reinhard Lorenz,
and Peter C. Weber. Raven Press, Ltd.,
New York © 1995.

# 23

# Expression of Adhesion Molecules in Human Heart, Liver, and Lung Transplants

Gustav Steinhoff

*Department of Cardiovascular Surgery, Christian Albrechts University of Kiel, 24105 Kiel, Germany*

Endothelial cells in vascularized organ allografts are the regulatory interface with the recipient's immune-reactive cells. They remain of donor type in the long-term course, expressing donor class I and class II MHC molecules (1). In recent years it has become clear that both allospecific MHC-directed and other inflammatory reactions of leukocytes are regulated by the receptor–ligand interaction of a number of cell adhesion molecules (2–6). The presence of certain adhesion molecules on the endothelial cell surface is essential for the regulation of leukocyte–endothelium interactions (cell–cell) and the interaction with basal membrane matrix components (cell–matrix). The cellular expression of both types of molecules is influenced by cytokines. However, a stream bed-specific difference of adhesion ligand molecules on arterial, venous, and capillary endothelia has been described in human liver grafts (7,8). Similar organ-specific differences exist in heart and lung tissue (9). Local differences in leukocyte ligand molecules on endothelia could explain the preferential sites of transplant infiltration and organ-specific peculiarities in susceptibility to the rejection response.

## RESULTS

The results are given as a survey of the expression patterns of adhesion molecules found in human liver, lung, and heart transplant biopsy specimens by standard immunohistology on cryostat sections (6–9). The monoclonal antibodies (MAbs) used are specified in Table 1.

One hundred thirty-four liver transplant biopsy specimens from 36 patients who received 55 transplants were studied. The post-transplant biop-

**TABLE 1.** *Adhesion molecules studied and monoclonal antibodies used*

| Adhesion molecule | Ligand structure | Monoclonal antibody | Producer |
|---|---|---|---|
| Immunoglobulin superfamily | | | |
| CD2 | LFA-3 | 6F10.3;8E6B3 | Dianova[a] |
| LFA-3 (CD58) | CD2 | G26.1;TS2/9 | T. Springer[b] |
| ICAM-I (CD54) | CD11a/b | 84H10;RR1/1;6.5B5 | Rothlein[c]; Haskard[d] |
| ICAM-2 | CD11a | CBR/IC2/1 | Bender Med/Serva[e] |
| VCAM-1 | VLA-4 | 1.4C3 | Dianova[a] |
| NCAM (CD56) | NCAM | T199 | Dianova[a] |
| PECAM-1 (CD31) | ? | 5.E.6 | Dianova[a] |
| Integrin family | | | |
| VLA-1 | Laminin, collagen | TS2/7 | Biermann/T-Cell Sciences[f] |
| VLA-2 | Laminin, collagen | Gi9 | Dianova[a] |
| VLA-3 | Fibronectin, laminin, collagen | P1B5 | Dianova[a] |
| VLA-4 | VCAM-1, fibronectin | HP2/1 | Dianova[a] |
| VLA-5 | Fibronectin | SAM-1 | Dianova[a] |
| VLA-6 | Laminin | GoH3 | Dianova[a] |
| CD51 | Vitronectin, fibrinogen vWF, thrombospondin, fibronectin | AMF/7 | Dianova[a] |
| Selectin family | | | |
| ELAM-1 | CD15, sialyl Lewis$^x$ | 1.2B6;HP18 | Dianova[a]; Becton–Dickinson[g] |
| CD62 | Sialyl Lewis$^x$ | CBL.thromb6 | Dianova |
| Unclassified | | | |
| CD44 | Hyaluronic acid | SBU24-32,F10-44-2<br>Mem-85 | McKenzie[h]; Dalchau[i]<br>Horeysi[j] |
| HECA452 | ? | HECA-452 | Duivestijn[k] |

Source of monoclonal antibodies: [a] Hamburg, Germany; [b,c] Boston, MA, U.S.A.; [d,i] London, England; [e,g] Heidelberg, Germany; [f] Bad Nauheim, Germany; [h] Melbourne, Australia; [j] Prague, Czechoslovakia; [k] Maastricht, The Netherlands.

sies showed various types of rejection- and nonrejection-related pathology. Lung transplant biopsy material was studied in four patients requiring retransplantation (days 11 to 463 post-transplantation) due to acute ($n = 1$) or chronic ($n = 3$) rejection. Heart transplant biopsies ($n = 304$) were studied in 24 patients after orthotopic heart transplantation with different degrees of acute and chronic rejection activity.

The results concerning the expression of adhesion molecules on the endothelial and parenchymal cell types are summarized in Table 2 (immunoglobulin superfamily), Table 3 (integrin matrix receptors), and Table 4 (selectin, CD44, and HECA452). Table 5 summarizes the adhesion receptor patterns found on graft-infiltrating lymphocytes in liver and lung allografts.

## DISCUSSION

The study of liver, heart, and lung transplants showed a cell type- and organ-specific expression of various immune adhesion molecules. In liver, heart, and lung tissue the main immunocompetent cell types, such as interstitial dendritic cells, Kupffer cells, and macrophages express a broad, high-density pattern of adhesion molecules (8). The relative interaction of endothelial and parenchymal cell types in the transplanted organ with cells of the immune system is most likely dependent on their respective expression of intercellular adhesion molecules. The endothelia in different parts of the vascular stream bed inside the liver, heart, and lung display such differences in vascular adhesion molecules, both in basic expression (high in the liver sinusoid and lung capillaries) and in differential inducibility (low in liver sinusoid and lung capillaries). It is conceivable that these differences in immunocompetence result in a fine regulation of leukocyte reactivity inside the vascular stream bed. Furthermore, specific molecule configurations may regulate homing of leukocyte subpopulations to certain vascular compartments. This is particularly possible for the interaction of memory T lymphocytes (CD45RO+) and NK cells that are specifically accumulated in the liver sinusoidal and lung capillaries (10,11).

Organ-specific differences in immunologic susceptibility can be explained by the patterns of basal or induced expression of intercellular adhesion molecules on the different organ cell types. The expression of MHC (1,12, 13) and other adhesion molecules, including cell–matrix receptors is clearly cell type-specific. Despite major differences in the patterns of ligand molecules and their inducibility by cytokines, it can be generally postulated that almost all cell types can be immunologically recognized by immune cells after induction of the main ligand molecules, allo-MHC (class I) and ICAM-1. A relative resistance against immunologic recognition can be caused by deficient basal expression (e.g., on hepatocytes and myocytes) or by a relative resistance against cytokine-mediated induction (myocytes). The composition of such differing immunocompetent cell types in an organ transplant may determine its susceptibility to immune destruction. In general, a minor basal expression of adhesion ligand molecules or an incomplete ligand pattern can be overcome by the action of locally released cytokines from infiltrating leukocytes or tissue macrophages. The susceptibility to cytokine stimulation in the different cell types may also determine patterns of adhesion ligand molecules induced in a specific organ. Therefore, the regulation of the anti-alloantigenic immune response and its manifestation may depend on organ-specific intravascular stimulation (in vascularized organ transplants). On the other hand, the localization and mode of interstitial infiltration may depend on the organ-specific composition of adhesion ligand and cell matrix molecules, leading to differences in stimulation and manifestation of immunoreactivity. The local cellular sensitivity to immune recognition,

**TABLE 2.** *Expression of immunoglobulin superfamily cell–cell ligand*

| | Liver transplants | | | | | | Lung | | |
|---|---|---|---|---|---|---|---|---|---|
| | pa | pv | se | cv | hep | bd | a | ao | c |
| LFA-3 | | | | | | | | | |
| Normal | – | – | + | – | – | – | ++ | ++ | +++ |
| Rejection | + | + | + | + | ++ | ++ | ++ | ++ | +++ |
| ICAM-1 | | | | | | | | | |
| Normal | – | – | ++ | – | – | – | + | +++ | +++ |
| Rejection | + | ++ | +++ | – | ++ | + | ++ | +++ | +++ |
| ICAM-2 | | | | | | | | | |
| Normal | + | – | + | – | – | – | – | +++ | +++ |
| Rejection | ++ | ++ | ++ | ++ | – | – | +++ | +++ | +++ |
| VCAM-1 | | | | | | | | | |
| Normal | – | (+) | – | – | – | – | +++ | – | – |
| Rejection | ++ | + | (+) | + | – | – | +++ | +++ | – |
| NCAM (CD56) | | | | | | | | | |
| Normal | – | – | – | – | – | – | – | – | – |
| Rejection | – | – | + | – | – | + | (+) | – | – |
| PECAM (CD31) | | | | | | | | | |
| Normal | ++ | ++ | ++ | ++ | – | – | +++ | +++ | +++ |
| Rejection | +++ | +++ | +++ | +++ | – | – | +++ | +++ | +++ |

Liver: portal artery, pa; portal vein, pv; sinusoidal endothelium, se; central vein, cv; hepatocyte; hep; bile duct epithelium, bd; Lung: large pulmonary artery, a; arteriole, ao; capillary, c; pulmonary vein, v; pneumocyte, pn; bronchial epithelium, be; Heart: arteriole, ao; capillary, c; vein, v; endocardium, end; myocyte, myo; n.d., not determined.

mainly interpreted by MHC and other adhesion ligand molecule expression, may determine organ-specific differences in cytokine-mediated effector mechanisms and the extent of organ destruction by the rejection response.

The tissue-specific immunogenicity of an organ transplant can be determined by differing patterns of adhesion ligand molecules on endothelial cells in liver, heart, or lung transplants. This could explain differing vascular adhesiveness and activation of leukocytes inside the vascular bed. A special regulation of lymphocyte adhesion must be assumed for lung capillaries and the liver sinusoid. Differences in susceptibility of epithelial and mesenchymal cell types to the immune response can be explained by the degree of basal expression and cytokine-dependent inducibility of immunologic ligand molecules. Organ-specific peculiarities exist in the intravascular or perivascular distribution of cells possessing major immunocompetence by expression of adhesion molecules and other cell functions (interstitial dendritic cells, Kupffer cells, lung macrophages).

## Adhesion Molecules in Ischemia–Reperfusion Injury

An important mechanism in the deterioration of transplant function is the ischemic damage that occurs during organ preservation and the postreper-

*adhesion molecules in human liver, lung, and heart transplants*

| Transplants | | | Heart transplants | | | | |
|---|---|---|---|---|---|---|---|
| v | pn | be | ao | c | v | end | myo |
| ++ | − | + | ++ | + | ++ | + | ++ |
| ++ | − | + | ++ | ++ | ++ | ++ | +++ |
| ++ | +++ | − | ++ | + | ++ | (+) | − |
| +++ | +++ | − | +++ | +++ | ++ | ++ | + |
| ++ | − | − | ++ | + | ++ | − | − |
| +++ | − | − | ++ | ++ | ++ | − | − |
| − | − | − | ++ | − | (+) | (+) | − |
| ++ | ++ | − | +++ | ++ | ++ | ++ | − |
| − | − | − | − | − | − | − | (+) |
| − | (+) | − | − | + | − | − | ++ |
| +++ | − | − | +++ | ++ | +++ | + | − |
| +++ | − | − | +++ | +++ | +++ | + | − |

fusion inflammatory response affecting graft function in the first days after transplantation (14). This implies the possibility of modifying this inflammatory response, which is mainly mediated by leukocytes (14). The immunologic changes that lead to intraoperative induction of adhesion molecules and MHC (7,8,12) in the transplant probably have a major influence on the primary immune reaction against transplantation antigens. The early inducibility of E- and P-selectin as well as VCAM-1 points to an endothelial activation that then may induce leukocyte adhesion. Further steps of firm leukocyte adhesion and infiltration may then be initiated depending on the extent of the post-ischemic reaction and the generation of a cytokine cascade reaction by injured and activated endothelial cells. The extent of adhesion molecule induction may very well relate to the metabolic impairment of the transplant. However, the endothelial activation and injury that result in a clinical disturbance of coagulation factor synthesis, especially in liver transplantation, may induce thrombogenesis in the transplant by direct adhesion ligand induction. Furthermore, inhibitory mechanisms of leukocyte and thrombocyte adhesion, similar to the synthesis of antithrombin III (15), may be disturbed. First, arteriolar or capillary leukocyte stasis induced by the endothelial reperfusion reaction may cause regional microcirculatory perfusion defects (16). Their additional ischemic consequences may potentiate the tissue damage, even after 1 or 2 days, if additional ligand molecules such as ICAM-1 and MHC are induced. Second, the first-passage sensitization of alloreactive T lymphocytes and the induction of rejection activity may very well be influenced by the initial post-ischemic inflammatory reaction. This has been observed for the early induction of class I and class II MHC

**TABLE 3.** *Expression of integrin cell matrix adhesion*

| | Liver transplants | | | | | | Lung | |
|---|---|---|---|---|---|---|---|---|
| | pa | pv | se | cv | hep | bd | a | ao |
| VLA-1 | | | | | | | | |
| Normal | ++ | ++ | ++ | ++ | + | − | +++ | +++ |
| Rejection | +++ | +++ | +++ | +++ | ++ | − | +++ | +++ |
| VLA-2 | | | | | | | | |
| Normal | (+) | (+) | − | − | − | + | − | − |
| Rejection | + | + | (+) | + | (+) | ++ | + | ++ |
| VLA-3 | | | | | | | | |
| Normal | + | + | (+) | − | − | + | ++ | ++ |
| Rejection | ++ | ++ | ++ | − | (+) | ++ | ++ | ++ |
| VLA-4 | | | | | | | | |
| Normal | − | − | (+) | − | − | − | − | − |
| Rejection | − | − | + | − | − | − | − | − |
| VLA-5 | | | | | | | | |
| Normal | ++ | ++ | ++ | ++ | + | − | +++ | +++ |
| Rejection | +++ | +++ | +++ | +++ | ++ | − | +++ | +++ |
| VLA-6 | | | | | | | | |
| Normal | ++ | ++ | − | − | − | + | − | ++ |
| Rejection | ++ | ++ | ++ | − | + | ++ | +++ | +++ |
| CD51 | | | | | | | | |
| Normal | ++ | ++ | ++ | − | − | + | +++ | ++ |
| Rejection | +++ | +++ | +++ | ++ | ++ | ++ | +++ | +++ |

Abbreviations as in Table 2.

molecules (17) and may also relate to the coinduction of other adhesion ligand molecules. Clinical observations of an increased rejection rate in initial poorly functioning grafts point in this direction (18). In experimental research, the post-ischemic reperfusion damage could be inhibited by anti-CD11/CD18 MAbs (19,20) and by prostaglandin E1 (21). This may open possibilities to modify the reaction clinically by the use of such tools. It can be assumed, however, that the selectin molecules have central importance in the induction phase of inflammation.

## Adhesion Molecules in Transplant Rejection

The understanding of the pathophysiology of transplant rejection has undergone major changes since the analysis of molecular alterations in transplants. Until recently the molecular mechanisms of transplant infiltration were somewhat unclear, and morphologic and functional knowledge of the contributing effector cell populations remained in the foreground of interest (22). It was not then known that on transplant cells, especially on endothelial and parenchymatous cell types, immunologic reactions may occur to the extent now recognized. Starting with the observations of Lampert et al. (23) and De Waal et al. (24), changes in MHC expression on parenchymatous *molecules in human liver, lung, and heart transplants*

| Transplants | | | | Heart transplants | | | |
|---|---|---|---|---|---|---|---|
| c | v | pn | be | a | c | v | myo |
| +++ | +++ | − | − | − | ++ | ++ | − |
| +++ | +++ | − | − | − | +++ | +++ | − |
| ++ | − | − | ++ | ++ | − | (+) | − |
| ++ | (+) | ++ | ++ | ++ | − | + | − |
| +++ | + | +++ | +++ | +++ | − | ++ | − |
| ++ | ++ | ++ | +++ | ++ | + | +++ | − |
| − | − | − | − | − | − | (+) | − |
| − | − | − | − | − | − | + | − |
| +++ | +++ | − | − | − | − | ++ | − |
| +++ | +++ | ++ | +++ | +++ | ++ | +++ | − |
| +++ | ++ | − | ++ | ++ | ++ | ++ | − |
| +++ | +++ | − | + | +++ | +++ | +++ | − |
| ++ | ++ | ++ | +++ | n.d. | | | |
| ++ | ++ | | ++ | n.d. | | | |

transplant cells were found. Although this appeared at first to explain the effector mechanism of reactive T lymphocytes on MHC-incompatible transplant cells, the basal and normal expression of class I and class II MHC molecules, especially on endothelial cells in a transplant without major rejection activity, pointed to the existence of additional binding molecules. Furthermore, the unspecific co-infiltration of other leukocyte subpopulations in cases of transplant rejection could not be unexplained on the molecular level. By identification of molecular families of accessory adhesion molecules and their functions in both specific and unspecific lymphocyte–leukocyte reactivity, various mechanisms in the recognition of cells and the genesis of tissue infiltration have been unraveled. Although at present the exact sequence of adhesive cellular interactions is not completely clarified in vivo, it is possible to draw conclusions concerning their functional order on the basis of their sequential changes in distribution pattern during transplant rejection.

A central event in the antigen-specific reaction of $CD4^+$ T-helper lymphocytes is the counterexpression of class II (HLA-DR) MHC molecules. These are present or may be induced by cytokines on many cells types in liver, heart, or lung transplants. Stimulation of T lymphocytes by the various cell types, however, may have different effects. This could have a basis in the fact that additional co-stimulatory ligand molecules, such as B7 (BB7 ligand to CD28), LFA-3, ICAM 1 and 2, as VCAM-1, are variably expressed

**TABLE 4.** *Expression of selectin and unclassified adhesion molecules in human lung and heart transplants*

| | Liver transplants | | | | | | Lung transplants | | | | | | Heart transplants | | | |
|---|---|---|---|---|---|---|---|---|---|---|---|---|---|---|---|---|
| | pa | pv | se | cv | hep | bd | a | ao | c | v | pn | be | a | v | c | myo |
| E-selectin (ELAM-1) | | | | | | | | | | | | | | | | |
| Normal | − | + | − | − | − | − | (+) | ++ | − | − | − | − | (+) | − | − | − |
| Rejection | + | + | − | (+) | − | − | − | − | − | − | − | − | ++ | + | ++ | − |
| P-selectin (CD62, GMP140) | | | | | | | | | | | | | | | | |
| Normal | (+) | + | − | − | − | − | ++ | − | − | ++ | − | − | (+) | − | − | − |
| Rejection | + | + | − | + | − | − | +++ | +++ | − | ++ | − | − | ++ | − | + | − |
| CD44 | | | | | | | | | | | | | | | | |
| Normal | − | − | − | − | − | − | + | +++ | +++ | +++ | +++ | +++ | − | − | − | − |
| Rejection | − | (+) | (+) | (+) | − | − | + | +++ | +++ | +++ | +++ | +++ | − | (+) | − | − |
| HECA452 | | | | | | | | | | | | | | | | |
| Normal | − | − | − | − | + | (+) | − | − | − | − | − | − | n.d. | | | |
| Rejection | − | + | + | − | ++ | ++ | − | − | − | ++ | +++ | +++ | n.d. | | | |

Abbreviations as in Table 2.

**TABLE 5.** *Intravascular and interstitial expression of lymphocyte adhesion receptors in human liver and lung transplants*

| Adhesion molecule | Intravascular[a] Normal | Rejection | Others[b] | Tissue infiltrate Normal | Rejection | Others |
|---|---|---|---|---|---|---|
| CD2 | + | + | + | + | + | + |
| LFA-3 | − | + | + | − | + | + |
| ICAM-1 | − | + | + | + | + | + |
| ICAM-2 | − | − | − | − | − | − |
| VCAM-1 | − | − | − | − | − | − |
| NCAM | +[c] | + | + | + | + | + |
| CD-31(PECAM) | (+) | + | + | − | (+) | − |
| LFA-1 | + | + | + | + | + | + |
| CD11b | − | − | − | − | − | − |
| CD11c | − | − | − | − | − | − |
| VLA-1 | −[d] | − | − | − | +[e] | − |
| VLA-2 | − | − | − | − | +[f] | − |
| VLA-3 | − | − | − | − | − | − |
| VLA-4 | + | + | + | + | + | + |
| VLA-5 | − | − | − | − | + | + |
| VLA-6 | − | − | − | − | − | − |
| CD51 | − | − | − | − | + | − |
| LECAM/Leu8 | (+) | + | (+) | − | + | (+) |
| ELAM-1 | − | − | − | − | (+) | − |
| CD62 | − | − | − | − | − | − |
| CD44 | − | + | + | − | + | + |
| HECA452 | − | (+) | (+) | − | + | − |

[a] Intravascular, endothelial adherent leukocytes; interstitial, portal infiltrating leukocytes.
[b] Cholangitis, sepsis, viral infection.
[c] NK cells.
[d] No expression found.
[e] Only found in late/advanced rejection.
[f] Only found in lung transplant rejection.

on different cell types and are coexpressed in high density mainly on the immunocompetent interstitial dendritic cells and monocytes–macrophages. Therefore, the order of response of different cell types to T lymphocytes may depend on the presence of a panel of adhesion ligand molecules. In addition, these cell types can produce stimulatory cytokines such as interleukin 1 and γ-interferon on binding of T lymphocytes. Moreover, intracellular functions, such as cleavage and presentation of antigen-peptides with MHC molecules, determine their immunologic interactivity. For the target cell interaction and cytolytic process of $CD8^+$ cytotoxic T lymphocytes, the presence of the target antigen, alloantigenic MHC molecules, is also a prerequisite. However, additional binding structures such as ICAM-1 and LFA-3 may also be required to induce a cascade reaction leading to cytolysis. Therefore, the presence, or rather the inducibility, of certain immune ligand molecules in an organ transplant is a prerequisite for the genesis of alloantigen-directed T-lymphocyte sensitization and the manifestation of an effector reaction.

## Immunosuppressive Potential of Adhesion Molecules

For the specific and unspecific immunosuppression of the anti-allogenic immune response after transplantation, a number of new aspects are appearing. Clinical immunosuppression after liver, heart, and lung transplantation consists of two main components: specific inhibition of T-lymphocyte activation by abrogation of interleukin-2–mediated stimulation (cyclosporin A, FK506, Rapamycin) or a short-term elimination of lymphocytes by antilymphocyte sera or MAbs (anti-CD3, anti-IL2rec). B-Lymphocyte activity is suppressed by azathioprine. Further unspecific leukocyte activation and adhesion are suppressed by corticosteroids. These are likely to interact with adhesion molecule-dependent processes of leukocyte interaction. The rapid effect on all leukocyte subpopulations by the classical steroid bolus rejection treatment supports this assumption. Thus far, however, it is unclear which molecular mechanisms are interrupted by corticosteroid therapy. Experimental studies using anti-ICAM-1 (25,26), anti-CD11/CD18 MAbs (19,20,27, 28), and anti–VLA-4 (29,30) support the central role of these molecular interactions in the infiltration process. For clinical purposes in the modification of immunosuppression, however, the use of such MAbs may be of lesser future importance than the identification of anti-adhesive capacities of peptides, oligosaccharides (31), and other drug agents.

Of particular relevance for the treatment of chronic transplant rejection are potential antiinflammatory therapies (32). Analysis of adhesion molecules in chronic liver (8), heart (33), and lung transplant rejection (9) reveals endothelial activation with increased expression of adhesion ligand molecules such as ICAM-1, VCAM-1, and MHC class II. Similar findings have been observed in renal transplants (34,35). The triggering mechanisms are speculative but may relate to antibody binding to endothelial alloantigens and local cytokine release of reactive memory T lymphocytes (36). Furthermore, general effects of altered organ homeostasis, such as vascular effects of flow-related shear stress on endothelial cells in a denervated organ (heart) and opportunistic viral infections, must be considered as possible triggering events. Although T-lymphocyte activation is usually sufficiently (immuno-)suppressed, endothelial activation in the transplant itself can recruit other leukocytes and thus maintain a chronic inflammatory reaction. The increased expression of adhesion molecules with endothelial activation and the induced intravascular adhesiveness in the transplant may play central roles in this pathomechanism (37). Although closer analysis of the pathomechanisms is still awaited, a main target in the treatment of chronic rejection is the inhibition of endothelial activation and increased adhesiveness. A number of possibilities are undergoing experimental and clinical evaluation (32).

A special aspect of new approaches to immunosuppression and immunomodulation arises from the biologic models of soluble forms of adhesion molecules such as ICAM-1, LFA-3, E- and P-selectin and MHC (38–40). These molecules, in addition to cytokines, may have major functions in the intravascular regulation of the immune response. The processes of intravascular leukocyte activation and depression may be influenced by the release and interaction of adhesion ligand peptide molecules. It is possible that the intravascular binding of soluble ICAM-1 or LFA-3 peptides to their respective receptors on leukocytes leads to a change in expression density or receptor conformation, either increasing or decreasing the state of intercellular affinity (leukocyte–thrombocyte, leukocyte–leukocyte, leukocyte–endothelium). It is also possible that intravascular or interstitial blockade of interactive adhesion receptors by soluble adhesion ligands downregulates target cell binding. In this context, the identification of various molecular isoforms (38,40) may point to a differentiated use of these for regulation of intercellular contacts. Similar effects have been postulated for the binding of soluble alloantigenic MHC to the T-cell receptor. For alloantigenic organ transplants, effects of soluble MHC in previous blood transfusions or simultaneous organ transplants (liver–kidney, heart–lung) may also modify the T-lymphocyte response. The use of soluble adhesion molecules or special blocking peptides is likely to open new avenues for manipulation of the immune response. This not only may involve organ transplant-related questions, such as the induction of tolerance or modification of chronic rejection, but also may have relevance to the treatment of several nontransplant inflammatory diseases and post-ischemic reactions. In this regard, the human liver may be of special interest as a main producer of soluble adhesion molecules in the bloodstream (ICAM-1, MHC/$\beta_2$ microglobulin). These may exert physiologic functions in the acute-phase reaction similar to those of complement and coagulation factor synthesis.

## Adhesion Molecules and Transplant Inflammation

Our knowledge of the membrane-bound local regulation of leukocyte and lymphocyte reactivity by activation of the transplant endothelium and of the possible systemic immune activity by the secretion of soluble adhesion molecules (MHC, ICAM-1, ICAM-2, ELAM-1) and cytokines (IL-1, TNF$\alpha$) by the transplant endothelium indicates that immunologic changes within an organ transplant are at the center of the regulation of the immune response. The endothelial cell and, in the liver, the Kupffer cell probably play an active and previously unrecognized role in the modification of immune reactions. These functions probably are only partially influenced by immunosuppression therapy. Organ-specific and general conclusions about the regulation of rejection and other inflammatory reactions can be drawn with new insight. An important factor in the generation of tissue inflammation is the key role of the binding of LFA-1/MAC-1 to ICAM-1 and of VLA-4 to VCAM-1. This could explain the synergistic inflammatory actions of different leukocyte subpopulations. Furthermore, it is of importance that different pathologic stimuli, e.g., viral infection (of leukocytes or transplant cells) or rejection of the transplant, may influence each other by way of identical inflammatory pathways. This may be of particular relevance to chronic inflammatory transplant disease of different etiologies.

Analysis of adhesion molecules in the transplant may be of clinical diagnostic interest. The agreement of induction phenomena in different rejection- and nonrejection–related inflammatory reactions reveals general mechanisms of the inflammatory response. Therefore, additional information about the kind of pathologic stimulus cannot be expected from such an analysis. However, information about the inflammatory state of the transplant can be obtained. An example of this is the expression and induction of vascular adhesion ligand molecules. These may point to subclinical and clinical pathologic activation states. The single activation of sinusoidal endothelial cells, for example, without coactivation of hepatocytes, may point to a systemic stimulus. Such information may help in the assessment of the risk factors for a given organ transplant as a result of endothelial activation (e.g., viral infection, chronic rejection). This may influence the choice of an immunosuppressive and anti-inflammatory drug regimen. A prospect for anti-inflammatory intervention is a direct approach to adhesion molecules by oligosaccharide or peptide-receptor blockade. In the near future, a close connection between the diagnosis of cellular changes in adhesion molecules or their excretion in soluble form in the serum and therapeutic modulation can be expected. A further prospect can be deduced from knowledge about the induction of cell matrix integrin receptors on transplant cells during inflammation. This could provide a way for diagnosis and intervention when fibrogenetic changes are present in the transplant. At present it is unclear which cytokines or inducers change integrin-receptor expression on the different cell types and how this relates to fibrogenetic inflammatory activity. Further investigation, however, may reveal the fibrogenetic activity of different cell types such as hepatocytes, myocytes, and pneumocytes. Further in vitro studies may clarify the interaction of cytokine-induced fibrogenesis and membrane or intracellular changes in different transplant cell types mediated by cell matrix integrin receptors.

## REFERENCES


1. Steinhoff G, Wonigeit K, Schäfers HJ, Haverich A. Sequential analysis of monomorphic and polymorphic major histocompatibility complex antigen expression in human heart allograft biopsy specimens. *J Heart Transplant* 1989;8:360–70.
2. Springer TA. Adhesion receptors of the immune system. *Nature* 1990;346:425–34.
3. Hogg N, Harvey J, Cabanas C, Landis RC. Control of leukocyte integrin activation. *Am Rev Respir Dis* 1993;148:55–9.
4. Springer TA, Lasky TA. Sticky sugars for selectins. *Nature* 1991;349:196–7.

5. Hynes RO. Integrins: versatility, modulation, and signaling in cell adhesion. *Cell* 1992;69: 11–25.
6. Steinhoff G. *Cell adhesion molecules in human organ transplants*. Austin, TX; RG Landes, 1993:1–110.
7. Steinhoff G, Behrend M, Schrader B, Duijvestijn AM, Wonigeit K. Expression patterns of leucocyte adhesion ligand molecules on human liver endothelia: lack of ELAM-1 and CD62 inducibility on sinusoidal endothelia and distinct distribution of VCAM-1, ICAM-1, ICAM-2, and LFA-3. *Am J Pathol* 1993;142:481–8.
8. Steinhoff G, Behrend M, Schrader B, Pichlmayr R. Intercellular immune adhesion molecules in human liver transplants—overview on expression patterns of leukocyte receptor and ligand molecules. *Hepatology* 1993;18:440–53.
9. Steinhoff G, Behrend M, Richter N, Schlitt HJ, Cremer J, Haverich A. Distinct expression of cell-cell and cell-matrix adhesion molecules on endothelial cells in human heart and lung transplants. *J Heart Lung Transplant [in press]*.
10. Volpes R, van den Oord JJ, Desmet VJ. Memory T-cells are the predominant lymphocyte subset in acute and chronic liver inflammation. *Hepatology* 1990;12:826–9.
11. Schlitt HJ, Raddatz G, Steinhoff G, Wonigeit K, Pichlmayr R. Passenger lymphocytes in human liver allografts and their potential role after transplantation. *Transplantation* 1993; 56:951.
12. Steinhoff G, Wonigeit K, Pichlmayr R. Analysis of sequential changes in major histocompatibility complex expression in human liver grafts after transplantation. *Transplantation* 1988;45:394–401.
13. Taylor PM, Rose ML, Yacoub MH. Expression of MHC antigens in normal human lung and transplanted lungs with obliterative bronchiolitis. *Transplantation* 1989;48:506–10.
14. Clavien PA, Harvey PRC, Strasberg SM. Preservation and reperfusion injuries in liver allografts. *Transplantation* 1992;53:957–78.
15. Absher E, Labarrere CA, Carter C, Haag B, Page Faulk W. The endothelial heparan sulfate-antithrombin III natural anticoagulant pathway in normal and transplanted human kidneys. *Transplantation* 1992;53:828–34.
16. Marzi I, Walcher F, Menger M, Bühren V, Harbauer G, Trentz O. Microcirculatory disturbances and leucocyte adherence in transplanted livers after cold storage in Euro-Collins, UW and HTK solutions. *Transplant Int* 1991;4:45–50.
17. Shackleton CR, Ettinger SL, McLoughlin MG, Scudamore CH, Miller R, Keown PA. Effect of recovery from ischemic injury on class I and class II MHC antigen expression. *Transplantation* 1990;49:641–4.
18. Howard TK, Klintmalm GB, Cofer JB, Huberg BS, Goldstein RM, Gonwa TA. The influence of preservation injury on rejection in the hepatic transplant recipient. *Transplantation* 1990;49:103–7.
19. Horgan MJ, Wright SD, Malik AB. Antibody against leukocyte integrin (CD18) prevents reperfusion-induced lung vascular injury. *Am J Physiol* 1990;259:315–9.
20. Simpson P, Todd JRF III, Fantone JC, Mickelson JM, Griffin JD, Lucchesi BR. Reduction of experimental canine myocardial reperfusion injury by a monoclonal antibody (anti-Mo1, anti-CD11b) that inhibits leukocyte adhesion. *J Clin Invest* 1988;81:624–9.
21. Simpson PJ, Michalson J, Fatane JC, Gallagher KP, Lucchesi BR. Reduction of experimental canine myocardial infarct size with prostaglandin E1: inhibition of neutrophil migration and activation. *J Pharmacol Exp Ther* 1988;244:619–24.
22. Häyry P. Intragraft events in allograft destruction. *Transplantation* 1984;38:1–6.
23. Lampert IA, Suitters AJ, Chilsom PM. Expression of Ia antigen in epidermal keratinocytes in graft-versus-host disease. *Nature* 1981;293:149–50.
24. De Waal RMW, Bogman MJJ, Maass CN, et al. Variable expression of Ia antigens on the vascular endothelium of mouse skin allografts. *Nature* 1983;303:426–9.
25. Cosimi AB, Conti D, Delmonico FL, et al. In vivo effects of monoclonal antibody to ICAM-1 (CD54) in non-human primates with renal allografts. *J Immunol* 1990;144:4604–12.
26. Flavin T, Ivens K, Rothlein R, et al. Monoclonal antibodies against intercellular adhesion molecule 1 prolong cardiac allograft survival in cynomolgus monkeys. *Transplant Proc* 1991;23:533–4.
27. Heagy W, Waltenbaugh C, Martz E. Potent ability of anti LFA-1 monoclonal antibody to prolong allograft survival. *Transplantation* 1984;37:520–3.

28. Kaslovsky RA, Horgan MJ, Lum H, et al. Pulmonary edema induced by phagocytosing neutrophils—protective effect of monoclonal antibody against phagocyte CD18 integrin. *Circ Res* 1990;67:795–802.
29. Paul LC, Davidoff A, Paul DW, Bendiktsson H, Issekutz TB. Monoclonal antibodies against LFA-1 and VLA-4 inhibit graft vasculitis in rat cardiac allografts. *Transplant Proc* 1993;25:813–4.
30. Orosz CG, Ohye RG, Pelletier RP, et al. Treatment with anti-vascular cell adhesion molecule 1 monoclonal antibody induces long-term murine cardiac allograft acceptance. *Transplantation* 1993;56:453–60.
31. Mulligan MS, Paulson JC, De Frees S, Zheng ZL, Lowe JB, Ward PA. Protective effects of oligosaccharides in P-selectin-dependent lung injury. *Nature* 1993;364:149–51.
32. Paul LC, Fellström B. Chronic vascular rejection of the heart and the kidney—have rational treatment options emerged? *Transplantation* 1993;53:1169–79.
33. Steinhoff G, Behrend M, Haverich A. Signs of endothelial activation in human heart allograft biopsies. *Eur Heart J* 1991;12:141–3.
34. Brockmeyer C, Ulbrecht M, Schendel DJ, et al. Distribution of cell adhesion molecules (ICAM-1, VCAM-1, ELAM-1) in renal tissue during allograft rejection. *Transplantation* 1993;55:610–15.
35. Fuggle SV, Sanderson JB, Gray DWR, Richardson A, Morris PJ. Variation in expression of endothelial adhesion molecules in pretransplant and transplanted kidneys—correlation with intragraft events. *Transplantation* 1993;55:117–23.
36. Damle NK, Eberhardt C, Van der Vieren M. Direct interaction with primed CD4$^+$ CD45RO$^+$ memory T-lymphocytes induces expression of endothelial leukocyte adhesion molecule-1 and vascular cell adhesion molecule-1 on the surface of vascular endothelial cells. *Eur J Immunol* 1991;21:2915–23.
37. Duijvestijn AM, Van Breda Vriesman PJC. Chronic renal allograft rejection: selective involvement of the glomerular endothelium in humoral immune reactivity and intravascular coagulation. *Transplantation* 1991;52:195–202.
38. Seth R, Raymond FD, Makgoba MW. Circulating ICAM-1 isoforms: diagnostic prospects for inflammatory and immune disorders. *Lancet* 1991;338:83–4.
39. Davies HS, Pollard SG, Calne RY. Soluble HLA antigens in the circulation of liver graft recipients. *Transplantation* 1989;47:524–7.
40. Haynes BF, Telen MJ, Hale JP, Denning SM. CD44—a molecule involved in leukocyte adherence and T-cell activation. *Immunol Today* 1989;10:423–8.

*Topics in Molecular Medicine, Volume 1,*
edited by Wolfgang Siess, Reinhard Lorenz,
and Peter C. Weber. Raven Press, Ltd.,
New York © 1995.

# 24

# Induction of Specific Immunological Tolerance to Transplanted Organs by Blocking Cell Adhesion

Mitsuaki Isobe and Jun-ichi Suzuki

*The First Department of Internal Medicine, Shinshu University School of Medicine, Matsumoto 390, Japan*

Burnet and Medawar were awarded the 1960 Nobel Prize for their discovery of immunologic tolerance. They showed that the injection of allogeneic cells into a newborn mouse induces lifelong immunologic tolerance for the donor's tissues and organs (1). Since then, a great deal of investigation has attempted to achieve a similar effect in adult animals and in humans. It is crucial to understand the mechanism of the immune system that distinguishes between "self" and "non-self." This knowledge would enable us to develop effective and specific immunologic tolerance, which is necessary in treating autoimmune diseases and managing organ rejection. Therefore, one of the ultimate goals of transplantation immunobiology is understanding of the mechanism of immunologic response to allogeneic antigens and development of ways to induce specific immunologic tolerance to transplanted organs in humans.

It is obvious that cellular adhesion molecules play significant roles in biologic reactions, since blockade of cell adhesion by monoclonal antibodies (MAbs) sometimes produces drastic biologic responses. A great deal of progress has been made in elucidating the roles of cellular adhesion molecules in eliciting immune responses. One of the most interesting targets of these investigations is transplantation immunology. We have found that temporal blockade of cell adhesion leads to specific immunologic tolerance in murine models (2). Results obtained from our animal experiments show that this tolerance induction is unique in many aspects and is potentially applicable to clinical immunosuppression.

## INDUCTION OF CELL ADHESION MOLECULES IN ORGAN REJECTION

Several adhesion molecules, such as intercellular adhesion molecule-1 (ICAM-1), vascular cell adhesion molecule-1 (VCAM-1), and E-selectin

(ELAM-1) (3), are induced in association with acute rejection of transplanted organs. ICAM-1 is normally expressed on vascular endothelial cells, fibroblasts, epithelial cells, macrophages, and activated lymphocytes. In addition, there is limited expression of VCAM-1 in the vascular endothelium of normal cardiac tissue. However, immunohistochemical studies of transplanted allografts revealed that ICAM-1 and VCAM-1 are strongly induced on the vascular endothelium and on cardiac myocytes from the early stages of graft rejection in human transplants or the animal model of transplantation (4–9), as shown in Fig. 1. Therefore, it is obvious that ICAM-1/LFA-1 or VCAM-1/VLA-4 interaction participates in the pathophysiology of allograft rejection.

Induction of the ICAM-1 molecule can be monitored and quantitated by noninvasive methods. In addition to measurement of the soluble form of

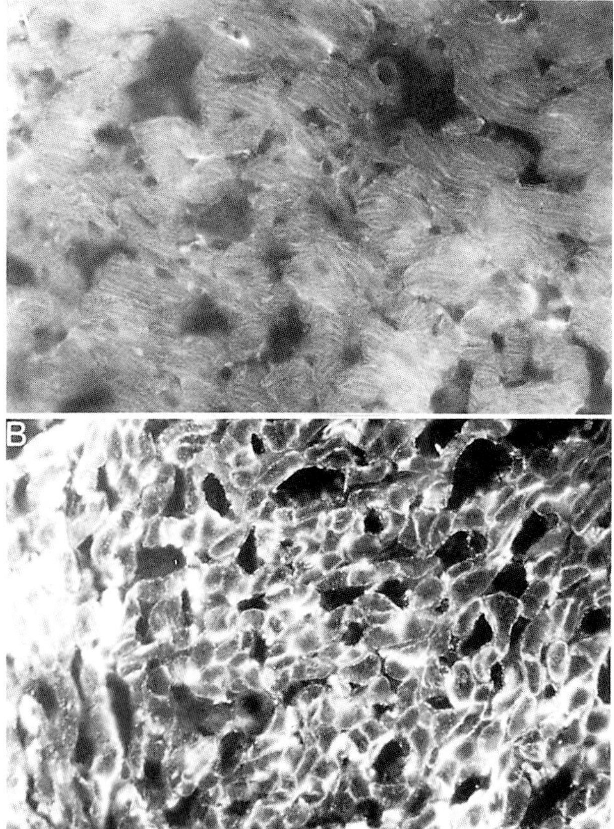


**FIG. 1.** Immunohistochemistry of rat cardiac allografts stained with a monoclonal antibody to the rat ICAM-1. **A:** Isografted rat heart 7 days after transplantation. Positive ICAM-1 expression is observed mainly on the vascular endothelium. **B:** Allografted rat heart 6 days after transplantation. Note the ICAM-1 on the cardiac myocytes.

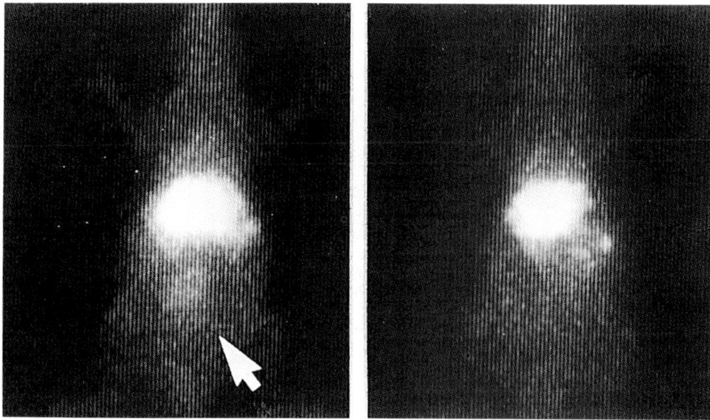

**FIG. 2.** Radioimmune scintigraphy using [$^{111}$In]DTPA anti-rat ICAM-1 MAb. One hundred μCi of radiotracer was injected 1 day before data acquisition. **Left:** Nontreated allograft at day 6. Arrow indicates heterotopic cardiac allograft with acute rejection. **Right:** Cardiac allograft in rat treated with FK506. Cardiac allograft without rejection could not be identified on the image.

ICAM-1, radioimmune scintigraphy targeting ICAM-1 induction has been reported (10,11). Rats transplanted with cardiac allografts were injected with $^{111}$In-labeled anti-rat ICAM-1 MAb (1A29) and γ-scintigraphy was performed 1 day after the injection. As shown in Fig. 2, rejecting allografts showed increased radiotracer uptake and could be identified on the images as early as 5 days after transplantation. In contrast, nonrejecting cardiac allografts and isografts did not show specific uptake. Mildly rejecting allografts with mononuclear cell infiltration could be scintigraphically identified, and the level of radiotracer uptake reflected the histologic severity of rejection. Therefore, radioimmune scintigraphy is a sensitive method for early detection and assessment of cardiac allograft rejection. This new technique could also be potentially effective in detecting various immunologic disorders accompanying ICAM-1 induction, such as autoimmune diseases and inflammatory diseases.

## TOLERANCE INDUCTION TO CARDIAC ALLOGRAFTS

### Experimental Protocol

Heterotopic heart transplantation was performed between C3H/He (H-$2^k$) donors and BALB/c (H-$2^d$) recipients (12,13). Because these two murine strains are fully immune-incompatible, donor hearts are rejected within 10 days in nontreated recipients. The MAbs used in this study were anti-mouse ICAM-1 (YN1/1.7, $IgG_{2b}$) (14), anti-mouse LFA-1 (KBA, $IgG_{2a}$) (15), and anti-mouse CD18 (M18/2, $IgG_{2a}$) (16). YN1/1.7 or KBA inhibits cell adhesion

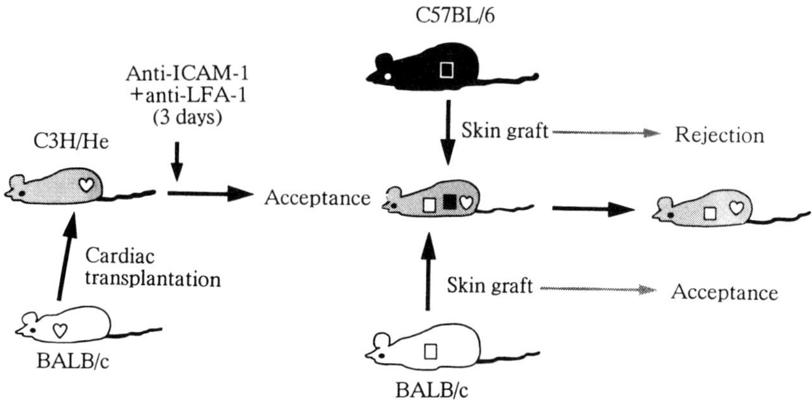


**FIG. 3.** Outline of the animal experiment.

in vitro. M18/2 was used as a control antibody because this antibody does not block cell adhesion in vitro (non-neutralizing antibody). Antibodies were injected intraperitoneally from immediately after cardiac transplantation until the third day after transplantation (Fig. 3).

### Survival of Cardiac Grafts and Challenge Test with Skin Grafts

As shown in Fig. 4, cardiac allografts treated with 50 μg each of anti–ICAM-1 and anti–LFA-1 MAbs were accepted indefinitely without excep-

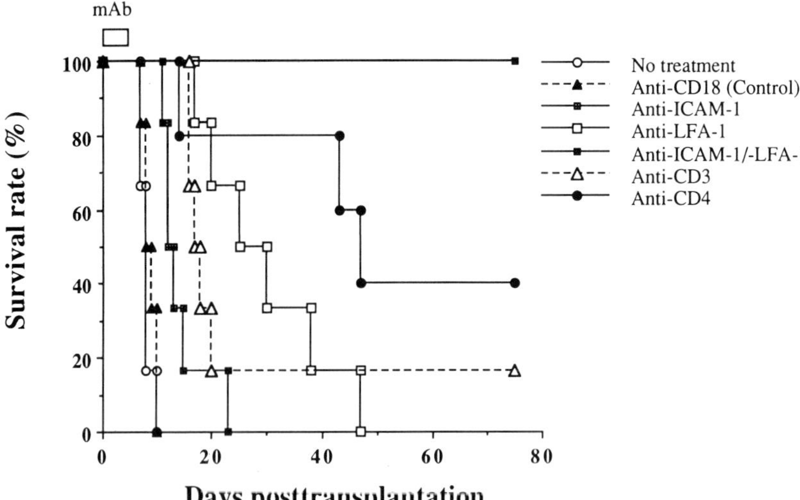


**FIG. 4.** Survival of cardiac allografts after treatment with monoclonal antibodies. One hundred micrograms of MAb was administered starting just after transplantation until day 5 after operation. All cardiac allografts treated with a combination of anti–ICAM-1 and anti–LFA-1 were accepted indefinitely ($n$ = 20).

tion, in contrast to mice without treatment, in which cardiac allografts were rejected within 10 days. This effect in prolonging graft survival was better than for anti-CD3 or anti-CD4 MAb given in the same protocol. Another group of mice received the same dose of MAbs from the fifth day after transplantation. Even though rejection with myocyte necrosis is usually observed at this stage, more than half of these mice accepted cardiac allografts. Therefore, these MAbs are able to reverse ongoing rejection. Histopathologic examination of cardiac allografts at 75 days after transplantation showed scattered areas of mononuclear cell infiltration but myocytes were essentially free from necrosis (Fig. 5). The grafted

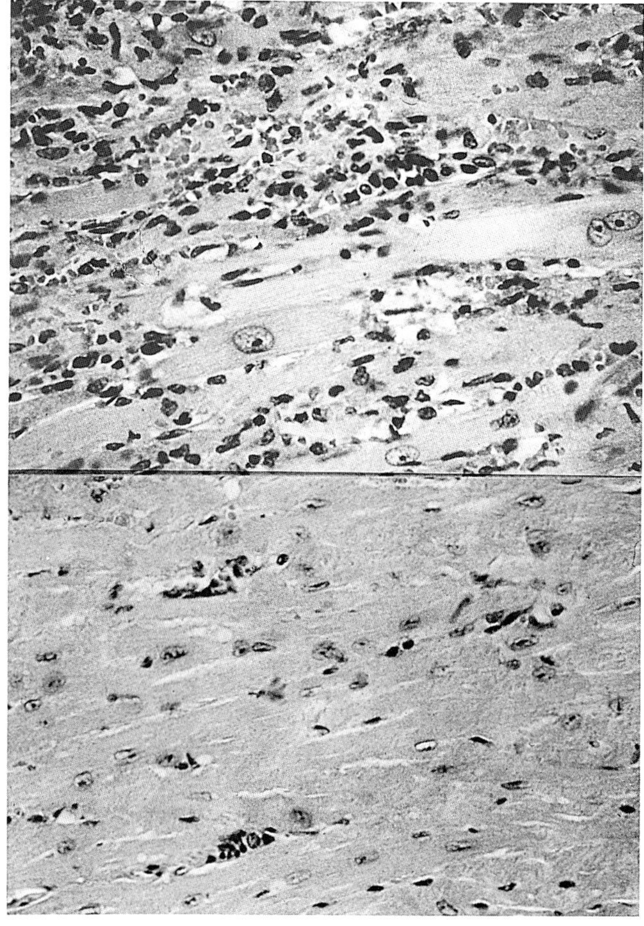

**FIG. 5.** Histologic analysis of heart allograft (HE staining). **Upper panel:** A control allograft 7 days after transplantation with no immunosuppressive treatment. Massive infiltration of leukocytes with myocyte necrosis and interstitial hemorrhage are noted. **Lower panel:** An allograft 75 days after transplantation from a recipient treated with MAbs to ICAM-1 and LFA-1. No sign of active rejection is seen. Magnification ×200.

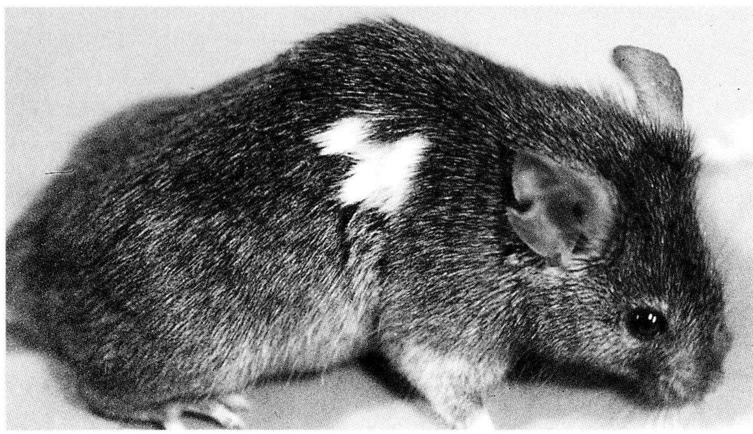

**FIG. 6.** A mouse with a cardiac allograft and 5-day course of MAbs to ICAM-1 and LFA-1, showing acceptance of the second skin graft from the donor strain.

hearts kept beating as long as the recipient mouse survived (more than 365 days).

A definitive way to prove the existence of specific immunologic tolerance is a challenge test. Mice that accepted cardiac allografts for more than 65 days were challenged with two skin grafts, one from donor-strain skin (BALB/c) and one from third-party strain skin (C57BL/6). Third-party skin was rejected acutely, as expected, within 14 days, whereas donor-strain skin was accepted indefinitely without exception (Fig. 6). In addition, the graft heartbeat was not affected by skin grafting. This experiment clearly demonstrates that short-term administration of MAbs to ICAM-1 and LFA-1 in association with cardiac allograft transplantation leads to donor alloantigen-specific immunologic tolerance.

## T-Cell Function Tests and FACS Analysis

Alloantigen-specific killer T cells are activated in mice with rejecting allografts, and the activity can be detected by cytotoxic T-lymphocyte (CTL) assay. BALB/c mice transplanted with C3H/He hearts were sacrificed at 7 or 40 days after operation and CTL activity of splenocytes against donor alloantigen was determined by a standard CTL assay.

Splenocytes from mice treated with MAbs did not show donor-specific CTL activity at 7 days or at 40 days, in contrast to those from mice with rejecting allografts, which showed significant CTL activity at 7 days after transplantation (Fig. 7, left panel). Interestingly enough, splenocytes of mice treated with MAbs responded to in vitro allostimulation and were cytotoxic against the donor alloantigens (Fig. 7, right panel) (17). These results suggest

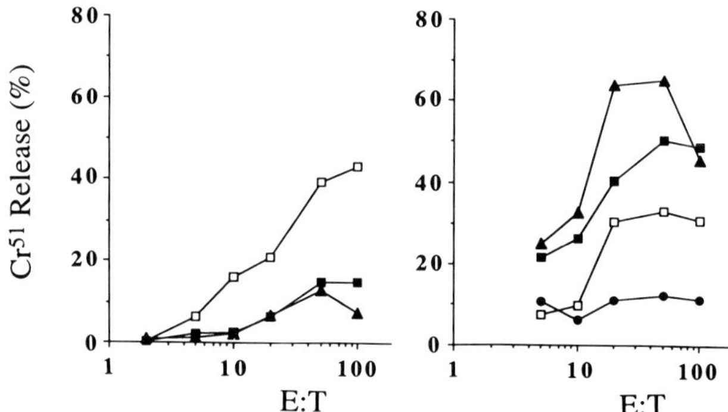


**FIG. 7.** Cytotoxic T-lymphocyte assay (CTL). Recipient BALB/c mice were killed 7 days after transplantation of C3H/He hearts. Fresh spleen cells from the recipients were used for a standard 4-h CTL assay **(left)** using RDM4 cells as targets. CTL activity is suppressed in splenocytes taken from mice treated with anti-ICAM-1 and anti-LFA-1 (open triangles) compared with those taken from an allografted mouse without treatment (closed squares). Closed squares denote CTL activity of splenocytes from a normal control mouse. The fresh splenocytes were cultured with irradiated C3H/He splenocytes for 4 days, and subjected to CTL assay **(right).** The splenocytes from an antibody-treated mouse showed CTL activity after the mixed culture with donor syngeneic splenocytes. Closed circles denote CTL activity of fresh splenocytes from a normal mouse.

that T cells are potentially active against alloantigen even after the induction of tolerance. Although the mechanism of this process requires further study, this finding would lead to the belief that clonal anergy may at least partly account for the induction of this tolerance, as discussed later.

Flow cytometry (indirect immunofluorescence) analysis of LFA-1 and ICAM-1 expression on splenocytes of allografted mice indicated that the MAb treatment led to significant reduction of LFA-1– and ICAM-1–positive cells 7 days after transplantation (Fig. 8). This downregulation of the cell surface antigens may account for the inability to detect alloreactive CTL activity at this time and could be responsible for the induction of tolerance against alloantigens. The expression of LFA-1 and ICAM-1 became normal 40 and 75 days after transplantation, although alloreactive CTL activity was still not detected. Therefore, the unresponsiveness is probably maintained by some mechanism other than downmodulation of LFA-1 and ICAM-1 molecules on allo-responding cells.

It would be of interest to determine whether or not this immunosuppression by MAbs to T-cell surface antigen is accompanied by T-cell depletion. A 6-day course of anti–ICAM-1/anti–LFA-1 treatment of ungrafted C3H/He mice did not reduce the circulating leukocyte count (Fig. 9) or the yield of leukocytes per spleen, and the Thy1.2$^+$, CD4$^+$, and CD8$^+$ subpopulations of splenocytes were not reduced at day 7 by treatment with the MAbs. This result was distinctly different from that with the anti-CD4 MAb GK1.5 treatment, which depleted CD4$^+$ cells from the circulation.

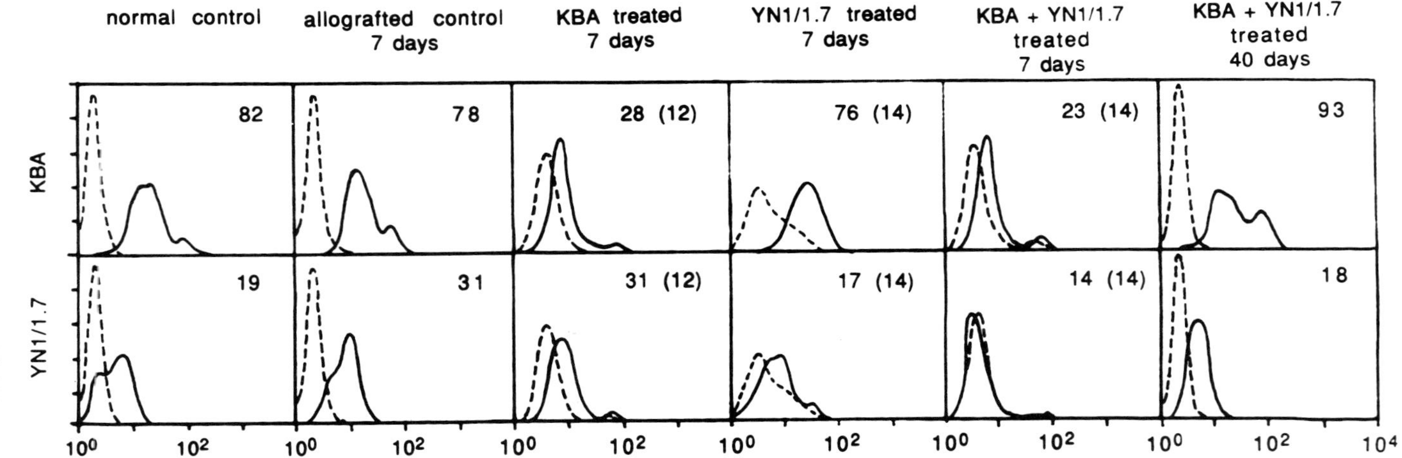


**FIG. 8.** FACS analysis of LFA-1 and ICAM-1 expression on splenocytes from allografted mice. Indirect immunofluorescent staining was performed. Background staining with FITC-conjugated goat antibodies to rat Ig alone (dotted lines) was less than 1% unless otherwise noted in parentheses. Profiles of stained cells are shown by solid lines and percentages of positively stained cells are expressed by numbers. Mice were treated daily with the indicated MAbs for the first 6 days after transplantation, except for the normal and allografted controls.

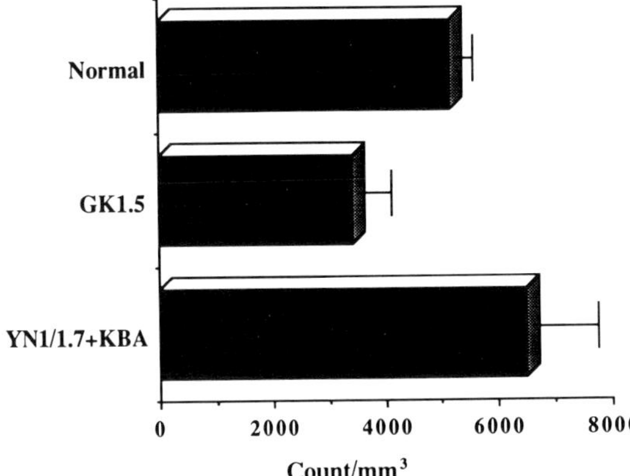


**FIG. 9.** Circulating white blood cell count after treatment with MAbs to GK1.5 (a depleting antibody) or YN1/1.7 plus KBA. One hundred micrograms of MAbs were injected into normal mice for 6 consecutive days and the blood samples were collected on the following day. The anti-YN1/1.7 and KBA MAbs did not deplete circulating white blood cells.

## VCAM-1/VLA-4

VLA-4 can also exert a co-stimulatory function on binding with VCAM-1 (18–20). The effects of MAbs against VCAM-1 and VLA-4 on cardiac allograft survival were investigated in the same model described above (21). MAbs (100 μg/day) were administered i.p. for the first 6 days. Graft survival in mice treated with M/K-2 (anti–VCAM-1) (median survival 20 days; $n = 6$) and those treated with PS/2 (anti–VLA-4) (30 days; $n = 6$) was greater than that in control mice (8 days; $n = 7$). Eight of 18 mice treated with both M/K-2 and PS/2 accepted the grafts for over 65 days, and five of them accepted the grafts for over 100 days. Mice with long-surviving cardiac grafts were challenged with skin grafts from donor and third-party (C57BL/6) strains. Survival of the donor-type skin was significantly greater than that of the third-party skin. These results indicate that in vivo administration of anti–VCAM-1 and anti–VLA-4 MAbs induces specific immunologic unresponsiveness to cardiac allografts.

ICAM-1 and VCAM-1 play reciprocal roles in the binding of resting T cells to resting and activated endothelial cells, respectively (22). In addition, the normal distribution of these molecules is different. The expression of ICAM-1 is more prevalent than VCAM-1. The immunosuppressive effect of anti–VCAM-1 and anti–VLA-4 MAbs was not as prominent in the present study compared with our previous observations using anti–ICAM-1 and LFA-1 MAbs given in the same protocol. The differences in the function and distribution of these molecules may explain the difference in the immunosuppressive effects.

It is of interest that the blockade of distinct co-stimulatory pathways, such as ICAM-1/LFA-1, CD28/B7, or VCAM-1/VLA-4, leads to a similar consequence, because some, if not all, of these co-stimulatory adhesion molecules may be simultaneously expressed on the surfaces of antigen-presenting cells

(23). An interesting question is whether or not these adhesion molecules provide the same kind of co-stimulatory signal or independently transmit distinct signals required for efficient T-cell activation. Differential participation of ICAM-1/VCAM-1 and B7 as the predominant co-stimulators at early and late stages of immune responses, respectively, has been demonstrated in vitro. It appears to be more likely that these adhesion molecules provide independent co-stimulatory signals, all of which are required for effectively eliciting immune responses in vivo.

## IMMUNOLOGIC CHARACTERIZATION OF THE TOLERANCE

### Skin Graft Transplantation

Because skin is one of the most immunogenic tissues, skin allografts are difficult to be accepted in incompatible combinations, even with treatment with conventional immunosuppressants. The effects of anti-adhesion therapy on skin graft transplantation were tested.

C57BL/6 (B6) mice and their mutant forms bm1 and bm12, were used for this experiment. B6 and bm1 are different in class I molecules and B6 and bm12 are different in class II molecules (24,25). The results are shown in Fig. 10. Nine of 15 bm12 grafts and 3 of 7 bm1 grafts were accepted for more than 100 days after treatment with anti–ICAM-1 and anti–LFA-1 for 20 days. However, there was no statistical difference in the survival days between the two groups. Rejection of bm1 skin in B6 recipients is mediated by CD8 T cells and that of bm12 in B6 recipients is mediated by CD4 T cells. These results suggest that tolerance induced by anti–ICAM-1 and anti–LFA-1 is not biased to either CD4- or CD8-mediated rejection.

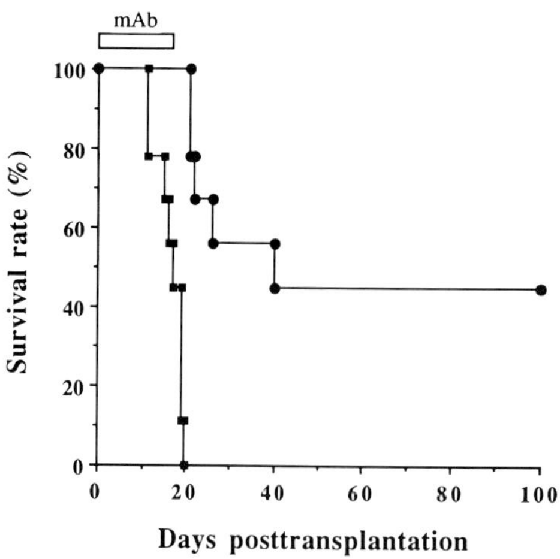


**FIG. 10.** Survival curve of bm12 skin graft transplanted into B6 recipients. Mice were treated with 0.5 mg each of KBA and YN1/1.7 every other day from 2 days before through 20 days after the operation. They showed significant prolongation of skin grafts (circles) compared with skin grafts in mice without treatment (squares). About half of the treated mice accepted the skin grafts indefinitely.

## Stability of the Tolerance

Approximately 2 weeks after MAb therapy is stopped, the MAbs are no longer detectable in the circulation and cellular adhesion molecules are not coated or modulated. In addition, the peripheral immune system is reconstituted with new thymic emigrants. Therefore, acceptance of the second skin grafts cannot be explained by coating or downmodulation of cell surface antigens or lack of presentation of foreign antigens by the graft to the recipient. This maintenance of tolerance might be due to mechanisms not related to MAb therapy. It has been stated that clonal deletion is a way of achieving tolerance to self antigens during T-cell development and to non-self antigens in the adult (26–28). To determine the role of the thymus and of thymic T-cell development, the following experiments were performed.

## Tolerance Induction in Thymectomized Mice

The role(s) of the thymus in this tolerance induction was evaluated. The thymus was removed from C3H/He mice at 4 weeks after birth. Four to 6 weeks after thymectomy, BALB/c hearts were transplanted, and anti–ICAM-1 and anti–LFA-1 MAbs were administered for 4 days. All recipients without treatment rejected the cardiac allografts within 23 days, but mice treated with MAbs accepted the allografts without exception. These data suggest that T-cell maturation in the thymus may not play a principal role in tolerance induction, supporting the idea that this tolerance is taking place in the periphery.

## Injection of Naive and Primed T Cells

The possibility that the tolerant donor-specific T cells were involved in resisting the ability of naive T cells to reject allografts was evaluated. Mature naive T cells were taken from normal C3H/He mice and injected into tolerant mice. Fifty to 100 million naive C3H/He splenocytes were unable to induce rejection of the BALB/c cardiac allograft. In addition, the same number of splenocytes were removed from a C3H/He mouse that was allografted with a BALB/c mouse heart 7 to 10 days previously. These primed T cells also could not abrogate the tolerance in BALB/c mice transplanted with C3H/He hearts. Therefore, the induction of tolerance leaves the mice in a state in which they are capable of resisting the ability of normal mature peripheral T cells that are syngeneic with the recipient. The tolerant donor-specific T cells may be involved in resisting the naive T cells, as suggested in anti-CD4/8 MAb-induced tolerance reported by other authors (29,30).

### Unresponsiveness to Soluble Antigens

The same regimen of treatment is capable of inducing tolerance to soluble antigens. C3H/He mice were immunized with heat-aggregated human γ-globulin (HGG). Mice were injected with either saline, 100 μg of M18/2 control MAb, PS/2 (anti–VLA-4), or 50 μg each of YN1/1.7 (anti–ICAM-1) and KBA (anti–LFA-1) at the time of immunization. Booster immunization was performed 3 weeks after the initial immunization without MAb therapy, and mice were bled 1 week after the booster injection. Antibody titers against HGG were measured by an enzyme-linked immunosorbent assay. Mice injected with saline or control M18/2 MAb produced antibodies to HGG, whereas antibody production was significantly suppressed in mice treated with PS/2 or a combination of YN1/1.7 and KBA. Furthermore, plates were coated with KBA MAb (rat $IgG_{2a}$) and antibody titers to rat $IgG_{2a}$ were compared with mice treated with saline, M18/2 (rat $IgG_{2a}$), and KBA. Antibody production to rat $IgG_{2a}$ was also greatly suppressed in mice injected with KBA but not in mice injected with M18/2, which showed antibody to $IgG_{2a}$ or antibody itself. Therefore, MAbs to LFA-1 and ICAM-1 are capable of inducing immunologic unresponsiveness not only to alloantigens but also to soluble antigens. The mechanism for this suppression in B-cell immunity should be investigated by further experiments.

## MECHANISM UNDERLYING TOLERANCE INDUCTION

### T-Cell Clonal Anergy

A cell that is anergic is one that is unable to proliferate on recognition of antigen by its antigen receptor. Clonal anergy is believed to be an alternative means by which T cells become tolerant to self antigens. There is evidence that it takes place during development both in the thymus (31) and in the periphery (32). It is not yet understood what drives a T cell to anergy rather than to activation or deletion.

Mueller et al. (33), and other investigators (34), reported that optimal activation of T cells requires two signals, one from processed antigen presented on the major histocompatibility complex via the T-cell receptor and another co-stimulatory signal from another molecule distinctive from the T-cell receptor (Fig. 11). These authors also proposed that recognition of antigen by the T-cell receptor in the absence of a co-stimulatory signal from the antigen-presenting cells might lead to T-cell inactivation. Tolerance to self antigen might be established through these mechanisms in the periphery. Furthermore, anergic T cells respond by activation in the presence of exogenously administered IL-2. Therefore, suppression of IL-2 production might be involved in the induction of anergy.

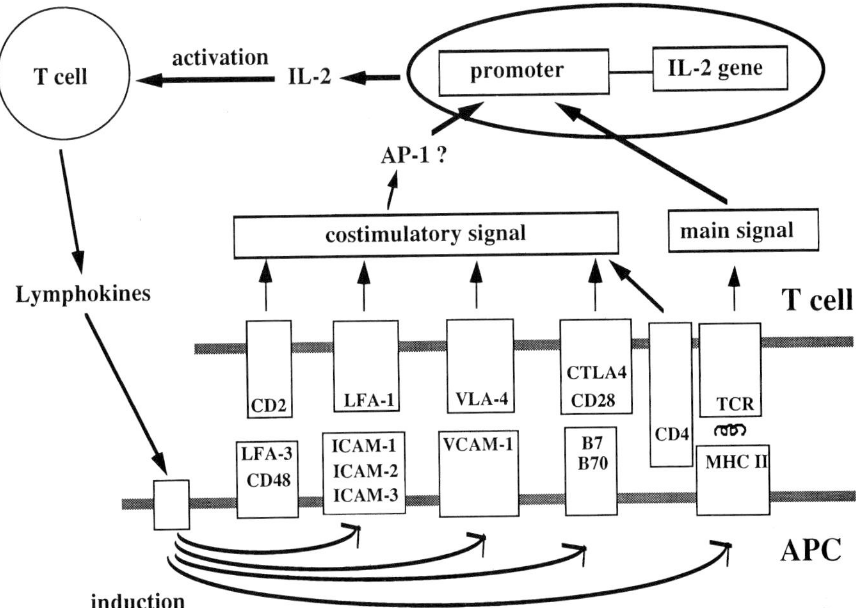


**FIG. 11.** A hypothesis on the activation and inactivation of T cells based on the proposal by R. Schwartz et al. (See text and 33–36)

Van Seventer (35) reported that LFA-1 produces this kind of co-stimulatory signal in the T-cell antigen-presenting cell interaction on binding with ICAM-1, a counter-receptor for LFA-1. Other adhesion molecules, such as CD28 (36) and VLA-4 (18–20), have been shown to produce co-stimulatory signals as well. Although the real nature of co-stimulation is not yet known, it is of interest that several different molecules can transmit similar signals.

### Alteration of the IL-2 Pathway

It is possible that tolerance by blockade of ICAM-1 and LFA-1 is induced by clonal anergy, resulting from absence of a co-stimulatory signal from the LFA-1 molecule in the presence of stimulation by alloantigens on the cardiac graft. To test this hypothesis, the production of IL-2 in tolerant mice was evaluated. rIL-2 was administered intraperitoneally with an osmotic pump for 7 days. This pump allows the release of rIL-2 over 7 days at a constant rate. Seven mice were injected with $10 \times 10^4$ units of IL-2. Five of the seven mice rejected the cardiac allograft within 25 days despite treatment with MAbs to ICAM-1 and LFA-1. Only two of the seven mice accepted the cardiac allografts indefinitely. The same dose of rIL-2 did not affect the isograft heartbeat in all mice tested. These results indicate that the induction of tolerance was prevented by administration of exogenous rIL-2. However, $10 \times 10^4$ units of rIL-2 did not abrogate tolerance at 50 days after operation. Therefore, induction and maintenance of tolerance are suggested to be regulated by different mechanisms with regard to IL-2 production. Furthermore, expression of mRNA for IL-2 in the cardiac allografts treated with MAbs was significantly suppressed compared with that in mice without treatment. Some deficits in the IL-2 pathway may be involved in the induction of the tolerance. Dallman et al. (37) reported a similar suppression of IL-2 production in the peripheral immunologic tolerance induced by donor-specific transfusion and kidney allograft transplantation. The same or a similar mechanism might be involved in the induction of different types of peripheral tolerance.

### Synergism of Two Monoclonal Antibodies

Another question in this tolerance induction is the synergism of the two antibodies. Tolerance was induced only when two MAbs were administered. All mice injected with only anti–ICAM-1 or anti–LFA-1 MAb eventually rejected the cardiac graft. We have no explanation for this phenomenon. However, redundancy of adhesion pairs may account for this synergism. ICAM-1 can bind to Mac-1 (38,39) and CD43 (40) molecules in addition to LFA-1 that are present on lymphocytes. Also, LFA-1 has at least three ICAMs, ICAM-1, -2 (41), and -3 (42). Although the physiologic role of this redundancy is not yet known, blockade of one side of the adhesion pair may not be enough to block cell adhesion or to suppress co-stimulatory signals. In fact, our in vitro experiments showed that anti–ICAM-1 and anti–LFA-1 MAbs synergized to suppress the mixed lymphocyte reaction (data not shown).

## CONCLUSIONS

Rejection is a complicated, poorly understood immunologic process. The pathophysiologic role(s) of ICAM-1/LFA-1 or VCAM-1/VLA-4 adhesion in acute rejection is still obscure. However, it is clear from our observations that cellular adhesion plays a crucial role in the rejection process. The molecular mechanism for the unresponsiveness observed in this study should be a target for further investigation. Knowledge of the preferential use of distinct co-stimulatory pathways in eliciting humoral and cellular immune responses to particular antigens may have important clinical implications, since the unique tolerance-inducing ability achieved by manipulating these adhesion molecules would enable us to develop a new mode of immunosuppression for clinical treatment of graft rejection, allergy, and autoimmune diseases.

## ACKNOWLEDGMENT

The authors acknowledge Professors Yoshio Yazaki (University of Tokyo), Morie Sekiguchi (Shinshu University), Ko Okumura (Juntendo University), Masayuki Miyasaka (Osaka University), and Edgar Haber and David H. Sachs (Harvard University) for their invaluable discussions. The monoclonal antibodies used were kind gifts from Professor Ko Okumura and Professor Masayuki Miyasaka. The bm1 and bm12 mice were a generous gift from Professor Michio Fujiwara (University of Tokyo). This investigation was supported in part by a grant-in-aid from The Ministry of Welfare and Health of Japan, the Terumo Life Science Foundation, and the Sankyo Foundation of Life Science.

## REFERENCES

1. Billingham RE, Brent L, Medawar PB. 'Actively acquired tolerance' of foreign cells. *Nature* 1953;172:603.
2. Isobe M, Yagita H, Okumura K, Ihara A. Specific acceptance of cardiac allograft after treatment with anti-ICAM-1 and anti-LFA-1. *Science* 1992;255:1125–7.
3. Peuchmaur FM, Desruennes M, Ghoussoub JJ, et al. Implications of de novo ELAM-1 and VCAM-1 expression in human cardiac allograft rejection. *Transplantation* 1993;55:605–9.
4. Cosimi AB, Conti D, Delmonico FL, et al. In vivo effects of monoclonal antibody to ICAM-1 (CD54) in nonhuman primates with renal allografts. *J Immunol* 1990;144:4604–12.
5. Dustin ML, Rothlein R, Bhan AK, Dinarello CA, Springer TA. Induction by IL 1 and interferon-gamma: tissue distribution, biochemistry, and function of a natural adherence molecule (ICAM-1). *J Immunol* 1986;137:245–54.
6. Pober JS, MA GJ, Lapierre LA, et al. Overlapping patterns of activation of human endothelial cells by interleukin 1, tumor necrosis factor and immune interferon. *J Immunol* 1986;137:1893–6.
7. Taylor PM, Rose ML, Yacoub MH, Pigott R. Induction of vascular adhesion molecules during rejection of human cardiac allografts. *Transplantation* 1992;54:451–7.
8. Carlos T, Gordon D, Fishbein D, et al. Vascular cell adhesion molecule-1 is induced on endothelium during acute rejection in human cardiac allografts. *J Heart Lung Transplant* 1992;11:1103–9.
9. Pelletier R, Ohye RP, Kincade P, Ferguson R, Orosz C. Monoclonal antibody to anti-VCAM-1 interferes with murine cardiac allograft rejection. *Transplant Proc* 1993;25:839–41.
10. Isobe M, Ohtani H, Yagita H, Okumura K, Strauss HW, Yazaki Y. Detection of cardiac rejection in mice by radioimmune scintigraphy using $^{123}$iodine-labeled anti-ICAM-1 monoclonal antibody. *Acta Cardiol* 1993;XLVIII:235–43.
11. Ohtani H, Strauss HW, Southern JF, Tamatani T, Miyasaka M, Isobe M. Imaging of intercellular adhesion molecule-1 induction in rejecting heart: a new scintigraphic approach to detect early allograft rejection. *Transplant Proc* 1993;25:867–9.
12. Isobe M, Haber E, Khaw BA. Early detection of rejection and assessment of cyclosporine therapy by indium-111 antimyosin imaging in mouse heart allografts. *Circulation* 1991;84:1246–55.
13. Isobe M, Narula J, Strauss HW, Khaw BA, Haber E. Imaging the rejecting heart: in vivo detection of MHC class II antigen induction. *Circulation* 1992;85:738–46.
14. Takei F. Inhibition of mixed lymphocyte response by a rat monoclonal antibody to a novel murine lymphocyte activation antigen (MALA-2). *J Immunol* 1985;134:1403–7.
15. Nishimura T, Yagi H, Yagita H, Uchiyama Y, Hashimoto Y. Lymphokine-activated cell-associated antigen involved in broad-reactive killer cell-mediated cytotoxicity. *Cell Immunol* 1985;94:122–32.

16. Sanchez MF, Simon P, Thompson S, Springer TA. Mapping of antigenic and functional epitopes on the alpha- and beta-subunits of two related mouse glycoproteins involved in cell interactions, LFA-1 and Mac-1. *J Exp Med* 1983;158:586–602.
17. Isobe M, Ihara A. Tolerance induction against cardiac allograft by anti-ICAM-1 and anti-LFA-1 treatment: T cells respond to in vitro allostimulation. *Transplant Proc* 1993;25:1079–80.
18. Shimizu Y, van Seventer GA, Horgan KJ, Shaw S. Costimulation of proliferative responses of resting CD4$^+$ T cells by the interaction of VLA-4 and VLA-5 with fibronectin or VLA-6 with laminin. *J Immunol* 1990;145:59–67.
19. Shimizu Y, Van Seventer GA, Horgan KJ, Shaw S. Roles of adhesion molecules in T-cell recognition: fundamental similarities between four integrins on resting human T cells (LFA-1, VLA-4, VLA-5, VLA-6) in expression, binding, and costimulation. *Immunol Rev* 1990; 114:109–43.
20. Nojima Y, Humphries MJ, Mould AP, et al. VLA-4 mediates CD3-dependent CD4$^+$ T cell activation via the CS1 alternatively spliced domain of fibronectin. *J Exp Med* 1990;172: 1185–92.
21. Isobe M, Suzuki J, Yagita H, et al. Immunosuppression to cardiac allographs and soluble antigens by anti-VCAM-1 and anti-VLA-4 monoclonal antibodies. *J Immunol* 1994;153:5810–8.
22. Oppenheimer-Marks N, Davis LS, Bogue DT, Ramberg J, Lipsky PE. Differential migration of human T lymphocytes. *J Immunol* 1991;147:2913–21.
23. Damle NK, Klussman K, Linsley PS, Aruffo A. Differential costimulatory effects of adhesion molecules B7, ICAM-1, LFA-3, and VCAM-1 on resting and antigen-primed CD4$^+$ T lymphocytes. *J Immunol* 1992;148:1985–92.
24. Rosenberg AS, Mizuochi T, Singer A. Analysis of T-cell subsets in rejection of Kb mutant skin allografts differing at class I MHC. *Nature* 1986;322:829–31.
25. Sprent J, Schaefer M, Lo D, Korngold R. Properties of purified T cell subsets. II. In vivo responses to class I vs. class II H-2 differences. *J Exp Med* 1986;163:998–1011.
26. Kisielow P, Bluthmann H, Staerz UD, et al. Tolerance in T cell receptor transgenic mice involves deletion of non-mature CD4$^+$8$^-$ thymocytes. *Nature* 1988;333:742–6.
27. Kappler JW, Roehm N, Marrack P. T cell tolerance by clonal elimination in the thymus. *Cell* 1987;49:273–80.
28. Webb S, Morris C, Sprent J. Extrathymic tolerance of mature T cells: clonal elimination as a consequence of immunity. *Cell* 1990;63:1249–56.
29. Qin S, Wise M, Cobbold SP, et al. Induction of tolerance in peripheral T cells with monoclonal antibodies. *Eur J Immunol* 1990;20:2737–45.
30. Qin S, Cobbold SP, Pope H, et al. "Infectious" transplantation tolerance. *Science* 1993; 259:974–7.
31. Ramsdell F, Lantz T, Fowlkes BJ. A non-deletional mechanism of thymic self tolerance. *Science* 1989;246:1038–41.
32. Burkly LC, Lo D, Kanagawa O, Flavell RA. T-cell tolerance by clonal anergy in transgenic mice with nonlymphoid expression of MHC class II I-E. *Nature* 1989;342:564–6.
33. Mueller D, Jenkins M, Schwartz R. Clonal expansion versus functional clonal inactivation: a costimulatory outcome of T cell antigen receptor occupancy. *Annu Rev Immunol* 1989; 7:445–80.
34. Geppert TD, Davis LS, Gur H, Wacholtz MC, Lipsky PE. Accessory cell signals involved in T-cell activation. *Immunol Rev* 1990;117:5–66.
35. van Seventer GA, Shimizu Y, Horgan KJ, Shaw S. The LFA-1 ligand ICAM-1 provides an important costimulatory signal for T cell receptor-mediated activation of resting T cells. *J Immunol* 1990;144:4579–86.
36. Harding FA, McAurthur JG, Gross JA, Raulet DH, Allison JP. CD28-mediated signaling co-stimulates murine T cells and prevents induction of anergy in T-cell clones. *Nature* 1992;356:607–9.
37. Dallman MJ, Shiho O, Page TH, Wood KJ, Morris PJ. Peripheral tolerance to alloantigen results from altered regulation of the interleukin 2 pathway. *J Exp Med* 1991;173:79–87.
38. Diamond MS, Staunton DE, de Fougerolles SA, et al. ICAM-1 (CD54): a counter-receptor for Mac-1 (CD11b/CD18). *J Cell Biol* 1990;111:3129–39.
39. Diamond MS, Staunton DE, Marlin SD, Springer TA. Binding of the integrin Mac-1

(CD11b/CD18) to the third immunoglobulin-like domain of ICAM-1 (CD54) and its regulation by glycosylation. *Cell* 1991;65:961–71.

40. Rosenstein Y, Park JK, Hahn WC, Rosen FS, Bierer BE, Burakoff SJ. CD43, a molecule defective in Wiskott-Aldrich syndrome, binds ICAM-1. *Nature* 1991;354:233–5.
41. Staunton DE, Dustin ML, Springer TA. Functional cloning of ICAM-2, a cell adhesion ligand for LFA-1 homologous to ICAM-1. *Nature* 1989;339:61–4.
42. Fawcett J, Holness LL, Needham LA, et al. Molecular cloning of ICAM-3, a third ligand for LFA-1, constitutively expressed on resting leukocytes. *Nature* 1992;360:481–4.

*Topics in Molecular Medicine, Volume 1,*
edited by Wolfgang Siess, Reinhard Lorenz,
and Peter C. Weber. Raven Press, Ltd.,
New York © 1995.

# 25

# Adhesion Molecule Expression in Renal Allograft Rejection

*†H. E. Feucht, ‡H. Schneeberger, *U. Bechtel, †S. Lederer, †C. Brockmeyer, §G. Hillebrand, §K. Burkhardt, ‡W. Land, and †J. Johnson

**Department of Internal Medicine, Klinikum Innenstadt, University of Munich; †Institute of Immunology, University of Munich; D-80336 Munich; ‡Department of Transplant Surgery, Klinikum Grosshadern; and §Nephrology Division, Department of Internal Medicine I, Klinikum Grosshadern, University of Munich, D-81366 Munich, Germany*

Acute and chronic rejections are the principal causes of allograft failure in recipients. Although both humoral and cell-mediated immune mechanisms are believed to cooperate, many studies suggest that rejection is in essence a T cell-mediated process directed against foreign MHC determinants. However, relatively few infiltrating cells appear to be actually MHC-specific, the majority being rather "unspecific" bystander cells recruited by other mediators of inflammation (1).

In renal transplants, the major targets of alloreactive T cells are graft endothelial cells and tubule epithelial cells. Acute cell-mediated rejections are characterized by extensive infiltration of peritubule areas and the invasion of tubules by lymphocytes. As a prerequisite to tissue infiltration, alloreactive mononuclear cells must leave the vascular space. Cell emigration in renal grafts usually takes place across peritubular capillaries (2). In the relatively rare instances of vascular rejection, lymphocytes also enter the vessel walls of arteries and arterioles. It has recently become clear that the process of leukocyte extravasation is controlled by different sets of adhesion molecules. Hence, the adhesive interactions of circulating leukocytes with graft cells were expected to be of importance also in the complex process of allograft rejection. That such interactions are indeed operational during transplantation was recently proven in clinical studies when monoclonal antibodies (MAbs) against ICAM-1 prevented rejection episodes (3).

In the present study, the differential expression of ICAM-1, VCAM-1, and

E-selectin in renal allografts with rejection was investigated. The question was also addressed of whether particular changes in tissue distribution were associated with different graft survival rates.

## PATIENTS AND METHODS

### Patients and Graft Biopsies

In this study, 108 recipients of cadaver kidney grafts, including 32 diabetic patients who simultaneously received a pancreas graft, were investigated. All patients underwent transplantation at the Division of Transplant Surgery between 1987 and 1993. Graft biopsies were performed at various time intervals after transplantation when rejection was clinically suspected. According to pathologic records, all patients had mild-to-severe acute cell-mediated rejections. The tissue distribution of cell adhesion molecules was assessed in parallel in these specimens. Normal renal tissue (control) was obtained from 15 tumor nephrectomies.

### Indirect Immunoperoxidase Staining and Monoclonal Antibodies

Indirect immunoperoxidase staining of cryostat sections prepared from normal kidney tissues and renal graft biopsies was performed as described in detail previously (4).

The following mouse MAbs were used: H18/7 (gift from Dr. M. A. Gimbrone, Boston, MA) and BBA 1 (British Biotechnology, Ltd., Oxford, U.K.) directed against ELAM-1; P3.58BA-14 directed against ICAM-1 (5); 4B9 ($IgG_1$) (gift from Dr. T. M. Carlos, Seattle, WA) and BBA 5 (British Biotechnology) against VCAM-1; T29/33 against common leucocyte antigen (CD45) (Hybritech, Inc., San Diego, CA); mouse $IgG_1$, $IgG_{2a}$ and $IgG_{2b}$ antibodies against irrelevant antigens were used as controls.

### Immunosuppressive Therapy

Immunosuppressive therapy in recipients was started either with a triple-drug regimen (cyclosporin, steroids, azathioprine) or, in recipients with immunologic risk factors, with a quadruple-drug regimen (supplemented with polyvalent or monoclonal anti-lymphocyte antibodies). Rejections were treated with high-dose steroids for 3 days and, if unresponsive, with anti-lymphocyte antibodies.

### Data Monitoring and Statistical Analysis

Patient data were monitored with computer assistance using the ENABLE data processing tools (Markt & Technik, Haar, Germany). Graft survival rates were computed by the Cutler–Ederer method.

## RESULTS

### Tissue Distribution of Cell Adhesion Molecules in Normal Kidneys and During Rejection

#### *Intercellular Adhesion Molecule-1*

Comparable immunohistologic results with MAb P3.58-BA14, reactive against ICAM-1, were obtained with 15 normal kidney specimens. Therefore, ICAM-1 was present on glomerular endothelial cells, in peritubule capillaries, in large vessels, and on some parietal epithelial cells of Bowman's capsule (Fig. 1A). During acute rejection, focally distributed expres-

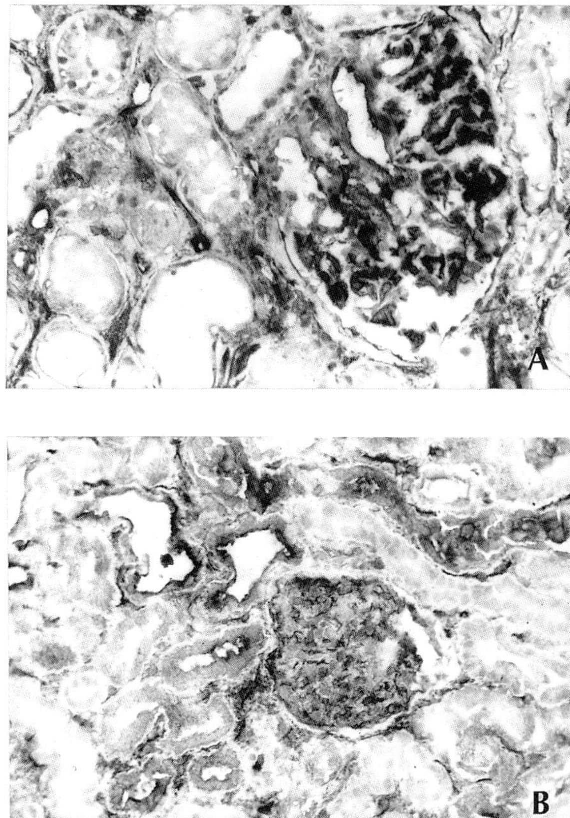


**FIG. 1.** Expression of ICAM-1. **A:** Normal kidney: vascular endothelial cells in glomerulus and peritubular capillaries. **B:** Transplanted kidney with rejection: capillaries and tubule cells.

sion of ICAM-1, predominantly in proximal tubules, was observed in 54 of 108 graft biopsies (Fig. 1B). In 35 biopsies, ICAM-1 was also detectable on graft-infiltrating leukocytes.

### *Vascular Cell Adhesion Molecule-1*

In 15 normal kidneys, MAbs 4B9 and BBA 5 against VCAM-1 reacted with parietal epithelial cells of Bowman's capsule and with a few cells in several proximal tubules, whereas vessels were devoid of VCAM-1 (Fig. 2A). During rejection, increased tubule staining (referring to both the number of tubules and the intensity of staining) with VCAM-1–specific monoclonal reagents was observed in 102 of 108 grafted kidneys. VCAM-1 stain-

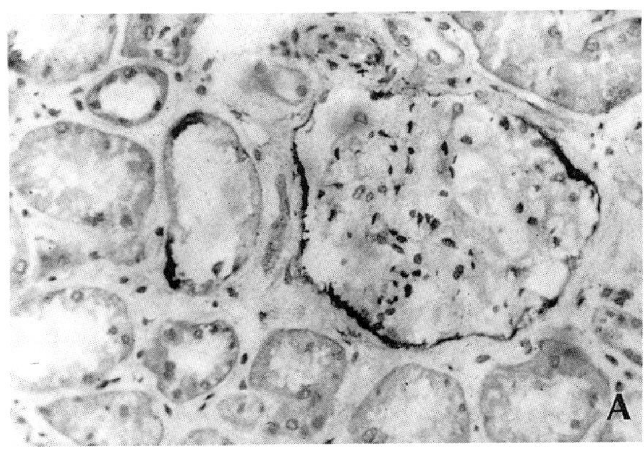

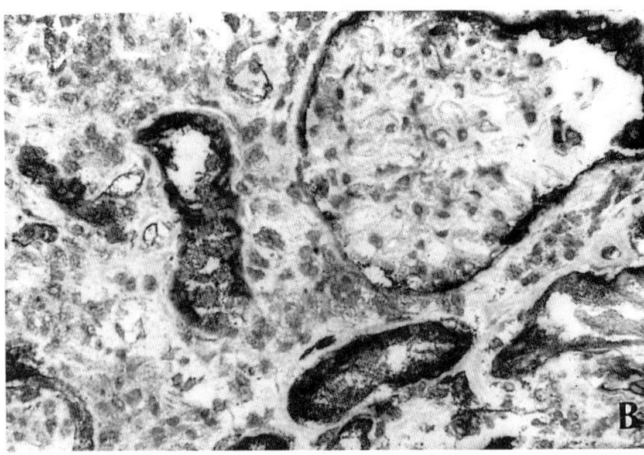

**FIG. 2.** Expression of VCAM-1. **A:** Normal kidney: epithelial cells of Bowman's capsule and a few tubule cells. **B:** Kidney transplant with rejection: enhanced staining of tubules and staining of several capillaries.

ing was focally distributed and included distal tubule segments as well. In 78 graft biopsies, VCAM-1 appeared also in variable numbers of peritubule capillaries, ranging from single vessels to approximately 30% of all vessels present in one specimen (Fig. 2B). However, glomerular capillaries were consistently devoid of VCAM-1. In 19 biopsies, infiltrating leukocytes also carried VCAM-1.

### *Endothelial Leukocyte Adhesion Molecule-1*

No reactivity of MAbs H18/7 and BBA 1 against ELAM-1 was observed in normal kidneys (not shown). During rejection, ELAM-1 could be detected only in a few capillaries in 61 of 108 biopsies (Fig. 3). ELAM-1 could not be detected in glomerular capillaries, in tubules, or in cellular infiltrates.

The immunohistologic results in normal kidneys and in graft biopsies are summarized in Figs. 4A,B.

## Tissue Distribution of ICAM-1, VCAM-1, ELAM-1, and Graft Survival Rates

The 1- through 5-year survival rates of all transplanted and biopsied grafts ($n$ = 108) were 86% and 61%, respectively. When these patients were further subdivided according to the de novo expression of ICAM-1 in tubules, consistent differences appeared. Grafts that showed tubule expression of ICAM-1 ($n$ = 54) had survival rates of only 79% at 1 year and 51% at 5 years. Grafts without tubule ICAM-1 expression ($n$ = 54) had an excellent outcome, with 93% survival at 1 year and 70% survival at 5 years (Fig. 5). No

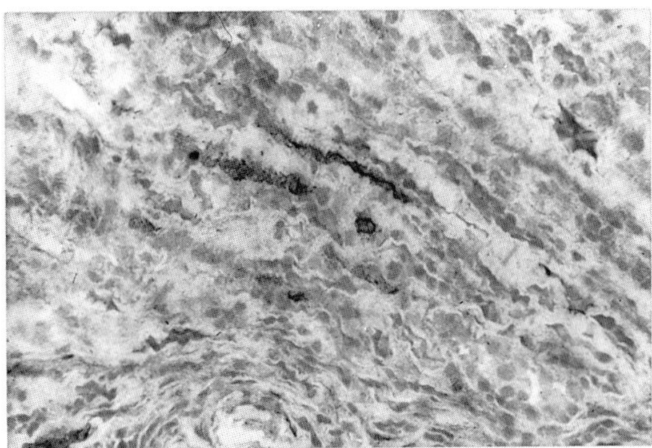

**FIG. 3.** ELAM-1–positive capillaries in kidney transplant with rejection.

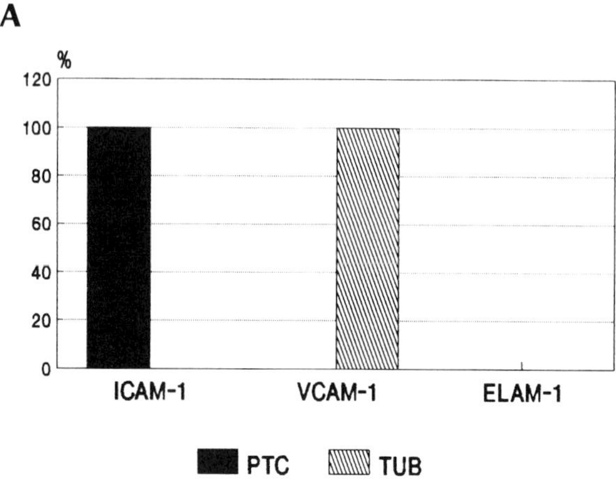


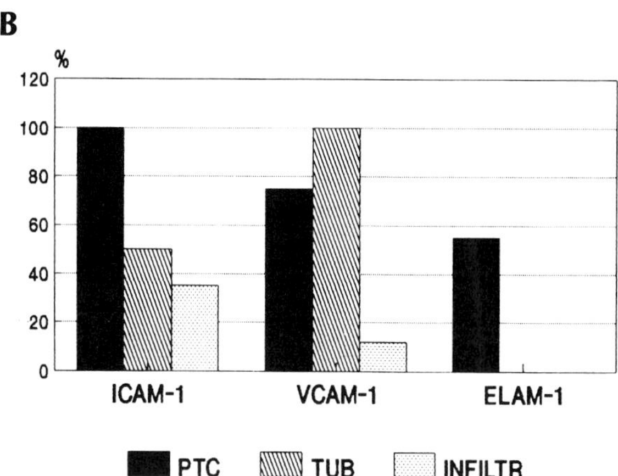


**FIG. 4.** Distribution of ICAM-1, VCAM-1 and ELAM-1 in 15 normal kidneys **(A)** and in 108 kidney transplants with rejection **(B).** PTC, peritubule capillaries; Tub, tubules; Infiltr, cellular infiltrates.

differences in survival rates emerged whether cell infiltrates within the graft were ICAM-1 positive or negative (data not shown).

Only moderate differences in graft survival were apparent when the endothelial distribution of VCAM-1 was analyzed. Grafts with capillary expression of VCAM-1 ($n$ = 78) had 82% survival at 1 year and 58% survival at 5 years. Grafts that were devoid of capillary VCAM-1 ($n$ = 30) had an improved survival (93%) at 1 year only, whereas the 5-year survival of 66% was comparable to that of total study group (Fig. 6). No differences were observed whether graft-infiltrating cells carried VCAM-1 or not (data not shown).

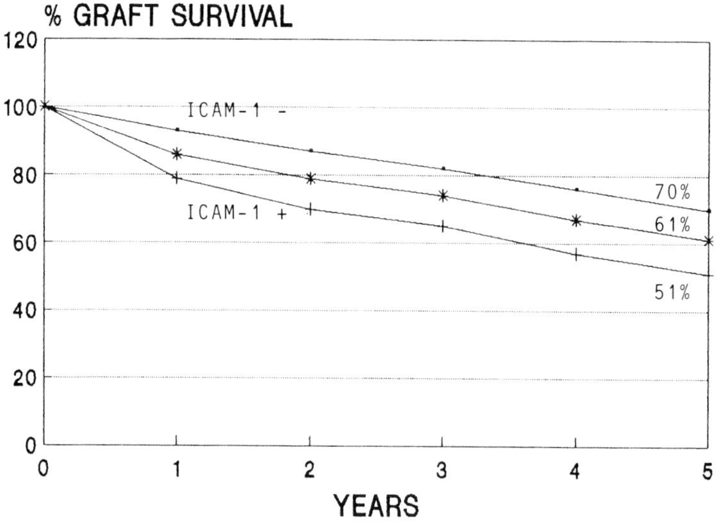


**FIG. 5.** Survival rates of all biopsied kidney grafts (*) and of grafts with ($n$ = 54) and without ($n$ = 54) ICAM-1 in tubules.

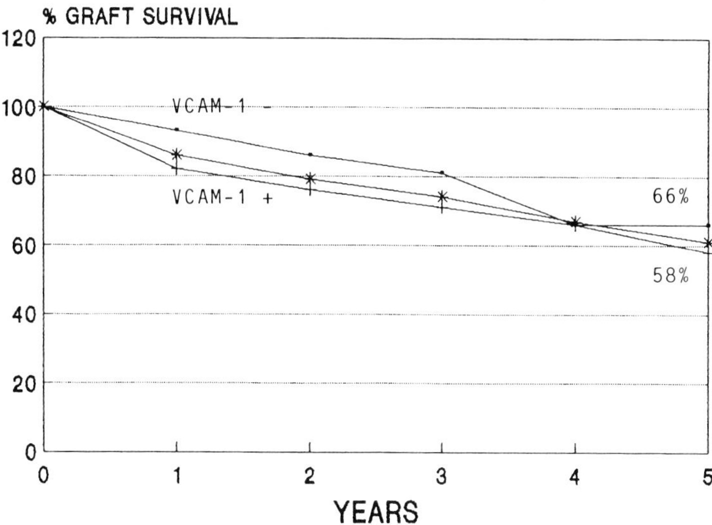


**FIG. 6.** Survival rates of all biopsied kidney grafts (*) with ($n$ = 78) and without ($n$ = 30) capillary expression of VCAM-1.

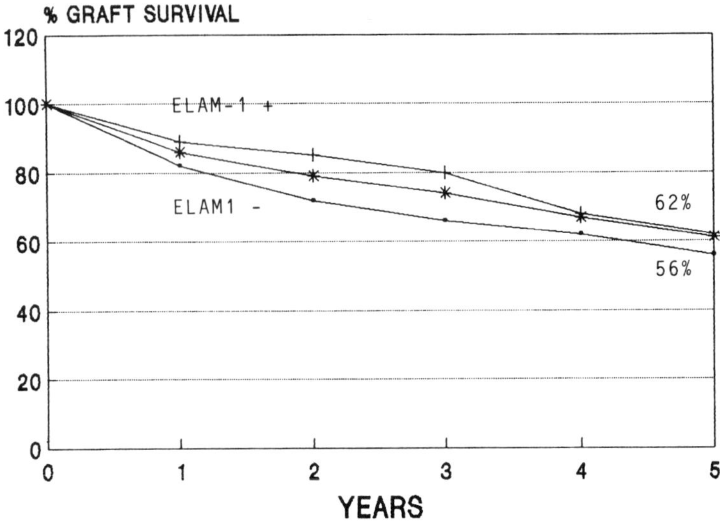


**FIG. 7.** Survival rates of all biopsied kidney grafts (*) with ($n$ = 61) and without ($n$ = 47) capillary expression of ELAM-1.

Analysis of the capillary de novo expression of ELAM-1 showed that ELAM-1–positive grafts ($n$ = 61) appeared to have slightly better survival rates than ELAM-1–negative grafts ($n$ = 47) (Fig. 7).

## DISCUSSION

Compared with normal kidneys, renal allografts show pronounced changes in the tissue distribution of cell adhesion molecules during acute rejection. It is shown in this study that ICAM-1, a member of the immunoglobulin gene superfamily that is constitutively expressed on endothelium, appears de novo on tubule cells in 50% of biopsied grafts. Conversely, VCAM-1, another member of the immunoglobulin superfamily, is constitutively expressed on few epithelial cells (Bowman's capsule, proximal tubule cells) but displays enhanced expression on tubule cells during rejection and appears de novo in capillaries in about 70% of biopsies. That the induced tubule expression of these molecules is not the result of glomerular filtration and tubule reabsorption but is the result of actual cellular synthesis could recently be demonstrated by Northern blot analysis using cultured tubular epithelial cells (4). It is of note that ICAM-1 and VCAM-1 apparently show some form of complementary distribution in normal and grafted kidneys. In contrast, ELAM-1, a member of the selectin family, is not expressed in extravascular sites but is restricted to single blood vessels in about half of renal grafts. The enhanced and/or de novo expression of these molecules in different tissue compartments is probably caused by locally acting lymphokines such as TNF (6).

It is particularly interesting that MHC molecules, also under the influence of cytokines, exhibit very similar changes with respect to the enhanced and de novo expression in grafted kidneys. MHC class II molecules, for example, are constitutively present on vascular endothelial cells and appear de novo in tubule cells (7). The combined display of MHC molecules and of adhesion receptors at vascular and nonvascular sites is likely to have several pathophysiologic consequences in the different phases of rejection. First, the assembly of alloreactive cells is promoted and cells are localized to given areas within the graft (8). The combined action of these molecules would not only stabilize MHC-directed effector–target interactions but would also amplify cell-mediated immune reactions by recruitment of leukocytes bearing suitable adhesion receptors. Therefore, different sets of adhesion molecules can modulate the composition of cellular infiltrates. Finally, in the effector phase, cytotoxic T lymphocytes are activated by a cascade of adhesion–signaling events (9). Furthermore, renal target cells expressing, e.g., surface ICAM-1, are more susceptible to lysis than target cells devoid of ICAM-1 (10).

Under these assumptions, analysis of graft survival rates was considered a first and direct approach to prove the clinical relevance of the immunohistologic findings reported above. Although the number of recipients is relatively small, grafts with de novo tubule expression of ICAM-1 consistently had worse short- and long-term survival rates than grafts without tubule ICAM-1. Hence, tubule ICAM-1 seems to reflect the severity of interstitial rejection. Conversely, the absence of tubule ICAM-1 identifies a subpopulation with a very good overall survival. Surprisingly, the de novo appearance of capillary VCAM-1 or ELAM-1 did not clearly distinguish among different subpopulations with respect to risk factors for graft survival. Although speculative, one possible explanation would be sampling error. Given the transient and limited expression of ELAM-1 in few vessels, probably many more, if not all, grafts could express ELAM-1 at a certain time in certain areas, which may not be represented in a single biopsy. On that basis, the appearance or absence of ELAM-1 would not really define a histological or clinical entity. The same could be true of endothelial VCAM-1. However, larger trials with more recipients are required for valid conclusions to be drawn.

It should be noted that the enhanced tubular expression of VCAM-1, the grading of which is sometimes difficult on purely histologic grounds, was not included in this type of analysis. It appeared, however, that the extent of tubule VCAM-1, in accordance with tubule ICAM-1, also correlated with the severity of rejection. This could be substantiated by measuring the concentrations of soluble ICAM-1 and VCAM-1 in urine samples from graft recipients. Thus, urinary levels of sICAM-1 and sVCAM-1 correlated well with immunohistologic findings and were reflective of the severity of acute and chronic rejection. Serial determinations of urinary sICAM-1 and sVCAM-1 proved to be most useful for the clinical monitoring of renal graft recipients (11).

In summary, the results further indicate that during rejection the constitutive and induced expression of adhesion molecules is also involved. In particular, the de novo appearance of ICAM-1 (and possibly of VCAM-1) in tubules appears to correlate with the severity of interstitial rejection and consequently with overall graft survival. Therefore, therapeutic strategies that would interfere with adhesive cellular interactions should improve graft outcome in the future.

## ACKNOWLEDGMENT

We are indebted to Prof. Dr. G. Thoenes (Department of Immunobiology, Med. Klinik) for providing cryostat sections of graft biopsies, to Dr. M. Weiss (Institute of Pathology) for communication of the histopathologic diagnoses, and to Ms. G. Schmid for expert technical assistance. This work was supported in part by the Deutsche Forschungsgemeinschaft (SFB 217:C8). H. E. Feucht is the recipient of a Hermann and Lilly Schilling professorship.

## REFERENCES

1. Hall BM. Cells mediating allograft rejection. *Transplantation* 1991;51:1141–51.
2. Turunen JP, Paavonen T, Majuri M-L, et al. Sialyl Lewis$^x$- and L-selectin-dependent site-specific lymphocyte extravasation into renal transplants during acute rejection. *Eur J Immunol* 1994;24:1130–6.
3. Haug CE, Colvin RB, Delmonico FL, et al. A phase I trial of immunosuppression with anti-ICAM-1 (CD 54) mAb in renal allograft recipients. *Transplantation* 1993;55:766–73.
4. Brockmeyer C, Ulbrecht M, Schendel DJ, et al. Distribution of cell adhesion molecules (ICAM-1, VCAM-1, ELAM-1) in renal tissue during allograft rejection. *Transplantation* 1993;55:610–5.
5. Johnson JP, Stade BG, Hupke U, Holzmann B, Riethmüller G. The melanoma progression-associated antigen P3.58 is identical to the intercellular adhesion molecule, ICAM-1. *Immunobiology* 1988;178:275–84.
6. Vandenbroecke C, Caillat-Zucman S, Legendre C, et al. Differential in situ expression of cytokines in renal allograft rejection. *Transplantation* 1991;51:602–9.
7. Hall BM, Bishop GA, Duggin GG, Horvath JS, Philips J, Tiller DJ. Increased expression of HLA-DR antigens on renal tubular cells in renal transplants: relevance to the rejection response. *Lancet* 1984;2:247–51.
8. Alpers CE, Hudkins KL, Davis CL, et al. Expression of vascular cell adhesion molecule-1 in kidney allograft rejection. *Kidney Int* 1993;44:805–16.
9. Ybarrondo B, O'Rourke AM, Brian AA, Mescher MF. Contribution of lymphocyte function-associated-1/intercellular adhesion molecule-1 binding to the adhesion/signaling cascade of cytotoxic T lymphocyte activation. *J Exp Med* 1994;179:359–63.
10. Suranyi MG, Bishop GA, Clayberger C, et al. Lymphocyte adhesion molecules in T cell-mediated lysis of kidney cells. *Kidney Int* 1991;39:312–9.
11. Bechtel U, Scheuer R, Landgraf R, König A, Feucht HE. Assessment of soluble cell adhesion molecules (sICAM-1, sVCAM-1, sELAM-1) and complement cleavage products (C4d, C5b–9) in urine: clinical monitoring of renal graft recipients. *Transplantation* 1994;58:905–11.

*Topics in Molecular Medicine, Volume 1,*
edited by Wolfgang Siess, Reinhard Lorenz,
and Peter C. Weber. Raven Press, Ltd.,
New York © 1995.

# 26

# Circulating Adhesion Molecules in Experimental and Human Disease

A. J. H. Gearing

*British Bio-technology Plc, Cowley, OX4 5LY, England*

It should be clear from the other chapters in this book that cellular interactions mediated by adhesion molecules are essential components of normal leukocyte trafficking, immunity, inflammation, and tumor metastasis. Monitoring the expression of cell surface adhesion molecules can provide useful information on the activation state of tissues. Leukocyte adhesion molecules can be quantified by FACS analysis of peripheral blood samples. Expression on other cells requires histologic analysis of tissue biopsies. Immunohistology provides a detailed picture of adhesion molecule expression at a particular moment, but repeated biopsy is not a practical means of sampling patients to generate a kinetic picture of disease progression. In most diseases, blood is the only tissue that can be regularly sampled. Work in our and other laboratories over the last 3 to 4 years has demonstrated that many adhesion molecules can be shed from the surface of endothelial, leukocytic, and tumor cells, and that these soluble forms of the adhesion molecules can be detected in body fluids (1). This chapter reviews the evidence that monitoring the levels of soluble adhesion molecules can provide useful information on the progress of experimental and human diseases, with particular emphasis on inflammation and cancer.

There are now several clear families of adhesion molecules [the selectins, immunoglobulin superfamily adhesins (IgSF), integrins], in addition to a number of proteoglycans, glycoproteins, and assorted oligosaccharides (2).

There are three selectins, L-, P-, and E-selectin. L-selectin is constitutively expressed on leukocytes. P-Selectin is stored in the α granules of platelets and the Weibel–Palade bodies of endothelial cells. E-Selectin is expressed only by activated endothelial cells. All contain a lectin-like N-terminal domain, followed by an EGF-like domain and a variable number of complement regulatory protein-like repeats.

The IGSF molecules include a subset that participates in cell–cell adhesion. The cell surface expression of many of these, such as ICAM-1 and VCAM-1, is upregulated after activation during inflammatory responses, mediating both cell migration and activation. ICAM is expressed on endothelial, epithelial, fibroblast, leukocyte, and many tumor cells. VCAM is also widely distributed on endothelial, epithelial, macrophage, and dendritic cells. ICAM-1 and VCAM-1 on endothelial cells appear to be particularly important for the firm attachment and transendothelial migration of leukocytes.

The integrins are a large family of heterodimeric structures comprising at least 11 $\alpha$-chains combined with 8 $\beta$-chains. Integrins are ligands for IgSF adhesins and for many extracellular matrix proteins. Although integrins are vital components of cell migration and activation pathways, there are no reports of their existence as soluble forms, and therefore they are of no further consequence in this article.

There is accumulating evidence that many adhesion molecules can be found as soluble forms either in cell culture supernatants or circulating in biologic fluids, such as blood, cerebrospinal fluid, synovial fluid, and bronchiolar lavage. The demonstration of soluble forms of adhesion molecules begs two separate questions. What is the role of such molecules and what, if any, is the significance of monitoring their levels in disease? The first question cannot yet be fully answered. The clearance of cell surface adhesion molecules does obviously downregulate the adhesive interactions of a cell. This may be a critical part of normal extravasation: a cell that is permanently bound to a surface is not free to migrate. In this case, soluble adhesion molecules are merely an essential byproduct of normal adhesive interactions. However, functional roles for soluble adhesion molecules cannot be discounted. There is some evidence that binding of soluble adhesion molecules can result in signal transduction. sE-Selectin can upregulate neutrophil CD11-b integrin function and sP-selectin has been reported to inhibit neutrophil activation. Neutrophils adhering to P-selectin are refractory to the generation of superoxide anion induced by TNF$\alpha$ or fMLP. In addition, sP-selectin inhibits the adhesion of neutrophils via CD11-b to endothelium by blocking the TNF$\alpha$-mediated activation of the $\beta_2$ integrin complex. These effects are reversible by removal of the sP-selectin or the use of blocking antibodies. Some adhesion molecules, such as CD44 and endoglin, are capable of binding and displaying cytokines, such as the chemokines and transforming growth factor $\beta_1$, respectively. If their soluble forms also carry such mediators, they could conceivably act as carrier molecules for cytokines in the circulation. Evidence for the biologic role of soluble adhesins will only come from in vivo experiments.

In contrast to the lack of information about the function of soluble adhesion molecules, there is now a wealth of information on their levels in both experimental and human diseases. The remainder of this chapter reviews the evidence that suggests that measuring their levels can provide useful information about the nature and progression of such diseases. This subject has recently been reviewed (1), and I have therefore concentrated on referencing recent papers.

## SOLUBLE ADHESION MOLECULES IN INFLAMMATORY DISEASE

### Selectins

The discovery of circulating forms of the selectins was preceded by circumstantial evidence that such isoforms may exist. P-Selectin was shown to have an alternately spliced mRNA lacking a transmembrane domain, and it shows only transient expression on the cell surface. L-Selectin was shown to be rapidly lost from the surface of stimulated neutrophils by a proteolytic mechanism, and E-selectin was shown to be induced on activated endothelial cells within 2 to 3 h, but it was absent from the surface by 24 h.

#### *L-Selectin*

Soluble L-selectin is 3 to 5 kDa smaller than the corresponding cell surface form. It can be released from the cell surface within minutes of leukocyte activation. sL-Selectin, when added at a high enough concentration, can inhibit essentially all L-selectin–mediated adhesion. At physiologic concentrations, sL-selectin can cause approximately 15% to 20% inhibition of lymphocyte attachment to endothelium. The cleavage site in L-selectin has been identified as between Lys321 and Ser322 (3).

Studies of patients with sepsis or HIV infection show a two- to threefold increase in sL-selectin levels compared to normal subjects (4). sL-Selectin can also be detected in synovial fluid of rheumatoid arthritis patients (5).

#### *P-Selectin*

P-Selectin is found in the plasma of normal individuals (6,7), and it appears to be approximately 3 kDa smaller than native P-selectin. The majority of sP-selectin in blood is the alternatively spliced form lacking a transmembrane domain with the N-terminus intact. Because transcripts of P-selectin with or without the transmembrane form exist in both endothelial cells and megakaryocytes, it is likely that both cell types contribute to blood levels. It is not known whether there is a sufficient concentration of P-selectin in disease states to affect this adhesion pathway. Circulating P-selectin values in patients with hemolytic uremic syndrome and thrombotic thrombocytopenic purpura are elevated two- to threefold over normal values (8).

### *E-Selectin*

E-Selectin is expressed on only activated endothelium, in contrast to other adhesion molecules that are widely distributed. Activated endothelial cells have been shown to release soluble E-selectin in vitro (9). Demonstration of soluble E-selectin in blood is therefore conclusive evidence of endothelial activation. The mechanism of release of sE-selectin has not been established, but immunochemical evidence suggests proteolytic cleavage (10).

sE-Selectin can inhibit leukocyte adhesion, and either recombinant soluble or cell surface E-selectin can activate the PMN CD11-b integrin receptor (11,12). The levels required for these effects are not attained in blood. However, this may be possible at local sites.

E-Selectin was detected in the blood of patients with the vasculitides polyarteritis nodosum, giant-cell arteritis, and scleroderma, with higher levels seen in lupus patients. Overall, however, there was no correlation with disease activity and only a weak correlation with the degree of organ involvement (13). Elevated levels of sE-selectin have also been found in diabetic patients independent of hypertension, nephropathy, or renal failure (14), and during acute asthma attacks (15).

sE-Selectin levels are elevated up to 20-fold in sepsis (10). The levels of sE-selectin appear to correlate with disease severity and outcome (16). Higher levels or persistent elevation were associated with greater mortality. Patients with acute *Plasmodium falciparum* malaria have also shown elevated levels of E-selectin (17). sE-Selectin has also been found in bronchiolar lavage (BAL) fluids from patients with interstitial lung disease and in BAL fluids of allergic subjects after segmental antigen challenge (18).

## Immunoglobulin Superfamily Adhesins

The cell surface expression of IgSF adhesion molecules such as ICAM-1 and VCAM-1 is increased on multiple cell types, including endothelium, during immune and inflammatory responses. Soluble forms of these adhesion molecules can readily be found in the supernatants of activated endothelial cells, as well as a number of other cell types and tumor cells (9).

### *ICAM-1*

ICAM-1 has been the most widely studied soluble adhesion molecule, predominantly because of the availability of antibodies. Seth et al. (19) demonstrated soluble ICAM in human serum in 1991 by immunoblotting, and this was confirmed by ELISA (20). Circulating ICAM-1 is functional and contains most of the extracellular domains of cell surface ICAM-1 (20). ICAM is released from endothelial cells, hepatocytes, and leukocytes in culture (9,21), and from melanoma and ovarian carcinoma cells (22). The mechanism of release of ICAM-1 is not clear.

A number of studies have now been published that give estimates of the mean level of circulating ICAM in healthy individuals as between 102 and 450 ng/ml (1). ICAM levels have been reported to be elevated in inflammation, autoimmune disease, infection, HIV, and cancers (1,23,24).

There have been a limited number of longitudinal surveys of serum levels of ICAM-1. ICAM levels rise before death in idiopathic pulmonary fibrosis (25). ICAM-1 increases are associated with episodes of graft rejection (26) and idiopathic uveoretinitis (27), and levels are also elevated in systemic vasculitis and fall during remission (28). ICAM-1 has also been detected in the cerebrospinal fluid of patients with inflammatory neurologic diseases (29,30), in synovial fluids from patients with rheumatoid arthritis (31), and in the BAL fluid of patients with interstitial lung disease (25).

There is no evidence that the related molecules ICAM-2 on endothelium and ICAM-3 on leukocytes exist as soluble forms.

## *VCAM*

Soluble VCAM is released by activated endothelial cells in culture (9). The mechanism of release is unknown. Estimates of the mean level of VCAM in blood range from 431 to 504 ng/ml in normal individuals (1). VCAM in blood is capable of supporting the adhesion of T cells (32). VCAM levels are elevated in serum from patients with cancer and with inflammatory diseases (1). Longitudinal studies have shown increases in VCAM associated with renal allograft rejection episodes and also with cytomegalovirus infection (M. Rose, A. Rees, and D. Haskard, personal communications). VCAM has also been demonstrated in the synovial fluids of patients with rheumatoid arthritis (31). VCAM levels appear to show an excellent correlation with disease activity in systemic lupus erythematosus and in systemic sclerosis (33). In rheumatoid arthritis patients, VCAM levels appeared to correlate with erythrocyte sedimentation rate and C-reactive protein levels, whereas ICAM did not (31).

## *CD44*

CD44 has been shown to be released from activated leukocytes in culture, and soluble CD44 can be detected in the blood of normal individuals and in synovial fluid of rheumatoid arthritis patients. The release appears to be dependent on the action of a metalloproteinase enzyme (34).

## SOLUBLE ADHESION MOLECULES IN CANCER

Work on soluble adhesion molecules in cancer has, with few exceptions, been largely restricted to ICAM-1 (35). Circulating ICAM-1 has been shown in blood of cancer patients with higher levels being associated with metastasis (36–38). Human melanoma cells have been shown to secrete ICAM-1, and high levels of ICAM-1 were found in the sera of nude mice bearing human melanoma (22). In addition, patients with ovarian carcinoma have higher levels of soluble ICAM-1 in their serum and ascites fluids than serum from normal individuals or ascites from patients with cirrhosis (39). Human ovarian carcinoma cell lines grown as ascites tumors in nude mice release ICAM-1 into serum and ascites fluid. The levels of ICAM-1 in serum and ascites fluid correlated with tumor burden. The level of ICAM-1 released by the tumors could be increased by in vivo treatment with interferon-$\gamma$.

The expression and release of ICAM-1 may have profound implications in tumor progression, ascites formation, and metastasis. ICAM-1 is important in mediating tumor recognition by leukocytes. Administration of cytokines could increase ICAM-1 expression on tumor cells and recognition by host cells (40–42). Alternatively, release of adhesion molecules by tumor cells may allow their escape from immunosurveillance by decreasing ICAM-1–mediated recognition. Cytokines that can stimulate release could therefore act in a detrimental way. Further studies in nude mouse, human, or homologous murine tumor systems are needed to confirm the role played by soluble ICAM-1 in tumor progression.

There is currently a great deal of interest in CD44 as a contributory factor in cancer. There are multiple exon splice variants of CD44, one of which (containing the v6 exon) is implicated in the metastatic process (43). The level of CD44, measured by an ELISA that detects all isoforms, has been shown to be elevated in patients with gastric and colon cancer (44). Levels correlated with tumor burden and metastasis. Surgical removal of the tumor resulted in a decrease in CD44 levels. It is not yet clear if particular splice variants are the major contributors to the levels of CD44.

## SUMMARY

Soluble forms of several adhesion molecules can be demonstrated circulating in the bloodstream of normal healthy individuals. Levels of these adhesion molecules can be increased in infectious and autoimmune inflammatory diseases, as well as in many cancers. The functional consequences of elevated levels of these adhesion molecules are not yet clear. Although there is at present no definitive clinical test based on an adhesion molecule, the evidence to date suggests that measuring soluble adhesion molecules will eventually find a place in patient monitoring.

## REFERENCES

1. Gearing AJH, Newman W. Circulating adhesion molecules in disease. *Immunol Today* 1993;14:506–12.
2. Pigott R, Power C. *The adhesion molecule factsbook*. San Diego: Academic Press, 1993.
3. Kahn J, Ingraham RH, Shirley F, Migaki GI, Kishimoto TK. Membrane proximal cleavage site of L-selectin: identification of the cleavage site and a 6kD transmembrane peptide fragment of L-selectin. *J Cell Biol* 1994;125:461–70.
4. Spertini O, Schleiffenbaum B, White-Owen C, Ruiz P Jr, Tedder TJ. ELISA for quantitation of L-selectin shed from leucocytes *in vivo*. *J Immunol Methods* 1992;156:115–23.
5. Humbria A, Diaz-Gonzales F, Campanero MR, et al. Expression of L-selectin, CD43 and CD44 in synovial fluid neutrophils in patients with inflammatory joint disease. Evidence for a soluble form of L-selectin in synovial fluid. *Arthritis Rheumatol* 1994;37:342–8.
6. Katayama M, Handa M, Hironobu A, et al. A monoclonal anntibody based ELISA for GMP-140/P-selectin. *J Immunol Methods* 1992;153:41–8.
7. Dunlop LC, Skinner MP, Bendall LJ, et al. Characterisation of GMP-140 as a circulating plasma protein. *J Exp Med* 1992;175:1147–50.
8. Katayama M, Handa M, Araki Y, Kawai Y, Watanabe K, Ikeda Y. Soluble P-selectin is present in normal circulation and its plasma level is elevated in patients with thrombotic thrombocytopenic purpura and hemolytic uraemic syndrome. *Br J Hematol* 1993;84:702–10.
9. Pigott R, Dillon LP, Hemingway IK, Gearing AJH. Soluble forms of E-selectin, ICAM-1 and VCAM-1 are present in the supernatants of cytokine-activated cultured endothelial cells. *Biochem Biophys Res Common* 1992;187:584–9.
10. Newman W, Beall DB, Carson CW, et al. Soluble E-selectin is found in supernatants of activated endothelial cells and is elevated in the serum of patients with septic shock. *J Immunol* 1993;150:633–54.
11. Lo SK, Lee S, Ramos RA, et al. ELAM-1 stimulates the adhesive activity of leucocyte integrin CR3 (CD11b/CD18, Mac-1) on human neutrophils. *J Exp Med* 1991;173:1493–1500.
12. Kuijpers T, Hakkert BC, Hoogerwerf M, Leeuwenberg JFM, Roos D. Role of ELAM-1 and PAF in neutrophil adherence to ILI-prestimulated endothelial cells. *J Immunol* 1991;147: 1369–76.
13. Carson CW, Beall LD, Hunder GG, Johnson CM, Newman W. Serum ELAM-1 is increased in vasculitis, scleroderma and systemic lupus erythematosus. *J Rheumatol* 1993;20:809–14.
14. Gearing AJH, Hemingway I, Pigott R, Hughes J, Rees A, Cashman SJ. Soluble forms of vascular adhesion molecules E-selectin, ICAM-1 and VCAM-1: pathological significance. *Ann NY Acad Sci* 1992;667:324–31.
15. Kobayashi T, Hashimoto S, Imai K, et al. Elevation of serum soluble ICAM-1 and sE-selectin in bronchial asthma. *Clin Exp Immunol* 1994;96:110–5.
16. Cowley HC, Heney D, Gearing AJH, Hemingway I, Webster NR. Increased circulating adhesion molecule concentrations in patients with the systemic inflamatory response syndrome: a prospective cohort study. *Crit Care Med* 1994;22:651–7.
17. Hviid L, Theander TG, Elhassan IM, Jensen JB. *Immunol Lett* [*in press*].
18. Georas SN, Liu MC, Newman W, et al. Altered adhesion molecule expression and endothelial cell activation accompany the recruitment of human granulocytes to the lung after segmental antigen challenge. *Am J Respir Cell Mol Biol* 1992;7:261–9.
19. Seth R, Raymond FD, Makgoba MW. Circulating ICAM-1 isoforms: diagnostic prospects for inflammatory and immune disorders. *Lancet* 1991;338:83–4.
20. Rothlein R, Mainolfi EA, Czajkowski M, Marlin SD. A form of circulating ICAM-1 in human serum. *J Immunol* 1991;147:3788–93.
21. Thomson AW, Satoh S, Nussler AK, et al. Circulating ICAM-1 in autoimmune liver disease and evidence for the production of ICAM-1 by cytokine-stimulated human hepatocytes. *Clin Exp Immunol* 1994;95:83–90.
22. Giavazzi R, Chirivi RGS, Garofalo A, et al. Soluble intercellular adhesion molecule 1 is released by human melanoma cells and is associated with tumor growth in nude mice. *Cancer Res* 1992;52:2628–30.

23. Martin S, Lampeter EF, Kolb H. A physiological role for circulating adhesion molecules? *Immunol Today* 1994;15:141.
24. Lai CK, Wong KC, Chan CH, et al. Circulating adhesion molecules in tuberculosis. *Clin Exp Immunol* 1993;94:522–6.
25. Shijubo N, Imai K, Shigehara K, et al. Soluble ICAM-1 in sera and broncheolar lavage fluid of patients with idiopathic pulmonary fibrosis and sarcoidosis. *Clin Exp Immunol* 1994;95: 155–61.
26. Gordon JL, Edwards RM, Cashman SJ, Rees AJ, Gearing AJH. In: Catravas JD, Callow AD, Ryan US., eds. *Vascular endothelium: physiological basis of clinical problems II. Proceedings from the NATO Advanced Study Institute.* New York: Plenum Press, 1993:115–22.
27. Zaman AG, Edelsten C, Stanford MR, et al. Soluble ICAM-1 as a marker of disease relapse in idiopathic uveoretinitis. *Clin Exp Immunol* 1994;95:60–5.
28. Wang CR, Liu MF, Tsai RT, Chuang CY, Chen CY. Circulating ICAM-1 and autoantibodies including endothelial cell, anti-cardiolipin and anti-neutrophil cytoplasm antibodies in patients with vasculitis. *Clin Rheumatol* 1993;12:375–80.
29. Sharief MK, Noori MA, Ciardi M, Cirelli A, Thompson EJ. Increased levels of circulating ICAM-1 in serum and CSF of patients with active MS. Correlation with TNFa and blood brain barrier damage. *J Neuroimmunol* 1993;43:15–21.
30. Tsukada N, Matsuda M, Miyagi K, Yanagisawa N. Increased levels of ICAM-1 and TNF receptor in the CSF of patients with MS. *Neurology* 1993;43:2679–82.
31. Mason JC, Kapahi P, Haskard DO. Detection of increased levels of circulating ICAM-1 in some patients with rheumatoid arthritis but not in patients with systemic lupus erythematosus. *Arthritis Rheum* 1993;36:519–27.
32. Wellicome SM, Kapahi P, Mason JC, Lebranchu Y, Yarwood H, Haskard DO. Detection of a circulating form of VCAM-1: raised levels in rheumatoid arthritis and systemic lupus erythematosus. *Clin Exp Immunol* 1993;92:412–8.
33. Gordon C, Sheeran T, Luqmani R, et al. *Abstracts of the American College of Rheumatology Meeting,* 1992.
34. Bazil V, Strominger JL. Metalloprotease and serine protease are involved in the cleavage of CD43, CD44 and CD16 from stimulated human granulocytes. *J Immunol* 1994;152:1314–22.
35. Banks RE, Gearing AJH, Hemingway IK, Norfolk DR, Perren TJ, Selby PJ. Circulating intercellular adhesion molecule-1, E-selectin and vascular cell adhesion molecule-1 in human malignancies. *Br J Cancer* 1993;68:122–4.
36. Harning R, Mainolfi E, Bystryn JC, Henn M, Merluzzi VY, Rothlein R. Serum levels of circulating intercellular adhesion molecule 1 in human malignant melanoma. *Cancer Res* 1991;51:5003–5.
37. Tsujisaki M, Imai K, Hirata H, et al. Detection of circulating intercellular adhesion molecule-1 antigen in malignant diseases. *Clin Exp Immunol* 1991;85:3–8.
38. Hyodo I, Jinno K, Tanimizu M, et al. Detection of circulating ICAM-1 in hepatocellular carcinoma. *Int J Cancer* 1993;55:775–9.
39. Giavazzi R, Nicoletli MI, Chirivi RGS, et al. Soluble ICAM-1 is released by human ovarian carcinoma into the serum and ascites of patients and in nude mice bearing tumour xerographs. *Eur J Cancer [in press]*.
40. Webb DSA, Mostowski HS, Gerrard TL. Cytokine-induced enhancement of ICAM-1 expression results in increased vulnerability of tumour cells to monocyte-mediated lysis. *J Immunol* 1991;146:3682–6.
41. Matsui M, Yoshimura S, Nakanishi T, Ferrone S. Effect of tumor size on the enhancement by g-interferon of the localization of radiolabeled F (ab')2 fragments of anti-intercellular adhesion molecule-1 monoclonal antibodies in human colon carcinoma cells grafted in nude mice. *Cancer Res* 1992;52:1309–13.
42. Naganuma H, Kiessling R, Patarroyo M, Hansson M, Handgretinger R, Gronberg A. Increased susceptibility of IFN-g-treated neuroblastoma cells to lysis by lymphokine-activated killer cells: participation of ICAM-1 induction on target cells. *Int J Cancer* 1991; 47:527–32.
43. Gunthert U, Hofman M, Rudy W, et al. A new variant of glycoprotein CD44 confers metastatic potential to rat carcinoma cells. *Cell* 1991;65:13–24.
44. Guo Y, Liu G, Wang X, et al. Potential use of soluble CD44 in serum as indicator of tumour burden and metastasis in patients with gastric or colon cancer. *Cancer Res* 1994;54:422–6.

# Subject Index

# Subject Index